T0192121

LONDON MATHEMATICAL SOCIETY LECTURE NOTE SERIES

225 A mathematical introduction to string theory, S. ALBEVERIO *et al*
226 Novikov conjectures, index theorems and rigidity I, S. C. FERRY, A. RANICKI & J. ROSENBERG (eds)
227 Novikov conjectures, index theorems and rigidity II, S. C. FERRY, A. RANICKI & J. ROSENBERG (eds)
228 Ergodic theory of Z^d-actions, M. POLLICOTT & K. SCHMIDT (eds)
229 Ergodicity for infinite dimensional systems, G. DA PRATO & J. ZABCZYK
230 Prolegomena to a middlebrow arithmetic of curves of genus 2, J. W. S. CASSELS & E. V. FLYNN
231 Semigroup theory and its applications, K. H. HOFMANN & M. W. MISLOVE (eds)
232 The descriptive set theory of Polish group actions, H. BECKER & A. S. KECHRIS
233 Finite fields and applications, S. COHEN & H. NIEDERREITER (eds)
234 Introduction to subfactors, V. JONES & V. S. SUNDER
235 Number theory: Séminaire de théorie des nombres de Paris 1993–94, S. DAVID (ed)
236 The James forest, H. FETTER & B. GAMBOA DE BUEN
237 Sieve methods, exponential sums, and their applications in number theory, G. R. H. GREAVES *et al* (eds)
238 Representation theory and algebraic geometry, A. MARTSINKOVSKY & G. TODOROV (eds)
240 Stable groups, F. O. WAGNER
241 Surveys in combinatorics, 1997, R. A. BAILEY (ed)
242 Geometric Galois actions I, L. SCHNEPS & P. LOCHAK (eds)
243 Geometric Galois actions II, L. SCHNEPS & P. LOCHAK (eds)
244 Model theory of groups and automorphism groups, D. M. EVANS (ed)
245 Geometry, combinatorial designs and related structures, J. W. P. HIRSCHFELD *et al* (eds)
246 p-Automorphisms of finite p-groups, E. I. KHUKHRO
247 Analytic number theory, Y. MOTOHASHI (ed)
248 Tame topology and O-minimal structures, L. VAN DEN DRIES
249 The atlas of finite groups – Ten years on, R. T. CURTIS & R. A. WILSON (eds)
250 Characters and blocks of finite groups, G. NAVARRO
251 Gröbner bases and applications, B. BUCHBERGER & F. WINKLER (eds)
252 Geometry and cohomology in group theory, P. H. KROPHOLLER, G. A. NIBLO & R. STÖHR (eds)
253 The q-Schur algebra, S. DONKIN
254 Galois representations in arithmetic algebraic geometry, A. J. SCHOLL & R. L. TAYLOR (eds)
255 Symmetries and integrability of difference equations, P. A. CLARKSON & F. W. NIJHOFF (eds)
256 Aspects of Galois theory, H. VÖLKLEIN, J. G. THOMPSON, D. HARBATER & P. MÜLLER (eds)
257 An introduction to noncommutative differential geometry and its physical applications (2nd edition), J. MADORE
258 Sets and proofs, S. B. COOPER & J. K. TRUSS (eds)
259 Models and computability, S. B. COOPER & J. TRUSS (eds)
260 Groups St Andrews 1997 in Bath I, C. M. CAMPBELL *et al* (eds)
261 Groups St Andrews 1997 in Bath II, C. M. CAMPBELL *et al* (eds)
262 Analysis and logic, C. W. HENSON, J. IOVINO, A. S. KECHRIS & E. ODELL
263 Singularity theory, W. BRUCE & D. MOND (eds)
264 New trends in algebraic geometry, K. HULEK, F. CATANESE, C. PETERS & M. REID (eds)
265 Elliptic curves in cryptography, I. BLAKE, G. SEROUSSI & N. SMART
267 Surveys in combinatorics, 1999, J. D. LAMB & D. A. PREECE (eds)
268 Spectral asymptotics in the semi-classical limit, M. DIMASSI & J. SJÖSTRAND
269 Ergodic theory and topological dynamics of group actions on homogeneous spaces, M. B. BEKKA & M. MAYER
271 Singular perturbations of differential operators, S. ALBEVERIO & P. KURASOV
272 Character theory for the odd order theorem, T. PETERFALVI. Translated by R. SANDLING
273 Spectral theory and geometry, E. B. DAVIES & Y. SAFAROV (eds)
274 The Mandelbrot set, theme and variations, T. LEI (ed)
275 Descriptive set theory and dynamical systems, M. FOREMAN, A. S. KECHRIS, A. LOUVEAU & B. WEISS (eds)
276 Singularities of plane curves, E. CASAS-ALVERO
277 Computational and geometric aspects of modern algebra, M. ATKINSON *et al* (eds)
278 Global attractors in abstract parabolic problems, J. W. CHOLEWA & T. DLOTKO
279 Topics in symbolic dynamics and applications, F. BLANCHARD, A. MAASS & A. NOGUEIRA (eds)
280 Characters and automorphism groups of compact Riemann surfaces, T. BREUER
281 Explicit birational geometry of 3-folds, A. CORTI & M. REID (eds)
282 Auslander-Buchweitz approximations of equivariant modules, M. HASHIMOTO
283 Nonlinear elasticity, Y. B. FU & R. W. OGDEN (eds)
284 Foundations of computational mathematics, R. DEVORE, A. ISERLES & E. SÜLI (eds)
285 Rational points on curves over finite fields, H. NIEDERREITER & C. XING
286 Clifford algebras and spinors (2nd Edition), P. LOUNESTO
287 Topics on Riemann surfaces and Fuchsian groups, E. BUJALANCE, A. F. COSTA & E. MARTÍNEZ (eds)
288 Surveys in combinatorics, 2001, J. W. P. HIRSCHFELD (ed)
289 Aspects of Sobolev-type inequalities, L. SALOFF-COSTE
290 Quantum groups and Lie theory, A. PRESSLEY (ed)
291 Tits buildings and the model theory of groups, K. TENT (ed)
292 A quantum groups primer, S. MAJID
293 Second order partial differential equations in Hilbert spaces, G. DA PRATO & J. ZABCZYK
294 Introduction to operator space theory, G. PISIER
295 Geometry and integrability, L. MASON & Y. NUTKU (eds)
296 Lectures on invariant theory, I. DOLGACHEV
297 The homotopy category of simply connected 4-manifolds, H.-J. BAUES
298 Higher operads, higher categories, T. LEINSTER (ed)
299 Kleinian groups and hyperbolic 3-manifolds, Y. KOMORI, V. MARKOVIC & C. SERIES (eds)
300 Introduction to Möbius differential geometry, U. HERTRICH-JEROMIN
301 Stable modules and the D(2)-problem, F. E. A. JOHNSON
302 Discrete and continuous nonlinear Schrödinger systems, M. J. ABLOWITZ, B. PRINARI & A. D. TRUBATCH
303 Number theory and algebraic geometry, M. REID & A. SKOROBOGATOV (eds)
304 Groups St Andrews 2001 in Oxford I, C. M. CAMPBELL, E. F. ROBERTSON & G. C. SMITH (eds)
305 Groups St Andrews 2001 in Oxford II, C. M. CAMPBELL, E. F. ROBERTSON & G. C. SMITH (eds)
306 Geometric mechanics and symmetry, J. MONTALDI & T. RATIU (eds)

307 Surveys in combinatorics 2003, C. D. WENSLEY (ed.)
308 Topology, geometry and quantum field theory, U. L. TILLMANN (ed)
309 Corings and comodules, T. BRZEZINSKI & R. WISBAUER
310 Topics in dynamics and ergodic theory, S. BEZUGLYI & S. KOLYADA (eds)
311 Groups: topological, combinatorial and arithmetic aspects, T. W. MÜLLER (ed)
312 Foundations of computational mathematics, Minneapolis 2002, F. CUCKER *et al* (eds)
313 Transcendental aspects of algebraic cycles, S. MÜLLER-STACH & C. PETERS (eds)
314 Spectral generalizations of line graphs, D. CVETKOVIĆ, P. ROWLINSON & S. SIMIĆ
315 Structured ring spectra, A. BAKER & B. RICHTER (eds)
316 Linear logic in computer science, T. EHRHARD, P. RUET, J.-Y. GIRARD & P. SCOTT (eds)
317 Advances in elliptic curve cryptography, I. F. BLAKE, G. SEROUSSI & N. P. SMART (eds)
318 Perturbation of the boundary in boundary-value problems of partial differential equations, D. HENRY
319 Double affine Hecke algebras, I. CHEREDNIK
320 L-functions and Galois representations, D. BURNS, K. BUZZARD & J. NEKOVÁŘ (eds)
321 Surveys in modern mathematics, V. PRASOLOV & Y. ILYASHENKO (eds)
322 Recent perspectives in random matrix theory and number theory, F. MEZZADRI & N. C. SNAITH (eds)
323 Poisson geometry, deformation quantisation and group representations, S. GUTT *et al* (eds)
324 Singularities and computer algebra, C. LOSSEN & G. PFISTER (eds)
325 Lectures on the Ricci flow, P. TOPPING
326 Modular representations of finite groups of Lie type, J. E. HUMPHREYS
327 Surveys in combinatorics 2005, B. S. WEBB (ed)
328 Fundamentals of hyperbolic manifolds, R. CANARY, D. EPSTEIN & A. MARDEN (eds)
329 Spaces of Kleinian groups, Y. MINSKY, M. SAKUMA & C. SERIES (eds)
330 Noncommutative localization in algebra and topology, A. RANICKI (ed)
331 Foundations of computational mathematics, Santander 2005, L. M PARDO, A. PINKUS, E. SÜLI & M. J. TODD (eds)
332 Handbook of tilting theory, L. ANGELERI HÜGEL, D. HAPPEL & H. KRAUSE (eds)
333 Synthetic differential geometry (2nd Edition), A. KOCK
334 The Navier–Stokes equations, N. RILEY & P. DRAZIN
335 Lectures on the combinatorics of free probability, A. NICA & R. SPEICHER
336 Integral closure of ideals, rings, and modules, I. SWANSON & C. HUNEKE
337 Methods in Banach space theory, J. M. F. CASTILLO & W. B. JOHNSON (eds)
338 Surveys in geometry and number theory, N. YOUNG (ed)
339 Groups St Andrews 2005 I, C. M. CAMPBELL, M. R. QUICK, E. F. ROBERTSON & G. C. SMITH (eds)
340 Groups St Andrews 2005 II, C. M. CAMPBELL, M. R. QUICK, E. F. ROBERTSON & G. C. SMITH (eds)
341 Ranks of elliptic curves and random matrix theory, J. B. CONREY, D. W. FARMER, F. MEZZADRI & N. C. SNAITH (eds)
342 Elliptic cohomology, H. R. MILLER & D. C. RAVENEL (eds)
343 Algebraic cycles and motives I, J. NAGEL & C. PETERS (eds)
344 Algebraic cycles and motives II, J. NAGEL & C. PETERS (eds)
345 Algebraic and analytic geometry, A. NEEMAN
346 Surveys in combinatorics 2007, A. HILTON & J. TALBOT (eds)
347 Surveys in contemporary mathematics, N. YOUNG & Y. CHOI (eds)
348 Transcendental dynamics and complex analysis, P. J. RIPPON & G. M. STALLARD (eds)
349 Model theory with applications to algebra and analysis I, Z. CHATZIDAKIS, D. MACPHERSON, A. PILLAY & A. WILKIE (eds)
350 Model theory with applications to algebra and analysis II, Z. CHATZIDAKIS, D. MACPHERSON, A. PILLAY & A. WILKIE (eds)
351 Finite von Neumann algebras and masas, A. M. SINCLAIR & R. R. SMITH
352 Number theory and polynomials, J. MCKEE & C. SMYTH (eds)
353 Trends in stochastic analysis, J. BLATH, P. MÖRTERS & M. SCHEUTZOW (eds)
354 Groups and analysis, K. TENT (ed)
355 Non-equilibrium statistical mechanics and turbulence, J. CARDY, G. FALKOVICH & K. GAWEDZKI
356 Elliptic curves and big Galois representations, D. DELBOURGO
357 Algebraic theory of differential equations, M. A. H. MACCALLUM & A. V. MIKHAILOV (eds)
358 Geometric and cohomological methods in group theory, M. R. BRIDSON, P. H. KROPHOLLER & I. J. LEARY (eds)
359 Moduli spaces and vector bundles, L. BRAMBILA-PAZ, S. B. BRADLOW, O. GARCÍA-PRADA & S. RAMANAN (eds)
360 Zariski geometries, B. ZILBER
361 Words: Notes on verbal width in groups, D. SEGAL
362 Differential tensor algebras and their module categories, R. BAUTISTA, L. SALMERÓN & R. ZUAZUA
363 Foundations of computational mathematics, Hong Kong 2008, F. CUCKER, A. PINKUS & M. J. TODD (eds)
364 Partial differential equations and fluid mechanics, J. C. ROBINSON & J. L. RODRIGO (eds)
365 Surveys in combinatorics 2009, S. HUCZYNSKA, J. D. MITCHELL & C. M. RONEY-DOUGAL (eds)
366 Highly oscillatory problems, B. ENGQUIST, A. FOKAS, E. HAIRER & A. ISERLES (eds)
367 Random matrices: High dimensional phenomena, G. BLOWER
368 Geometry of Riemann surfaces, F. P. GARDINER, G. GONZÁLEZ-DIEZ & C. KOUROUNIOTIS (eds)
369 Epidemics and rumours in complex networks, M. DRAIEF & L. MASSOULIÉ
370 Theory of p-adic distributions, S. ALBEVERIO, A. YU. KHRENNIKOV & V. M. SHELKOVICH
371 Conformal fractals, F. PRZYTYCKI & M. URBAŃSKI
372 Moonshine: The first quarter century and beyond, J. LEPOWSKY, J. MCKAY & M. P. TUITE (eds)
373 Smoothness, regularity, and complete intersection, J. MAJADAS & A. RODICIO
374 Geometric analysis of hyperbolic differential equations: An introduction, S. ALINHAC
375 Triangulated categories, T. HOLM, P. JØRGENSEN & R. ROUQUIER (eds)
376 Permutation patterns, S. LINTON, N. RUŠKUC & V. VATTER (eds)
377 An introduction to Galois cohomology and its applications, G. BERHUY
378 Probability and mathematical genetics, N. H. BINGHAM & C. M. GOLDIE (eds)
379 Finite and algorithmic model theory, J. ESPARZA, C. MICHAUX & C. STEINHORN (eds)
380 Real and Complex Singularities, M. MANOEL, M. C. ROMERO FUSTER & C. T. L. WALLS (eds)

London Mathematical Society Lecture notes series: 377

An Introduction to Galois Cohomology and its Applications

GRÉGORY BERHUY

Université Joseph Fourier, Grenoble

CAMBRIDGE
UNIVERSITY PRESS

University Printing House, Cambridge CB2 8BS, United Kingdom

One Liberty Plaza, 20th Floor, New York, NY 10006, USA

477 Williamstown Road, Port Melbourne, VIC 3207, Australia

314-321, 3rd Floor, Plot 3, Splendor Forum, Jasola District Centre, New Delhi - 110025, India

103 Penang Road, #05-06/07, Visioncrest Commercial, Singapore 238467

Cambridge University Press is part of the University of Cambridge.

It furthers the University's mission by disseminating knowledge in the pursuit of education, learning and research at the highest international levels of excellence.

www.cambridge.org
Information on this title: www.cambridge.org/9780521738668

First published 2010

A catalogue record for this publication is available from the British Library

ISBN 978-0-521-73866-8 Paperback

To my dear friend Frédérique

Contents

Foreword

Like an idea whose time has come, nonabelian Galois cohomology burst into the world in the mid 50's. There had been harbingers, of course. Châtelet's *méthode galoisienne* for genus 1 curves and Weil's observations on homogeneous spaces had opened the way, and it was a small step to write down the basic operations so that they make sense in a noncommutative situation. Within a few years, several pioneers realized almost simultaneously that the formalism of Galois cohomology could be used to classify various algebraic structures and to illuminate the definition of some of their invariants. This simple and remarkably penetrating idea, soon popularized by Serre's famous monograph *Cohomologie galoisienne*, immediately took hold. Galois cohomology is indeed algebra at its best: a few formal basic operations with a broad spectrum of far-reaching applications.

Grégory Berhuy's monograph provides a very welcome introduction to Galois descent techniques and nonabelian Galois cohomology, aimed at people who are new to the subject. Beginners will find here a thorough discussion of the technical details that are usually left to the reader. Together with advanced readers, they will appreciate a tasteful tour of applications, including some to which the author, himself an avid cocyclist, has contributed. (Incidentally, the title of Section III.8.1 also offers a glimpse into his taste in movies.) As may be expected, the list of applications discussed here is far from exhaustive, and in the last chapters the exposition is more demanding. It strikes a nice balance between a thorough account and a survey, and it provides a unique introduction to several of the exciting developments of the last decade, such as essential dimension and new advances on rationality problems. The many who did not have the good fortune to take his course at the University of Southampton will be thankful to Grégory Berhuy for making available the text of his lectures. I trust it will give to a large audience an idea of the scope and beauty of the subject, and inspire many of them to contribute to it in their turn.

Jean-Pierre Tignol

Introduction

A recurrent problem arising in mathematics is to decide if two given mathematical structures defined over a field k are isomorphic. Quite often, it is easier to deal with this problem after scalar extension to a bigger field Ω containing k, for example an algebraic closure of k, or a finite Galois extension. In the case where the two structures happen to be isomorphic over Ω, this leads to the natural descent problem: if two k-structures are isomorphic over Ω, are they isomorphic over k? Of course, the answer is no in general. For example, consider the following matrices $M, M_0 \in \mathrm{M}_2(\mathbb{R})$:

$$M_0 = \begin{pmatrix} 0 & -2 \\ 1 & 0 \end{pmatrix}, M = \begin{pmatrix} 0 & 2 \\ -1 & 0 \end{pmatrix}.$$

It is easy to see that they are conjugate by an element of $\mathrm{GL}_2(\mathbb{C})$, since they have same eigenvalues $\pm i\sqrt{2}$, and therefore are both similar to $\begin{pmatrix} i\sqrt{2} & 0 \\ 0 & -i\sqrt{2} \end{pmatrix}$. In fact we have

$$\begin{pmatrix} i & 0 \\ 0 & -i \end{pmatrix} M \begin{pmatrix} i & 0 \\ 0 & -i \end{pmatrix}^{-1} = M_0,$$

so M and M_0 are even conjugate by an element of $\mathrm{SL}_2(\mathbb{C})$.

A classical result in linear algebra says that M and M_0 are already conjugate by an element of $\mathrm{GL}_2(\mathbb{R})$, but this is quite obvious here since the equality above rewrites

$$\begin{pmatrix} 1 & 0 \\ 0 & -1 \end{pmatrix} M \begin{pmatrix} 1 & 0 \\ 0 & -1 \end{pmatrix}^{-1} = M_0.$$

However, they are not conjugate by an element of $\mathrm{SL}_2(\mathbb{R})$. Indeed, it is easy to check that a matrix $P \in \mathrm{GL}_2(\mathbb{R})$ such that $PM = M_0P$ has the form

$$P = \begin{pmatrix} a & 2c \\ c & -a \end{pmatrix}.$$

Since $\det(P) = -(a^2 + 2c^2) < 0$, P cannot belong to $\mathrm{SL}_2(\mathbb{R})$. Therefore, M and M_0 are conjugate by an element of $\mathrm{SL}_2(\mathbb{C})$ but not by an element of $\mathrm{SL}_2(\mathbb{R})$.

Hence, the descent problem for conjugacy classes of matrices has a positive answer when we conjugate by elements of the general linear group, but has a negative one when we conjugate by elements of the special linear group. So, how could we explain the difference between these two cases? This is where Galois cohomology comes into play, and we would like now to give an insight of how this could be used to measure the obstruction to descent problems on the previous example. If k is a field, let us denote by $G(k)$ the group $\mathrm{GL}_2(k)$ or $\mathrm{SL}_2(k)$ indifferently.

Assume that $QMQ^{-1} = M_0$ for some $Q \in G(\mathbb{C})$. The idea is to measure how far is Q to have real coefficients, so it is natural to consider the difference $Q\overline{Q}^{-1}$, where $\overline{Q}$ is the matrix obtained from Q by letting the complex conjugation act coefficientwise. Indeed, we will have $Q \in G(\mathbb{R})$ if and only if $\overline{Q} = Q$, that is if and only if $Q\overline{Q}^{-1} = I_2$. Of course, if $Q\overline{Q}^{-1} = I_2$, then M and M_0 are conjugate by an element of $G(\mathbb{R})$, but this is not the only case when this happens to be true. Indeed, if we assume that $PMP^{-1} = M_0$ for some $P \in G(\mathbb{R})$, then we easily get that $QP^{-1} \in G(\mathbb{C})$ commutes with M_0. Therefore, there exists $C \in Z_G(M_0)(\mathbb{C}) = \{C \in G(\mathbb{C}) \mid CM_0 = M_0C\}$ such that $Q = CP$. We then easily have $\overline{Q} = \overline{C}\,\overline{P} = \overline{C}P$, and therefore

$$Q\overline{Q}^{-1} = C\overline{C}^{-1} \text{ for some } C \in Z_G(M_0)(\mathbb{C}).$$

Conversely, if the equality above holds then $P = C^{-1}Q$ is an element of $G(\mathbb{R})$ satisfying $PMP^{-1} = M_0$. Indeed, we have

$$\overline{P} = \overline{C}^{-1}\overline{Q} = C^{-1}Q = P,$$

so $P \in G(\mathbb{R})$, and

$$PMP^{-1} = C^{-1}QMQ^{-1}C = C^{-1}M_0C = M_0C^{-1}C = M_0.$$

Thus, M and M_0 will be congugate by an element of $G(\mathbb{R})$ if and only if

$$Q\overline{Q}^{-1} = C\overline{C}^{-1} \text{ for some } C \in Z_G(M_0)(\mathbb{C}).$$

Notice also for later use that $Q\overline{Q}^{-1} \in G(\mathbb{C})$ commutes with M_0, as we may check by applying complex conjugation on both sides of the equality $QMQ^{-1} = M_0$.

If we go back to our previous example, we have $Q = \begin{pmatrix} i & 0 \\ 0 & -i \end{pmatrix}$, and therefore $Q\overline{Q}^{-1} = -I_2$. Easy computations show that we have

$$Z_G(M_0)(\mathbb{C}) = \left\{ C \in G(\mathbb{C}) \mid C = \begin{pmatrix} z & -2z' \\ z' & z \end{pmatrix} \text{ for some } z, z' \in \mathbb{C} \right\}.$$

Therefore, we will have $C \in Z_G(M_0)(\mathbb{C})$ and $C\overline{C}^{-1} = Q\overline{Q}^{-1} = -I_2$ if and only if

$$C = \begin{pmatrix} iu & -2iv \\ iv & iu \end{pmatrix} \text{ for some } u, v \in \mathbb{R}, (u, v) \neq (0, 0).$$

Notice that the determinant of the matrix above is $-(u^2 + 2v^2) < 0$. Thus, if $G(\mathbb{C}) = \mathrm{GL}_2(\mathbb{C})$, one may take $u = 1$ and $v = 0$, but if $G(\mathbb{C}) = \mathrm{SL}_2(\mathbb{C})$, the equation $C\overline{C}^{-1} = -I_2 = Q\overline{Q}^{-1}$ has no solution in $Z_G(M_0)(\mathbb{C})$. This explains a bit more conceptually the difference between the two descent problems. In some sense, if $QMQ^{-1} = M_0$ for some $Q \in G(\mathbb{C})$, the matrix $Q\overline{Q}^{-1}$ measures how far is M to be conjugate to M_0 over $\mathbb{R}$.

Of course, all the results above remain valid if M and M_0 are square matrices of size n, and if $G(k) = \mathrm{GL}_n(k), \mathrm{SL}_n(k), \mathrm{O}_n(k)$ or even $\mathrm{Sp}_{2n}(k)$. If we have a closer look to the previous computations, we see that the reason why all this works is that $\mathbb{C}/\mathbb{R}$ is a Galois extension, whose Galois group is generated by complex conjugation.

Let us consider now a more general problem: let Ω/k be a finite Galois extension, and let $M, M_0 \in \mathrm{M}_n(k)$ be two matrices such that

$$QMQ^{-1} = M_0 \text{ for some } Q \in G(\Omega).$$

Does there exist $P \in G(k)$ such that $PMP^{-1} = M_0$?

Since Ω/k is a finite Galois extension, then for all $x \in \Omega$, we have $x \in k$ if and only if $\sigma(x) = x$ for all $\sigma \in \mathrm{Gal}(\Omega/k)$. If now $Q \in G(\Omega)$, then let us denote by $\sigma \cdot Q \in G(\Omega)$ the matrix obtained from Q by letting σ act coefficientwise. Then we have

$$\begin{aligned} Q \in G(k) &\iff \sigma \cdot Q = Q \text{ for all } \sigma \in \mathrm{Gal}(\Omega/k) \\ &\iff Q(\sigma \cdot Q)^{-1} = I_2 \text{ for all } \sigma \in \mathrm{Gal}(\Omega/k). \end{aligned}$$

As before, applying $\sigma \in \mathrm{Gal}(\Omega/k)$ to the equality $QMQ^{-1} = M_0$, we see that $Q(\sigma \cdot Q)^{-1} \in Z_G(M_0)(\Omega)$. We therefore get a map

$$\alpha^Q : \begin{array}{rcl} \mathrm{Gal}(\Omega/k) & \longrightarrow & Z_G(M_0)(\Omega) \\ \sigma & \longmapsto & Q(\sigma \cdot Q)^{-1}. \end{array}$$

Arguing as at the beginning of this introduction, one can show that M and M_0 will be conjugate by an element of $G(k)$ if and only if there exists $C \in Z_G(M_0)(\Omega)$ such that $\alpha^Q = \alpha^C$, that is if and only if there exists $C \in Z_G(M_0)(\Omega)$ such that

$$Q(\sigma \cdot Q)^{-1} = C(\sigma \cdot C)^{-1} \text{ for all } \sigma \in \mathrm{Gal}(\Omega/k).$$

To summarize, to any matrix $M \in \mathrm{M}_n(k)$ which is conjugate to M_0 by an element of $G(\Omega)$, we may associate a map $\alpha^Q : \mathrm{Gal}(\Omega/k) \longrightarrow Z_G(M_0)(\Omega)$, which measures how far is M to be conjugate to M_0 by an element of $G(k)$.

This has a kind of a converse: for any map

$$\alpha : \begin{array}{rcl} \mathrm{Gal}(\Omega/k) & \longrightarrow & Z_G(M_0)(\Omega) \\ \sigma & \longmapsto & \alpha_\sigma \end{array}$$

such that $\alpha = \alpha^Q$ for some $Q \in G(\Omega)$, one may associate a matrix of $M_n(k)$ which is conjugate to M_0 by an element of $G(k)$ by setting $M_\alpha = Q^{-1} M_0 Q$. To see that M_α is indeed an element of $\mathrm{M}_n(k)$, notice first that we have

$$\sigma \cdot (CM'C^{-1}) = (\sigma \cdot C)(\sigma \cdot M')(\sigma \cdot C)^{-1}$$

for all $C \in G(\Omega), M' \in \mathrm{M}_n(\Omega), \sigma \in \mathrm{Gal}(\Omega/k)$. Thus, for all $\sigma \in$

$\mathrm{Gal}(\Omega/k)$, we have

$$\begin{aligned} \sigma\cdot M_\alpha &= (\sigma\cdot Q)^{-1}M_0(\sigma\cdot Q) \\ &= Q^{-1}Q(\sigma\cdot Q)^{-1}M_0(\sigma\cdot Q) \\ &= Q^{-1}M_0Q(\sigma\cdot Q)^{-1}(\sigma\cdot Q) \\ &= Q^{-1}M_0Q \\ &= M_\alpha, \end{aligned}$$

the third equality coming from the fact that $\alpha_\sigma = Q(\sigma\cdot Q)^{-1}$ lies in $Z_G(M_0)(\Omega)$.

Not all the maps $\alpha\colon \mathrm{Gal}(\Omega/k) \longrightarrow Z_G(M_0)(\Omega)$ may be written α^Q for some $Q \in G(\Omega)$. In fact, easy computations show that a necessary condition for this to hold is that α is a **cocycle**, that is

$$\alpha_{\sigma\tau} = \alpha_\sigma\, \sigma\cdot\alpha_\tau \text{ for all } \sigma,\tau \in \mathrm{Gal}(\Omega/k).$$

This condition is not sufficient in general. However, it happens to be the case if $G(\Omega) = \mathrm{GL}_n(\Omega)$ or $\mathrm{SL}_n(\Omega)$ (this will follow from Hilbert 90).

Notice that until now we picked a matrix $Q \in G(\Omega)$ which conjugates M into M_0, but this matrix Q is certainly not unique. We could therefore wonder what happens if we take another matrix $Q' \in G(\Omega)$ which conjugates M into M_0. Computations show that we have $Q'Q^{-1} \in Z_G(M_0)(\Omega)$. Therefore, there exists $C \in Z_G(M_0)(\Omega)$ such that $Q' = CQ$, and we easily get that

$$\alpha_\sigma^{Q'} = C\alpha_\sigma^Q(\sigma\cdot C)^{-1} \text{ for all } \sigma \in \mathrm{Gal}(\Omega/k).$$

Two cocycles $\alpha,\alpha'\colon \mathrm{Gal}(\Omega/k) \longrightarrow Z_G(M_0)(\Omega)$ such that

$$\alpha'_\sigma = C\alpha_\sigma(\sigma\cdot C)^{-1} \text{ for all } \sigma \in \mathrm{Gal}(\Omega/k)$$

for some $C \in Z_G(M_0)(\Omega)$ will be called **cohomologous**. Being cohomologous is an equivalence relation on the set of cocycles, and the set of equivalence classes is denoted by $H^1(\mathrm{Gal}(\Omega/k), Z_G(M_0)(\Omega))$. If α is a cocycle, we will denote by $[\alpha]$ the corresponding equivalence class. Therefore, to any matrix $M \in \mathrm{M}_n(k)$ which is conjugate to M_0 by an element of $G(\Omega)$, one may associate a well-defined cohomology class $[\alpha^Q]$, where $Q \in G(\Omega)$ is any matrix satisfying $QMQ^{-1} = M_0$.

It is important to notice that the class $[\alpha^Q]$ does not characterize M completely. Indeed, for every $P \in G(k)$, it is easy to check that $\alpha^{QP^{-1}} =$

α^Q. In particular, the cohomology classes associated to the matrices M and PMP^{-1} are equal, for all $P \in G(k)$.

Conversely, if $\alpha = \alpha^Q$ and $\alpha' = \alpha^{Q'}$ are cohomologous, it is not too difficult to see that $P = Q^{-1}C^{-1}Q' \in G(k)$, and that the corresponding matrices M_α and $M_{\alpha'}$ satisfy $PM_{\alpha'}P^{-1} = M_\alpha$.

Thus the previous considerations show that, in the case where every cocycle $\alpha\colon \mathrm{Gal}(\Omega/k) \longrightarrow Z_G(M_0)(\Omega)$ may be written $\alpha = \alpha^Q$ for some $Q \in G(\Omega)$, the set $H^1(\mathrm{Gal}(\Omega/k), Z_G(M_0)(\Omega))$ is in one-to-one correspondence with the set of $G(k)$-conjugacy classes of matrices $M \in \mathrm{M}_n(k)$ which are conjugate to M_0 by an element of $G(\Omega)$.

Many situations can be dealt with in a similar way. For example, reasoning as above and using Hilbert 90, one can show that the set of isomorphism classes of quadratic forms q which are isomorphic to the quadratic form $x_1^2 + \ldots + x_n^2$ over Ω is in one-to-one correspondence with $H^1(\mathrm{Gal}(\Omega/k), \mathrm{O}_n(\Omega))$.The case of k-algebras is a little bit more subtle, but one can show that the set of isomorphism classes of k-algebras which are isomorphic to a given k-algebra A over Ω is in one-to-one correspondence with $H^1(\mathrm{Gal}(\Omega/k), \mathrm{Aut}_{\Omega-\mathrm{alg}}(A \otimes_k \Omega))$.

Quite often, algebraic structures can be well understood over a separable closure k_s of k. In the best cases, they even become isomorphic over k_s. Therefore, it is useful to extend this setting to the case of infinite Galois field extensions. To do this, we will introduce the notion of a profinite group in Chapter 1, and recollect some facts on infinite Galois theory. Then in Chapter 2 we define the cohomology sets $H^i(\Gamma, A)$ for any profinite group Γ and any Γ-group A, and study their functorial properties and their behavior with respect to short exact sequences. We also introduce the cup-product, which is useful to construct higher cohomology classes. Chapter 3 deals with Galois cohomology and the central part of this chapter is devoted to formalize Galois descent and to give applications. We then come back to the conjugacy problem for matrices and compute the total obstruction in an example. In Chapter 4, we study Galois cohomology of quadratic forms and give a cohomological interpretation of some classical invariants attached to quadratic forms, such as the determinant or the Hasse invariant. In Chapter 5, we obtain an algebraic interpretation of Galois field extensions with Galois group G in terms of $H^1(\mathrm{Gal}(k_s/k), G)$. In Chapter 6, we give a cohomological obstruction of the following Galois embedding problem: given a group extension $1 \longrightarrow A \longrightarrow \tilde{G} \longrightarrow G \longrightarrow 1$, where A is a central subgroup of $\tilde{G}$, and given a Galois field extension E/k with Galois group G, does there exists a Galois field extension $\tilde{E}/k$ with Galois group $\tilde{G}$ such that $\tilde{E}^A = E$?

The next chapters describe various applications of Galois cohomology. Chapter 7 is devoted to the study of a certain Galois embedding problem with kernel $A = \mathbb{Z}/2\mathbb{Z}$. In this particular case we prove a formula of Serre which computes the obstruction in terms of the classical invariant of the trace form of E, and we give simple applications. We then study Galois cohomology of central simple algebras with or without involutions in Chapter 8. As an application of Galois cohomology techniques, we compute the Hasse invariant of certain quadratic forms attached to these algebras. In Chapter 9, we briefly introduce the notion of a G-torsor, which gives a geometric interpretation of Galois cohomology. We apply this point of view to derive some results on cohomological invariants of algebraic groups. In Chapter 10, we describe applications of Galois cohomology to the so-called Noether's problem: given a field k and a finite group G, is there a linear faithful representation V of G such that the field extension $k(V)^G/k$ is purely transcendental ? This is known to be true when G is abelian and $k \supset \mu_n$, but false for $G = \mathbb{Z}/8\mathbb{Z}$ and $k = \mathbb{Q}$. We will introduce the residue maps in Galois cohomology and use their properties to prove that Noether's problem has a negative solution when $G = \mathbb{Z}/2^m\mathbb{Z}, m \geq 3$ and $k = \mathbb{Q}$. To do so, we attach to each Galois extension of group G over a field $K \supset k$ a non-vanishing cohomological obstruction. In Chapter 11, we study another kind of rationality problem: given a linear algebraic group G over k, is the underlying variety rational? This is known to be true for classical groups when k is algebraically closed. We will show that the answer is negative in general when k is an arbitrary field. We will focus on the case where G is an automorphism group of some algebra with a symplectic involution. Once again, the answer will come from the existence of a non-zero cohomological obstruction. Finally in Chapter 12, we introduce the notion of essential dimension of a functor, which is an active research topic, for which substantial progress has been made recently. If G is a finite group, the essential dimension of the Galois cohomology functor $H^1(\_, G)$ will be the number of independent parameters needed to describe a Galois extension of group G.

This introduction to Galois cohomology does not pretend to be complete. For example, we are aware that an historical introduction to the subject is missing. The curious reader is referred to [30], p. 446-449, as well as [58] and [59] for more information and numerous references. Moreover, we tried to reduce the prerequisites necessary to read these notes to the minimum. Only some basic knowledge on Galois theory and algebra (definition of group, ring, field, k-algebra, notion of tensor product) is required. Also it was impossible to cover all the 'hot topics' (such as Serre's conjecture II, Hasse principle, Rost invariants) or applications of Galois cohomology. Once again, we refer to [30], [58] and [59]. More advanced material on Galois cohomology may be found in [25],[26], [30] or [58], each of these references focusing on a different aspect of the theory: cohomological invariants (including the construction of Rost invariants) and applications to Noether's problem in [25], Merkurjev-Suslin's theorem in [26], algebras with involution in [30] or cohomology of algebraic groups over fields of small cohomological dimension in [58].

This book is an extended version of notes of some postgraduate lectures on Galois cohomology that we gave at the University of Southampton, which included originally Chapters 1-7. The main goal of these lectures was to introduce enough material on Galois cohomology to fully understand the proof of Serre's formula [61] aiming at an audience having a minimal background in algebra, and to give applications to Galois embedding problems. The method we chose to establish this formula differs a bit from the original one. It was suggested as an alternative proof by Serre himself in [61]. Moreover, it was a good occasion to introduce classical tools such as exact sequences in cohomology, Galois descent, Hilbert 90 and some standard results such as Springer's cohomological interpretation of the Hasse invariant. Consequently, the material introduced in Part I is really basic, but is sufficient to obtain beautiful applications to inverse Galois theory or to the conjugacy problem. We also took a particular care to make the first half of this book self-contained, with an exception made for the section on infinite Galois theory and for Proposition III.7.23. Let us also mention the existence of lectures notes [2] presenting a shortened and simplified exposition of the material introduced in Chapters II and III (in these notes, all Galois extensions considered are finite, only the first cohomology set is presented and the functorial aspect of the theory is not treated). The second part of the book gives an insight of how Galois cohomology may be useful to solve some algebraic problems, and presents active research topics, such as rationality questions or essential dimension of algebraic groups and often requires more advanced material. Therefore, proofs of the most difficult results are skipped. We hope that these notes will help the reader willing to study more advanced books on this subject, such as those cited above.

This book could not have been written without the encouragements and the support of Gerhard Roerhle, and we would like to thank him warmly. We are also grateful to our colleagues and friends Vincent Beck, Jérôme Ducoat, Jean Fasel, Nicolas Grenier-Boley, Emmanuel Lequeu, Frédérique Oggier, Gerhard Roerhle and Jean-Pierre Tignol, who took time to read partly or integrally some earlier versions of the manuscript, despite the fact they certainly had better things to do. Their careful reading, judicious comments and remarks permitted to improve significantly the exposition and to detect many misprints or inaccuracies. The whole LaTeX support team of Cambridge University Press deserves a special mention for its efficiency and its patience. Finally, we would like to thank Roger Astley, Caroline Brown and Clare Dennison for their helpfulness in the whole editing process.

Part I

An introduction to Galois cohomology

I
Infinite Galois theory

In the introduction we explained how Galois cohomology could be used to classify mathematical structures defined over k which become isomorphic over a finite Galois extension, and why it would be useful to extend this setting to arbitrary Galois extensions (not necessarily finite). In this chapter, we would like to briefly recall some standard facts on infinite Galois theory. The reader may refer to [42] for details.

§I.1 Reminiscences on field theory

Definition I.1.1. Let k be a field. A **field extension** of k is a pair (K, ε), where K is a field and $\varepsilon : k \longrightarrow K$ is a ring morphism (necessarily injective since k is a field). In other words, K is an extension of k if it contains a subfield isomorphic to k. We will also say that K is an extension of k. We will denote it by K/k if ε is clear from the context.

A **morphism** $\iota : (K_1, \varepsilon_1) \longrightarrow (K_2, \varepsilon_2)$ **of field extensions** of k is a ring morphism $u : K_1 \longrightarrow K_2$ such that $u \circ \varepsilon_1 = \varepsilon_2$. We will also say that ι is a **k-embedding** of K_1 into K_2.

An **isomorphism** of field extensions is a morphism which is bijective. If $\iota : (K_1, \varepsilon_1) \longrightarrow (K_2, \varepsilon_2)$ is an isomorphism of field extensions of k, we will also say that ι is a **k-isomorphism** of K_1 onto K_2, or that K_1 and K_2 are k-isomorphic.

For the rest of this section, we will identify k with its image $\varepsilon(k)$, and therefore consider that we have an inclusion $k \subset K$.

Definition I.1.2. If K/k is a field extension, then K has a natural structure of a k-vector space. The **degree** of K/k is the dimension of K as a k-vector space, and is denoted by $[K : k]$.

Definition I.1.3. Let K/k be a field extension, and let A be a subset of K. The **subring of K generated by A** over k is the smallest subring of K containing A and k. It is denoted by $k[A]$. The **subextension of K generated by A** over k is the smallest subfield of K containing A and k. It is denoted by $k(A)$. If $A = \{\alpha_1, \dots, \alpha_n\}$, we denote them by $k[\alpha_1, \dots, \alpha_n]$ and $k(\alpha_1, \dots, \alpha_n)$ respectively.

Remark I.1.4. It is not difficult to see that

$$k[\alpha_1, \dots, \alpha_n] = \{P(\alpha_1, \dots, \alpha_n) \mid P \in k[X_1, \dots, X_n]\},$$

and that $k(\alpha_1, \dots, \alpha_n)$ is the field of fractions of $k[\alpha_1, \dots, \alpha_n]$.

Definition I.1.5. Let K/k be a field extension and let K_1, K_2 be two subfields of K containing k. The **compositum** of K_1 and K_2 is the subfield generated by $K_1 \cup K_2$. It is denoted by K_1K_2.

Definition I.1.6. Let K/k be a field extension, and let $\alpha \in K$. We say that α is **algebraic** over k if there exists a **non-zero** polynomial $P \in k[X]$ such that $P(\alpha) = 0$. We say that α is **transcendental** over k otherwise. A field extension K/k is called **algebraic** if every element of K is algebraic over k.

Proposition I.1.7. *Let K/k be a field extension, and let $\alpha \in K$. The set*

$$I_\alpha = \{P \in k[X] \mid P(\alpha) = 0\}$$

is an ideal of $k[X]$. It is a non-zero ideal if and only if α is algebraic over k. In this case, there exists a unique **monic irreducible** *polynomial $\mu_{\alpha,k}$ such that*

$$I_\alpha = (\mu_{\alpha,k}).$$

Definition I.1.8. The polynomial $\mu_{\alpha,k}$ is called the *minimal polynomial of α over k.*

Remark I.1.9. From the definition of the minimal polynomial, it follows that if $P \in k[X]$ satisfies $P(\alpha) = 0$, then $\mu_{\alpha,k} | P$; if moreover P is monic and irreducible then $P = \mu_{\alpha,k}$.

Theorem I.1.10. *Let K/k be a field extension. Then $\alpha \in K$ is algebraic over k if and only if $k(\alpha)/k$ has finite degree. In this case, a k-basis of $k(\alpha)$ is given by $1, \alpha, \dots, \alpha^{d-1}$, where $d = \deg(\mu_{\alpha,k})$. In particular, $k(\alpha) = k[\alpha]$ and we have the equality*

$$[k(\alpha) : k] = \deg(\mu_{\alpha,k}).$$

Remark I.1.11. It follows easily that if $\alpha_1, \ldots, \alpha_n \in K$ are algebraic over k, then $k(\alpha_1, \ldots, \alpha_n)/k$ has finite degree.

Definition I.1.12. We say that a field k is **algebraically closed** if every non-constant polynomial with coefficients in k has a root in k. An **algebraic closure** of a field k is an **algebraic** field extension k_{alg}/k such that k_{alg} is algebraically closed.

One can show that every field k has an algebraic closure, and that two algebraic closures are k-isomorphic.

Definition I.1.13. Let k be a field, and let k_{alg} be a fixed algebraic closure of k. A polynomial $f \in k[X]$ of degree n is **separable** over k if it has n distinct roots in k_{alg}. If K/k is a field extension, we say that $x \in K$ is **separable** over k if x is algebraic over k and its minimal polynomial over k is separable. Finally we say that K/k is **separable** if every element of K is separable over k.

Remark I.1.14. One can show that the compositum of two separable extensions is again separable and that $k(\alpha_1, \ldots, \alpha_n)/k$ is separable if and only if $\alpha_1, \ldots, \alpha_n$ are separable over k.

Definition I.1.15. The **separable closure** of k in k_{alg} is the maximal subfield k_s of k_{alg} such that k_s/k is separable. It is exactly the subfield of elements of k_{alg} xhich are separable over k.

We continue by stating some results on extensions of morphisms which will be useful in the sequel. First, we need a definition.

Definition I.1.16. Let K and K' be two fields, let L/K and L'/K' be two field extensions, and let $\iota : K \longrightarrow K'$ be a ring morphism. We say that a ring morphism $\varphi : L \longrightarrow L'$ is an **extension** of ι if the diagram

$$\begin{array}{ccc} L & \xrightarrow{\varphi} & L' \\ \uparrow & & \uparrow \\ K & \xrightarrow{\iota} & K' \end{array}$$

commutes.

Notation: Let K and K' be two fields, and let $\iota : K \longrightarrow K'$ be a ring morphism. If $P \in K[X]$, $P = a_n X^n + a_{n-1} X^{n-1} + \ldots + a_0$, we denote by $\iota(P)$ the element of $K'[X]$ defined by

$$\iota(P) = \iota(a_n) X^n + \iota(a_{n-1}) X^{n-1} + \ldots + \iota(a_0).$$

Lemma I.1.17. *Let K and K' be two fields, let L/K be a field extension, and let $\iota : K \longrightarrow K'$ be a ring morphism. Finally, let $\alpha \in L$. For every extension $\varphi : L \longrightarrow K'_{alg}$ of ι, $\varphi(\alpha)$ is a root of $\iota(\mu_{\alpha,K})$.*

Proof. Write $\mu_{\alpha,K} = X^n + a_{n-1}X^{n-1} + \ldots + a_0$. Since φ is an extension of ι, we have

$$\iota(\mu_{\alpha,K}) = X^n + \varphi(a_{n-1})X^{n-1} + \ldots + \varphi(a_0).$$

Since φ is a ring morphism, we then get

$$\iota(\mu_{\alpha,K})(\varphi(\alpha)) = \varphi(\mu_{\alpha,K}(\alpha)) = \varphi(0) = 0.$$

Hence $\varphi(\alpha)$ is a root of $\iota(\mu_{\alpha,K})$ as claimed. □

Proposition I.1.18. *Let K, K' be two fields, let $\iota : K \longrightarrow K'$ be a ring morphism, and let $\alpha \in K_{alg}$. If β is a root of $\iota(\mu_{\alpha,K})$ in K'_{alg}, then there exists a unique extension $\varphi : K(\alpha) \longrightarrow K'_{alg}$ of ι such that $\varphi(\alpha) = \beta$. In particular, the set of extensions of ι is in bijection with the set of roots of $\iota(\mu_{\alpha,K})$.*

Theorem I.1.19. *Let K be a field, let L/K be an algebraic field extension. Let E be an algebraically closed field, and let $\tau : K \longrightarrow E$ be a ring morphism. Then there exists a ring morphism $\sigma : L \longrightarrow E$ such that $\sigma_{|K} = \tau$. In other words, there exists an extension $\sigma : L \longrightarrow E$ of τ.*

Corollary I.1.20. *Let K and K' be two fields, and let $\iota : K \longrightarrow K'$ be a ring morphism. Then there exists an extension $\varphi : K_s \longrightarrow K'_s$ of ι.*

Proof. Let τ be the composition of ι with the inclusion $K' \subset K'_{alg}$. By Theorem I.1.19, there exists $\varphi : K_s \longrightarrow K'_{alg}$ such that $\varphi_{|K} = \tau$. Let $\alpha \in K_s$, so that $\mu_{\alpha,K}$ has no multiple roots. Let $\alpha = \alpha_1, \ldots, \alpha_n$ be its distinct roots in K_{alg}. Let $P = \tau(\mu_{\alpha,K})$. Since $\mu_{\alpha,K} = \mu_{\alpha_i,K}$ for all i, Lemma I.1.17 implies that $\varphi(\alpha_1), \ldots, \varphi(\alpha_n)$ are roots of P. Since P and $\mu_{\alpha,K}$ have the same degree and φ is injective, this implies that P is separable. Since $\varphi(\alpha)$ is a root of P, we have $\mu_{\varphi(\alpha),K'}|P$. Hence $\mu_{\varphi(\alpha),K'}$ has no multiple roots and $\varphi(\alpha)$ is separable over K'. In other words, $\varphi(K_s) \subset K'_s$; this concludes the proof. □

§I.2 Galois theory

I.2.1 Definitions and first examples

Definition I.2.1. We say that a field extension Ω/k (contained in k_{alg}) is a **Galois extension** if it is separable and for every k-linear embedding $\sigma : \Omega \longrightarrow k_{alg}$ we have $\sigma(\Omega) = \Omega$ (so σ is a k-automorphism of Ω).

In this case, the group $\mathrm{Gal}(\Omega/k)$ of all k-automorphisms of Ω/k is called the **Galois group** of Ω/k.

Notice that we did not assume Ω/k to be finite in the previous definition.

Example I.2.2. Let $\alpha \in k_{alg}$ be a separable element over k, and let $\Omega = k(\alpha_1, \ldots, \alpha_n)$, where $\alpha_1 = \alpha, \ldots, \alpha_n$ are the n distinct roots of $\mu_{\alpha,k}$ in k_{alg}. Then Ω/k is a finite Galois extension.

Indeed, since the α_i's have the same minimal polynomial, they are all separable over k, and $k(\alpha_1, \ldots, \alpha_n)/k$ is separable by Remark I.1.14. Now let $\sigma : \Omega \longrightarrow k_{alg}$ be a k-embedding. By Lemma I.1.17, $\sigma(\alpha_i)$ is one of the α_j's, and thus is an element of Ω. This proves that $\sigma(\Omega) \subset \Omega$. Since Ω is a finite dimensional k-vector space and $\sigma : \Omega \longrightarrow \Omega$ is k-linear and injective, then σ is bijective and we are done.

Lemma I.2.3. *Let Ω/k be a Galois extension. Let $\alpha \in \Omega$, and let $\alpha_1 = \alpha, \ldots, \alpha_n$ be the roots of $\mu_{\alpha,k}$ in k_{alg}. Then $\alpha_i \in \Omega$ for $i = 1, \ldots, n$.*

Proof. For $i = 1, \ldots, n$, there exists a unique k-embedding $\tau_i : k(\alpha) \longrightarrow k_{alg}$ satisfying $\tau_i(\alpha) = \alpha_i$ by Proposition I.1.18. Each τ_i extends to a k-embedding $\sigma_i : \Omega \longrightarrow k_{alg}$ by Theorem I.1.19. Since Ω/k is a Galois extension, $\sigma_i(\Omega) = \Omega$. In particular, $\sigma_i(\alpha) = \tau_i(\alpha) = \alpha_i \in \Omega$. □

The next lemma is very useful and will be frequently used without further reference.

Lemma I.2.4. *Let Ω/k be a Galois extension, and let $\alpha \in \Omega$. Then there exists a finite Galois subextension of Ω/k containing α.*

Proof. Let $\alpha_1 = \alpha, \ldots, \alpha_n$ be the roots of $\mu_{\alpha,k}$ in k_{alg}. The previous lemma implies that $k(\alpha_1, \ldots, \alpha_n) \subset \Omega$. Since $k(\alpha_1, \ldots, \alpha_n)/k$ is a finite Galois extension containing α by Example I.2.2, we are done. □

We now give an example of an infinite Galois extension.

Lemma I.2.5. *Let k_s be the separable closure of k in a fixed algebraic closure of k. Then the extension k_s/k is Galois.*

Proof. First, k_s/k is separable. Now let $\sigma : k_s \longrightarrow k_{alg}$ be a k-linear embedding. Let $x \in k_s$ and let L/k be a finite Galois extension containing x (which exists by Example I.2.2). Then $\sigma_{|L} : L \longrightarrow k_{alg}$ is a k-embedding of L into k_{alg}. Since L/k is a Galois extension, we have $\sigma(x) \in L$. In particular $\sigma(x)$ is separable, since L/k is separable. Therefore, we have proved that $\sigma(k_s) \subset k_s$. To prove the missing inclusion, let $x' \in k_s$ and let L'/k be a finite Galois extension containing x'. Since L'/k is a Galois extension, $\sigma_{|L'}$ is a k-automorphism of L' . Hence, there exists $x \in L' \subset k_s$ such that $\sigma_{|L'}(x) = x'$. Thus we have $x' = \sigma_{|L'}(x) = \sigma(x)$, and therefore, $\sigma(k_s) = k_s$. □

I.2.2 The Galois correspondence

We would like now to understand better the structure of the Galois group of a Galois extension Ω/k, not necessarily of finite degree over k. In particular, we would like to have a Galois correspondence between subfields of Ω and subgroups of $\mathrm{Gal}(\Omega/k)$ as in the case of finite Galois extensions. Unfortunately, the following example shows that this correspondence does not hold in the infinite case.

Example I.2.6. Let $\Omega = \mathbb{Q}(\sqrt{p}, p \text{ prime })$. Then $\Omega/\mathbb{Q}$ is a Galois extension, as the reader may check. For a prime number p, let σ_p the unique element of $\mathrm{Gal}(\Omega/\mathbb{Q})$ which fixes $\sqrt{p'}$ if $p' \neq p$, and which maps $\sqrt{p}$ onto $-\sqrt{p}$.

Now consider the subgroup H of $\mathrm{Gal}(\Omega/\mathbb{Q})$ generated by the σ_p's. Notice that $H \neq \mathrm{Gal}(\Omega/\mathbb{Q})$ since H does not contain the element $\sigma \in \mathrm{Gal}(\Omega/\mathbb{Q})$ which maps $\sqrt{p}$ onto $-\sqrt{p}$ for all prime numbers p. However, we have

$$\Omega^H = \Omega^{\mathrm{Gal}(\Omega/\mathbb{Q})} = \mathbb{Q}.$$

Indeed, any element $x \in \Omega$ is contained in some subfield E of the form $E = \mathbb{Q}(\sqrt{p_1}, \ldots, \sqrt{p_r})$. Notice that $E/\mathbb{Q}$ is a finite Galois extension. Now assume that $x \in \Omega^H$. Since $\sigma_{p_1}, \ldots, \sigma_{p_r} \in H$, and since they generate $\mathrm{Gal}(E/\mathbb{Q})$, we conclude that $x \in \mathbb{Q}$ by classical Galois theory.

In order to get a Galois correspondence, we define a topology on the Galois group of Ω/k.

Definition I.2.7. Let Ω/k be a Galois extension. The **Krull topology** on Ω/k is the unique topology such that for all $\sigma \in \mathrm{Gal}(\Omega/k)$, the familly of subsets

$$\{\sigma \mathrm{Gal}(\Omega/L) \mid \sigma \in \mathrm{Gal}(\Omega/k), L/k \text{ a finite Galois extension}, L \subset \Omega\}$$

is a basis of open neighbourhoods of σ.

We may now state the fundamental theorem of Galois theory.

Theorem I.2.8 (Fundamental theorem of Galois theory). *Let Ω/k be a Galois extension. Then there exist one-to-one correspondences between the following sets:*

(1) *The set of subfields K of Ω containing k and the set of closed subgroups of* $\mathrm{Gal}(\Omega/k)$.

(2) *The set of subfields K of Ω containing k such that $[K:k] < +\infty$ and the set of open subgroups of* $\mathrm{Gal}(\Omega/k)$.

(3) *The set of subfields K of Ω containing k such that K/k is a finite Galois extension and the set of open normal subgroups of* $\mathrm{Gal}(\Omega/k)$.

In all cases, the correspondence is given by

$$\begin{array}{ccc} K & \longmapsto & \mathrm{Gal}(\Omega/K) \\ \Omega^H & \longleftarrow & H. \end{array}$$

Moreover, if H is an open normal subgroup of $\mathrm{Gal}(\Omega/k)$, *then we have*

$$\mathrm{Gal}(\Omega^H/k) \simeq \mathrm{Gal}(\Omega/k)/H.$$

In particular, for any finite Galois subextension L/k of Ω/k, we have

$$\mathrm{Gal}(\Omega/k)/\mathrm{Gal}(\Omega/L) \simeq \mathrm{Gal}(L/k).$$

All these results and their proofs may be found in [42], Chapter IV. See also [26], Chapter 4.

I.2.3 Morphisms of Galois extensions

To continue this section on Galois theory, we would like to have a closer look at morphisms of Galois extensions.

Proposition I.2.9. *Let K and K' be two fields, let Ω/K and Ω'/K' be two Galois extensions (not necessarily finite), and let $\iota : K \longrightarrow K'$ be a ring morphism. Assume that there exist two extensions $\varphi_i : \Omega \longrightarrow \Omega'$ of $\iota, i = 1, 2$. Then for all $\tau' \in \mathrm{Gal}(\Omega'/K')$, there exists a unique $\tau \in \mathrm{Gal}(\Omega/K)$ such that*

$$\tau' \circ \varphi_1 = \varphi_2 \circ \tau.$$

In particular, there exists $\rho \in \mathrm{Gal}(\Omega/K)$ such that $\varphi_1 = \varphi_2 \circ \rho$.

Proof. Let φ_1, φ_2 as in the statement of the proposition, and let $\tau' \in \mathrm{Gal}(\Omega'/K')$. Let $x \in \Omega$. We have to show that there exists $y \in \Omega$ such that $\tau'(\varphi_1(x)) = \varphi_2(y)$. Notice that if such a y exists, it is unique since φ_2 is injective.

Let $x_1, \ldots, x_n$ be the n distinct roots of $\mu_{x,K}$ in K_{alg}. Since Ω/K is Galois, $x_i \in \Omega$ for $i = 1, \ldots, n$ by Lemma I.2.3. Since φ_1 is an extension of ι and τ' is K'-linear, we easily deduce that

$$\iota(\mu_{x,K})(\tau'(\varphi_1(x))) = \tau'(\varphi_1(\mu_{x,K}(x))) = 0.$$

Thus, $\tau'(\varphi_1(x))$ is a root of $\iota(\mu_{x,K})$ in K'_{alg}. Now since φ_2 is an extension of ι, $\varphi_2(x_i)$ is a root of $\iota(\mu_{x,K})$ by Lemma I.1.17. Since φ_2 is injective, it follows that $\varphi_2(x_1), \ldots, \varphi_2(x_n)$ are the n distinct roots of $\iota(\mu_{x,K})$. Therefore, $\tau'(\varphi_1(x)) = \varphi_2(x_i)$ for some i. We then set $y = x_i$.

Hence we have shown that there is a unique map $\tau : \Omega \longrightarrow \Omega$ such that $\tau' \circ \varphi_1 = \varphi_2 \circ \tau$. We have to check that $\tau \in \mathrm{Gal}(\Omega/K)$. If $x, x' \in \Omega$ and $\lambda \in K$, we have

$$\tau'(\varphi_1(\lambda x + x')) = \lambda \tau'(\varphi_1(x)) + \tau'(\varphi_1(x')) = \lambda \varphi_2(\tau(x)) + \varphi_2(\tau(x')).$$

Since φ_2 is K-linear, we get

$$\tau'(\varphi_1(\lambda x + x')) = \varphi_2(\lambda \tau(x) + \tau(x')).$$

But we also have $\tau'(\varphi_1(\lambda x + x')) = \varphi_2(\tau(\lambda x + x'))$. By injectivity of φ_2, we get

$$\tau(\lambda x + x') = \lambda \tau(x) + \tau(x').$$

Similarly, we can check that $\tau(xx') = \tau(x)\tau(x')$ and $\tau(1) = 1$.

It remains to show that τ is bijective, but this follows immediately from the fact that Ω/K is Galois. The last part of the proposition is an immediate application of the first one. □

Corollary I.2.10. *Let K and K' be two fields, let Ω/K and Ω'/K' be two Galois extensions, and let $\iota : K \longrightarrow K'$ be a ring morphism. Let $\varphi : \Omega \longrightarrow \Omega'$ be an extension of ι. For all $\tau' \in \mathrm{Gal}(\Omega'/K')$, let $\overline{\varphi}(\tau')$ be the unique element of $\mathrm{Gal}(\Omega/K)$ such that*

$$\tau' \circ \varphi = \varphi \circ \overline{\varphi}(\tau').$$

Then the map $\overline{\varphi} : \mathrm{Gal}(\Omega'/K') \longrightarrow \mathrm{Gal}(\Omega/K)$ is a continuous group morphism. Moreover, if φ' is another extension of ι, then there exists $\rho \in \mathrm{Gal}(\Omega/K)$ such that $\varphi = \varphi' \circ \rho$, and we have

$$\overline{\varphi}' = \mathrm{Int}(\rho) \circ \overline{\varphi}.$$

Proof. Let $\tau_1', \tau_2' \in \mathrm{Gal}(\Omega'/K)$. By definition of $\overline{\varphi}$, we have

$$(\tau_1' \circ \tau_2') \circ \varphi = \tau_1' \circ \varphi \circ (\overline{\varphi}(\tau_2')) = \varphi \circ (\overline{\varphi}(\tau_1') \circ \overline{\varphi}(\tau_2')).$$

Since $\overline{\varphi}(\tau_1' \circ \tau_2')$ is the unique element of $\mathrm{Gal}(\Omega/K)$ satisfying

$$(\tau_1' \circ \tau_2') \circ \varphi = \varphi \circ \overline{\varphi}(\tau_1' \circ \tau_2'),$$

we have $\overline{\varphi}(\tau_1' \circ \tau_2') = \overline{\varphi}(\tau_1') \circ \overline{\varphi}(\tau_2')$. This proves that $\overline{\varphi}$ is a group morphism. The continuity is left to the reader. Now, if φ' is another extension of ι, then by the previous proposition, there exists $\rho \in \mathrm{Gal}(\Omega/K)$ such that $\varphi = \varphi' \circ \rho$. Therefore, for every $\tau' \in \mathrm{Gal}(\Omega'/K')$, we have

$$\tau' \circ \varphi' = (\tau' \circ \varphi) \circ \rho^{-1} = \varphi \circ (\overline{\varphi}(\tau') \circ \rho^{-1}) = \varphi' \circ (\rho \circ \overline{\varphi}(\tau') \circ \rho^{-1}).$$

We conclude as before. □

I.2.4 The Galois group as a profinite group

Let Ω/k be a Galois extension. Lemma I.2.4 shows in particular that an element $\sigma \in \mathrm{Gal}(\Omega/k)$ is completely determined by its restrictions to finite Galois subextensions L/k of Ω/k. Intuitively, $\mathrm{Gal}(\Omega/k)$ then should be completely determined by the finite groups $\mathrm{Gal}(L/k)$. This is indeed the case, and in order to make this statement more precise, we need to introduce the concept of an inverse limit.

Definition I.2.11. A **directed set** is a partially ordered set $(I, \leq)$, such that for all $i, j \in I$, there exists $k \in I$ such that $i \leq k$ and $j \leq k$.

Examples I.2.12. The reader will easily convince himself that the following sets are examples of directed sets:

(1) The set $\mathbb{N}$ with the order relation $\leq$.

(2) The set $\mathbb{N}^* = \{1, 2, \ldots\}$ with the divisibility relation.

Definition I.2.13. A **projective system** of sets (groups, rings, etc) is a family of sets (groups, rings, etc) $(X_i)_{i \in I}$, indexed by a directed set I, together with maps (resp. group morphisms, ring morphisms, etc) $\pi_{ij} : X_j \longrightarrow X_i$ for any $i, j \in I, i \leq j$, satisfying the following properties:

(1) $\pi_{ii} = \mathrm{Id}_{X_i}$ for all $i \in I$.

(2) For all $i, j, k \in I, i \leq j \leq k$, we have $\pi_{ij} \circ \pi_{jk} = \pi_{ik}$.

Examples I.2.14. We now give some examples of projective systems indexed by the directed sets introduced above:

(1) Let p be a fixed prime number. For any $n \in \mathbb{N}$, let $X_n = \mathbb{Z}/p^n\mathbb{Z}$, and let $\pi_{mn} : \mathbb{Z}/p^n\mathbb{Z} \longrightarrow \mathbb{Z}/p^m\mathbb{Z}$ be the natural projection for $m \leq n$. Then we obtain a projective system of rings.

(2) For any $n \in \mathbb{N}^*$, let $X_n = \mathbb{Z}/n\mathbb{Z}$, and let $\pi_{mn} : \mathbb{Z}/n\mathbb{Z} \longrightarrow \mathbb{Z}/m\mathbb{Z}$ be again the natural projection for all $m|n$. Then we obtain once again a projective system of rings.

Definition I.2.15. If $((X_i)_{i\in I}, (\pi_{ij}))$ is a projective system of sets (groups, rings, etc), the **inverse limit** $\varprojlim_{i\in I} X_i$ is the subset (subgroup, subring, etc)

$$\varprojlim_{i\in I} X_i = \left\{ (x_i)_{i\in I} \in \prod_{i\in I} X_i \mid \pi_{ij}(x_j) = x_i \text{ for all } i \leq j \right\}.$$

Recall now a definition from topology.

Definition I.2.16. Let $(X_i)_{i\in I}$ be a family of topological spaces. The **product topology** on $\prod_{i\in I} X_i$ is the unique topology such that a basis of open neighbourboods of $(x_i)_{i\in I}$ consists of the subsets $\prod_{i\in I} U_i$, where $U_i \subset X_i$ is an open neighbourhood of x_i and $U_i = X_i$ for all but finitely many $i \in I$.

If $((X_i)_{i\in I}, (\pi_{ij}))$ is a projective system of topological spaces (groups, rings, etc), then the inverse limit is also a topological space (group, ring, etc) with respect to the topology induced by the product topology. In particular, if each X_i is finite, it may be endowed with the discrete topology, and in this case we get a natural structure of a topological space on the inverse limit $\varprojlim_{i\in I} X_i$.

Examples I.2.17. The projective systems introduced previously allow us to define two topological rings by taking the corresponding inverse limits:

(1) Let p be a fixed prime number. The topological ring

$$\mathbb{Z}_p = \varprojlim_{n\in\mathbb{N}} \mathbb{Z}/p^n\mathbb{Z}$$

is called the ring of p-adic integers. One can show that the ring $\mathbb{Z}_p$ defined above is homeomorphic to the completion of $\mathbb{Z}$ with respect to the p-adic valuation as a topological ring.

(2) The topological ring

$$\hat{\mathbb{Z}} = \varprojlim_{n \in \mathbb{N}^*} \mathbb{Z}/n\mathbb{Z}$$

is called the profinite completion of $\mathbb{Z}$.

Both of them play an important role in number theory.

We can now elucidate the structure of the Galois group as a topological group.

If k is a field, the set of all finite Galois subextensions of Ω/k with the partial order relation '$\subset$' is a directed set, since the compositum of two finite Galois extensions is a finite Galois extension. Moreover, for any finite Galois extension L/k, let $X_L = \mathrm{Gal}(L/k)$, and for any $L/k, L'/k$ such that $L \subset L'$, let $\pi_{L,L'}$ be the group morphism defined by

$$\pi_{L,L'} : \begin{array}{rcl} \mathrm{Gal}(L'/k) & \longrightarrow & \mathrm{Gal}(L/k) \\ \sigma & \longmapsto & \sigma_{|L}. \end{array}$$

We obtain in this way a projective system of groups. Therefore, the following statement makes sense:

Theorem I.2.18. *Let Ω/k be a Galois extension. Then we have an isomorphism of topological groups*

$$\mathrm{Gal}(\Omega/k) \simeq \varprojlim_{L} \mathrm{Gal}(L/k),$$

where L/k runs over all finite Galois subextensions of Ω/k.

Proof. Let us consider the map

$$\Theta : \begin{array}{rcl} \mathrm{Gal}(\Omega/k) & \longrightarrow & \varprojlim_{L} \mathrm{Gal}(L/k) \\ \sigma & \longmapsto & (\sigma_{|L})_L. \end{array}$$

This is clearly an abstract group morphism. Now assume that $\sigma \in \ker(\Theta)$, and let $x \in \Omega$. Pick any finite Galois subextension L/k of Ω/k containing x. By assumption $\sigma_{|L}$ is the identity, so we get $\sigma(x) = \sigma_{|L}(x) = x$. Hence $\sigma = \mathrm{Id}_\Omega$ and Θ is injective. Now let $(\sigma^{(L)})_L \in \varprojlim_{L} \mathrm{Gal}(L/k)$. For any $x \in \Omega$, we set $\sigma(x) = \sigma^{(L)}(x)$, where L/k is a finite Galois subextension of Ω/k containing x. We claim that the result does not depend on the choice of L. Indeed, assume that L_1, L_2 are two finite Galois subextensions of Ω/k containing x. Then $L = L_1L_2$ is a

finite Galois subextension of Ω/k containing x, and $L_i \subset L$. Therefore, by assumption, we have $\sigma^{(L)}(x) = (\sigma^{(L)})_{|L_i}(x) = \sigma^{(L_i)}(x)$.
Now it is clear that $\sigma \in \mathrm{Gal}(\Omega/k)$, and that it is a preimage of $(\sigma^{(L)})_L$ by Θ. Hence Θ is surjective as well. It is easy to check that Θ is bicontinuous (it follows from the definition of the Krull topology), so we are done. □

Definition I.2.19. A topological group Γ is **profinite** if it is isomorphic as a topological group to an inverse limit of finite groups (each of them being endowed with the discrete topology). In view of the previous result, the Galois group of an arbitrary Galois extension is profinite.

We now list some properties of profinite groups without proof. We refer the reader to [12] for more details.

A profinite group Γ is compact and totally disconnected (that is the only non-empty connected subsets are one-point subsets). In particular, one-point subsets are closed, every open subgroup is also closed and has finite index. Moreover, every neighbourhood of 1 contains an open normal subgroup (hence of finite index). If Γ' is a closed subgroup of Γ, then Γ' is profinite, and if moreover Γ' is normal, so is Γ/Γ'. Finally, if $\mathcal{N}$ denotes the set of **open** normal subgroups of Γ, the map

$$\theta : \begin{array}{ccc} \Gamma & \longrightarrow & \varprojlim_{U \in \mathcal{N}} \Gamma/U \\ g & \longmapsto & (gU)_U \end{array}$$

is an isomorphism of topological groups.

Exercises

1. Show that a finite group (endowed with the discrete topology) is profinite.

2. Show that the algebraic definition of $\mathbb{Z}_p$ given in this chapter coincides with the classical analytic definition.

3. Prove that there exists an isomorphism of topological groups

$$\hat{\mathbb{Z}} \simeq \prod_p \mathbb{Z}_p,$$

where p runs over the set of prime numbers.

4. Let $q = p^r, r \geq 1$, where p is a prime number, and let $k = \mathbb{F}_q$ the finite field with q elements. Show that $\mathrm{Gal}(k_s/k) \simeq \hat{\mathbb{Z}}$.

5. Let $\mathbb{Q}_{ab} \subset \mathbb{C}$ be the maximal abelian subextension of $\mathbb{Q}$. Determine $\mathrm{Gal}(\mathbb{Q}_{ab}/\mathbb{Q})$.

   Hint: Use the fact that every finite abelian extension of $\mathbb{Q}$ is contained in a cyclotomic extension (Kronecker-Weber's theorem).

II
Cohomology of profinite groups

In this chapter, we define the cohomology sets associated to a profinite group Γ, and establish fundamental properties which will be crucial when studying Galois cohomology. From now on, if Ω/k is a Galois extension, we will write $\mathcal{G}_\Omega$ for its Galois group whenever k is clear from the context.

§II.3 Cohomology sets: basic properties

II.3.1 Definitions

In the introduction, we 'solved' the descent problem for conjugacy classes of matrices associated to a finite Galois extension Ω/k of Galois group $\mathcal{G}_\Omega$. We would like now to investigate the case where Ω/k is an infinite Galois extension. The main idea is that the problem locally boils down to the previous case. Let us fix $M_0 \in \mathrm{M}_n(k)$ and let us consider a specific matrix $M \in \mathrm{M}_n(k)$ such that

$$QMQ^{-1} = M_0 \text{ for some } Q \in \mathrm{SL}_n(\Omega).$$

If L/k is any finite Galois subextension of Ω/k with Galois group $\mathcal{G}_L$ containing all the entries of Q, then $Q \in \mathrm{SL}_n(L)$ and the equality above may be read in $\mathrm{M}_n(L)$. Therefore, for this particular matrix M, the descent problem may be solved by examining the corresponding element in $H^1(\mathcal{G}_L, Z_{\mathrm{SL}_n}(M_0)(L))$. Now if we take another finite Galois subextension L'/k such that $M \in \mathrm{M}_n(L')$, we obtain an obstruction living in $H^1(\mathcal{G}_{L'}, Z_{\mathrm{SL}_n}(M_0)(L'))$. But the fact that M is conjugate or not to M_0 by an element of $\mathrm{SL}_n(k)$ is an intrinsic property of M and of the field k, and should certainly not depend on the chosen Galois extension L/k. Therefore, we need to find a way to patch these local obstructions together.

If we try to imitate the method followed in the introduction, we need first an appropriate action of $\mathcal{G}_\Omega$ on $\mathrm{M}_n(\Omega)$ and $\mathrm{SL}_n(\Omega)$. Since we want to patch together the local obstructions, we need this action to coincide with the local actions on the various sets $\mathrm{M}_n(L)$ and $\mathrm{SL}_n(L)$. Setting

$$\sigma\cdot(m_{ij}) = (\sigma(m_{ij}))$$

gives rise to an action on $\mathrm{M}_n(\Omega)$ and $\mathrm{SL}(\Omega)$ which satisfies the desired properties. Indeed, if $M \in \mathrm{M}_n(\Omega)$ (or $\mathrm{SL}_n(\Omega)$) and if L/k is a finite Galois extension containing all the entries of M, then the action of σ on M is nothing but the action of $\sigma_{|_L}$ on M, when M is viewed as an element of L.

The reason why this works here is that $\mathcal{G}_\Omega$ is a profinite group, isomorphic to the projective limit of the finite groups $\mathcal{G}_L$. In particular, an element $\sigma \in \mathcal{G}_\Omega$ is completely determined by its restrictions to finite Galois subextensions. Moreover, and maybe more importantly, the stabilizer of a given matrix $M \in \mathrm{M}_n(\Omega)$ for the action of $\mathcal{G}_\Omega$ is equal to the open subgroup $\mathrm{Gal}(\Omega/K)$, where K is the subfield of Ω generated by the entries of M. Indeed, $\sigma \in \mathcal{G}_\Omega$ will act trivially on M if and only if it acts trivially on the entries of M, that is if it restricts to the identity on K. Therefore, this stabilizer contains an open normal subgroup $\mathrm{Gal}(\Omega/L)$ (where L is a finite Galois subextension containing the entries of M), and consequently the action of $\mathcal{G}_\Omega$ induces an action on $\mathcal{G}_\Omega/\mathrm{Gal}(\Omega/L)$, and thus on $\mathcal{G}_L$ since these two groups are isomorphic by Theorem I.2.8.

These considerations generalize to arbitrary profinite groups, and lead to the following definitions:

Definition II.3.1. Let Γ be a profinite group. A left action of Γ on a discrete topological space A is called **continuous** if for all $a \in A$, the set

$$\mathrm{Stab}_\Gamma(a) = \{\sigma \in \Gamma \mid \sigma\cdot a = a\}$$

is an open subgroup of Γ.

One can show that this is equivalent to ask for the map

$$\begin{aligned} \Gamma \times A &\longrightarrow A \\ (\sigma, a) &\longmapsto \sigma\cdot a \end{aligned}$$

to be continuous.

Discrete topological spaces with a continuous left action of Γ are called

Γ**-sets**. A group A which is also a Γ-set is called a Γ**-group** if Γ acts by group morphisms, i.e.

$$\sigma\cdot(a_1a_2) = (\sigma\cdot a_1)(\sigma\cdot a_2) \text{ for } \sigma \in \Gamma, a_1, a_2 \in A.$$

A Γ-group which is commutative is called a Γ**-module**.

A **morphism** of Γ-sets (resp. Γ-groups, Γ-modules) is a map (resp. a group morphism) $f : A \longrightarrow A'$ satisfying the following property:

$$f(\sigma\cdot a) = \sigma\cdot f(a) \text{ for all } \sigma \in \Gamma \text{ and all } a \in A.$$

Examples II.3.2.

(1) Assume that Γ is a finite group. Then any discrete topological set A on which Γ acts on the left is a Γ-set. Indeed such an action is continuous since any finite set is open for the discrete topology.

(2) Any discrete topological set A on which Γ acts trivially is a Γ-set.

(3) Let Ω/k be a Galois extension of group $\mathcal{G}_\Omega$. Then the map

$$\begin{aligned} \mathcal{G}_\Omega \times \Omega &\longrightarrow \Omega \\ (\sigma, x) &\longmapsto \sigma(x) \end{aligned}$$

endows Ω with the structure of a $\mathcal{G}_\Omega$-module.

(4) Let V be a k-vector space, and let us denote by V_Ω the tensor product $V \otimes_k \Omega$. Then the action of $\mathcal{G}_\Omega$ on V_Ω defined on elementary tensors by

$$\sigma\cdot(v \otimes \lambda) = v \otimes (\sigma\cdot\lambda) \text{ for all } v \in V, \lambda \in \Omega$$

is continuous, and therefore endows V_Ω with the structure of a $\mathcal{G}_\Omega$-module.

(5) Let V and W be two k-vector spaces of dimension n and m respectively. If Ω/k is a Galois extension of group $\mathcal{G}_\Omega$, then $\mathcal{G}_\Omega$ acts on $\mathrm{Hom}_\Omega(V_\Omega, W_\Omega)$ as follows: for all $\sigma \in \mathcal{G}_\Omega$ and $f \in \mathrm{Hom}_\Omega(V_\Omega, W_\Omega)$, set

$$(\sigma\cdot f)(x) = \sigma\cdot(f(\sigma^{-1}\cdot v)) \text{ for all } x \in V_\Omega.$$

The choice of bases induces an isomorphism $\mathrm{Hom}_\Omega(V_\Omega, W_\Omega) \simeq \mathrm{M}_{m\times n}(\Omega)$, and the corresponding action of $\mathcal{G}_\Omega$ on $\mathrm{M}_{m\times n}(\Omega)$ is simply the action entrywise. Therefore, the action defined above is continuous. This remains true if we replace V and W by finite dimensional k-algebras, and if we consider morphisms of Ω-algebras.

(6) The action in the previous example also induces an action of $\mathcal{G}_\Omega$ on $\mathbf{GL}(V_\Omega)$. One may check easily that it is an action by group automorphisms, so that $\mathrm{GL}(V_\Omega)$ is a $\mathcal{G}_\Omega$-group. In particular, $\mathrm{GL}_n(\Omega)$ is a $\mathcal{G}_\Omega$-group. The same is true for other matrix groups such as $\mathrm{SL}_n(\Omega)$ or $\mathrm{O}_n(\Omega)$.

(7) Let $\mu_n(\Omega)$ be the group of nth roots of 1 in Ω. Then $\mu_n(\Omega)$ is a $\mathcal{G}_\Omega$-module.

As already observed, a matrix $M \in \mathrm{M}_n(\Omega)$ may be viewed as an element of $\mathrm{M}_n(L)$ for a suitable finite Galois subextension L/k of Ω/k. In other words,

$$\mathrm{M}_n(\Omega) = \bigcup_{L/k} \mathrm{M}_n(L) = \bigcup_{L/k} \mathrm{M}_n(\Omega)^{\mathrm{Gal}(\Omega/L)} = \bigcup_{U \in \mathcal{N}} \mathrm{M}_n(\Omega)^U,$$

where $\mathcal{N}$ is the subset of open normal subgroups of $\mathcal{G}_\Omega$.

This equality is not specific to $\mathrm{M}_n(\Omega)$. In fact, it characterizes more generally Γ-sets, where Γ is a profinite group.

Lemma II.3.3. *Let Γ be a profinite group, and let A be a discrete topological set on which Γ acts on the left. Then the action of Γ on A is continuous if and only if we have*

$$A = \bigcup_{U \in \mathcal{N}} A^U,$$

where $\mathcal{N}$ denotes the set of open normal subgroups of Γ.

Proof. Assume that the action of Γ is continuous, and let $a \in A$. Then $\mathrm{Stab}_\Gamma(a)$ is an open subgroup, which contains 1. Hence, it contains some $U \in \mathcal{N}$. In particular, $a \in A^U$. It follows that we have

$$A = \bigcup_{U \in \mathcal{N}} A^U.$$

Conversely, assume that the equality above holds, and let $a \in A$. By assumption, there exists $U \in \mathcal{N}$ such that $a \in A^U$. Therefore, for all $\sigma \in U$, we have $\sigma \cdot a = a$. If now $\tau \in \mathrm{Stab}_\Gamma(a)$, then

$$\tau\sigma \cdot a = \tau \cdot a = a \text{ for all } \sigma \in U.$$

Thus $\bigcup_{\tau \in \mathrm{Stab}_\Gamma(a)} \tau U \subset \mathrm{Stab}_\Gamma(a)$. Since $1 \in U$, the other inclusion holds as well, and we get

$$\mathrm{Stab}_\Gamma(a) = \bigcup_{\tau \in \mathrm{Stab}_\Gamma(a)} \tau U.$$

It follows that $\mathrm{Stab}_\Gamma(a)$ is open. This concludes the proof. □

At this point, we may define the 0th-cohomology set $H^0(\Gamma, A)$.

Definition II.3.4. For any Γ-set A, we set

$$H^0(\Gamma, A) = A^\Gamma.$$

If A is a Γ-group, this is a subgroup of A. The set $H^0(\Gamma, A)$ is called the **0th cohomology set of Γ with coefficients in A.**

Remark II.3.5. We will use this notation only episodically in this book, and will prefer the notation A^Γ.

We would like now to define the main object of this chapter, namely the first cohomology set $H^1(\Gamma, A)$. We first need an appropriate definition of a 1-cocycle. Let us go back to our conjugacy problem. Now that we have a suitable action of $\mathcal{G}_\Omega$ on $\mathrm{SL}_n(\Omega)$, we can mimick the reasoning made in the introduction and obtain a map

$$\alpha^Q : \begin{array}{rcl} \mathcal{G}_\Omega & \longrightarrow & Z_{\mathbf{SL}_n}(M_0)(\Omega) \\ \sigma & \longmapsto & Q(\sigma\cdot Q)^{-1} \end{array}$$

which measures the obstruction to the conjugacy problem for the pair of matrices M and M_0. This map satisfies the cocycle condition stated in the introduction, but has the extra property to contain all the local obstructions we wanted to patch together. Indeed, let L/k be a finite Galois subextension of Ω/k containing all the entries of Q. As already observed at the very beginning of this chapter, we have $\sigma\cdot Q = \sigma_{|L}\cdot Q$, so $\alpha^Q_\sigma \in Z_{\mathbf{SL}_n}(M_0)(L)$. Moreover let $\sigma, \sigma' \in \mathrm{Gal}(\Omega/L)$, and assume that $\sigma' = \sigma\tau$ for some $\tau \in \mathrm{Gal}(\Omega/L)$. Since L contains all the entries of Q, we have $\tau\cdot Q = Q$ and thus $\alpha^Q_{\sigma'} = \alpha^Q_\sigma$. Taking into account that we have a group isomorphism

$$\mathcal{G}_\Omega/\mathrm{Gal}(\Omega/L) \simeq \mathcal{G}_L$$

induced by restriction to L, we see that the map α^Q factors through a map

$$\begin{array}{rcl} \mathcal{G}_L & \longrightarrow & Z_{\mathbf{SL}_n}(M_0)(L) \\ \sigma & \longmapsto & Q(\sigma\cdot Q)^{-1} \end{array}$$

which is the local obstruction obtained when considering Q as an element of $\mathrm{SL}_n(L)$.

The crucial point here is that for all $\sigma \in \mathcal{G}_\Omega$, α^Q is constant on an open neighbourhood of σ. This is equivalent to say that α^Q is a continuous map, as the next proposition shows:

Proposition II.3.6. *Let Γ be a profinite group, let A be a Γ-set and let $n \geq 1$ be an integer. For any map $\alpha : \Gamma^n \longrightarrow A$, the following conditions are equivalent:*

(1) *α is continuous*

(2) *α is locally constant, that is for every $\mathbf{s} = (\sigma_1, \ldots, \sigma_n) \in \Gamma^n$, there exists an open neighbourhood of $\mathbf{s}$ on which α is constant*

(3) *There exist $U \in \mathcal{N}$ and a map $\alpha^{(U)} : (\Gamma/U)^n \longrightarrow A^U$ such that*

$$\alpha_{\sigma_1,\ldots,\sigma_n} = \alpha^{(U)}_{\overline{\sigma}_1,\ldots,\overline{\sigma}_n} \text{ for all } \sigma_1, \ldots, \sigma_n \in \Gamma.$$

Proof. For every $\mathbf{s} = (\sigma_1, \ldots, \sigma_n) \in \Gamma^n$, we will write $\alpha_{\mathbf{s}}$ instead of $\alpha_{\sigma_1,\ldots,\sigma_n}$.

(1) $\Rightarrow$ (2) Assume that α is continuous, and let $\mathbf{s} = (\sigma_1, \ldots, \sigma_n) \in \Gamma^n$. Then the set $U_{\mathbf{s}} = \alpha^{-1}(\{\alpha_{\mathbf{s}}\})$ is an open neighbourhood of $\mathbf{s}$, since $\{\alpha_{\mathbf{s}}\}$ is open in A and α is continuous. By definition, α is constant on $U_{\mathbf{s}}$.

(2) $\Rightarrow$ (3) Assume that α is locally constant. For all $\mathbf{s} = (\sigma_1, \ldots, \sigma_n) \in \Gamma^n$, let $U_{\mathbf{s}}$ be an open neighbourhood of $\mathbf{s}$ on which α is constant. By definition of the product topology, one may assume that

$$U_{\mathbf{s}} = V_{\mathbf{s}}^{(1)} \times \cdots \times V_{\mathbf{s}}^{(n)},$$

where $V_{\mathbf{s}}^{(i)}$ is an open neighbourhood of σ_i in Γ. The family $(U_{\mathbf{s}})_{\mathbf{s} \in \Gamma^n}$ is an open covering of Γ^n. Since Γ is compact, so is Γ^n, and thus there exists a finite subset T of Γ^n such that

$$\Gamma^n = \bigcup_{\mathbf{t} \in T} U_{\mathbf{t}}.$$

For every $\mathbf{t} = (\tau_1, \ldots, \tau_n) \in T$, notice that $U_{\mathbf{t}}^{(i)} = \tau_i^{-1} V_{\mathbf{t}}^{(i)}$ is an open neighbourhood of 1, and that we have

$$U_{\mathbf{t}} = \tau_1 U_{\mathbf{t}}^{(1)} \times \cdots \times \tau_n U_{\mathbf{t}}^{(n)}.$$

By Lemma II.3.3, for every $\mathbf{t} \in T$, there exists an open subgroup $U'_{\mathbf{t}}$ of Γ such that $\alpha_{\mathbf{t}} \in A^{U'_{\mathbf{t}}}$. Since T is finite, the set

$$U_0 = \bigcap_{i,\mathbf{t}} U_{\mathbf{t}}^{(i)} \cap \bigcap_{\mathbf{t}} U'_{\mathbf{t}}$$

is an open neighbourhood of 1 in Γ. Since Γ is a profinite group, there exists a normal open subgroup U of Γ contained in U_0. Now let $\mathbf{s} = (\sigma_1, \ldots, \sigma_n) \in \Gamma^n$. By choice of T, there exists $\mathbf{t} \in T$ such that $\mathbf{s} \in U_{\mathbf{t}}$, so for $i = 1, \ldots, n$ we may write $\sigma_i = \tau_i u_i$ for some $u_i \in U_{\mathbf{t}}^{(i)}$. We then get $\sigma_1 U \times \cdots \times \sigma_n U \subset U_{\mathbf{t}}$, since $U \subset U_{\mathbf{t}}^{(i)}$ for all i and $U_{\mathbf{t}}^{(i)}$ is a subgroup. Hence for all $\mathbf{s}' \in \sigma_1 U \times \cdots \times \sigma_n U$, we have

$$\alpha_{\mathbf{s}'} = \alpha_{\mathbf{t}} \in A^{U'_{\mathbf{t}}},$$

since α is constant on $U_{\mathbf{t}}$. Moreover, since $U \subset U_0 \subset U'_{\mathbf{t}}$, we have $A^{U'_{\mathbf{t}}} \subset A^U$. Therefore, the map

$$\alpha^{(U)} : \begin{array}{rcl} (\Gamma/U)^n & \longrightarrow & A^U \\ (\overline{\sigma}_1, \ldots, \overline{\sigma}_n) & \longmapsto & \alpha_{\sigma_1, \ldots, \sigma_n} \end{array}$$

is well-defined, and satisfies the required conditions.

(3) $\Rightarrow$ (1) Assume that we have an open normal subgroup U and a map $\alpha^{(U)} : (\Gamma/U)^n \longrightarrow A^U$ satisfying (3). Let V be an open subset of A. We have to prove that $\alpha^{-1}(V)$ is open in Γ. Since $\alpha^{-1}(V) = \bigcup_{v \in V} \alpha^{-1}(\{v\})$, it is enough to show that $\alpha^{-1}(\{v\})$ is open for every $v \in V$. If v does not lie in the image of α, this is obvious, so we can assume that $\alpha^{-1}(\{v\})$ is not empty. Let $\mathbf{s} = (\sigma_1, \ldots, \sigma_n) \in \alpha^{-1}(\{v\})$. The assumption implies that for all $\mathbf{t} \in \sigma_1 U \times \cdots \times \sigma_n U$, we have

$$\alpha_{\mathbf{t}} = \alpha^{(U)}_{\overline{\sigma}_1, \ldots, \overline{\sigma}_n} = \alpha_{\mathbf{s}} = v.$$

Thus $\alpha^{-1}(\{v\})$, contains an open neighbourhood of $\mathbf{s}$. Hence $\alpha^{-1}(\{v\})$ is open, and we are done. □

Taking the previous observations into consideration, it is natural to set the following definition:

Definition II.3.7. Let A be a Γ-group. A **1-cocycle** of Γ with values in A is a **continuous** map $\alpha : \Gamma \longrightarrow A$ such that

$$\alpha_{\sigma\tau} = \alpha_\sigma \, \sigma{\cdot}\alpha_\tau \text{ for } \sigma, \tau \in \Gamma.$$

We denote by $Z^1(\Gamma, A)$ the set of all 1-cocycles of Γ with values in A. The constant map

$$\begin{array}{c} \Gamma \longrightarrow A \\ \sigma \longmapsto 1 \end{array}$$

is an element of $Z^1(\Gamma, A)$, which is called the **trivial** 1-cocycle. Notice also that for any 1-cocycle α, we have $\alpha_1 = 1$.

Remark II.3.8. If Γ acts trivially on A, a 1-cocycle is just a continuous morphism $\alpha : \Gamma \longrightarrow A$.

In order to define the cohomology set $H^1(\Gamma, A)$, we need now an appropriate notion of cohomologous cocycles, which coincides with the one defined in the introduction in a particular case. This will be provided by the following lemma:

Lemma II.3.9. *Let Γ be a profinite group, let A be a Γ-group and let $\alpha : \Gamma \longrightarrow A$ be a 1-cocycle. Then for all $a \in A$, the map*

$$\alpha' : \begin{array}{l} \Gamma \longrightarrow A \\ \sigma \longmapsto a\alpha_\sigma\, \sigma \cdot a^{-1} \end{array}$$

is again a 1-cocycle.

Proof. Let $\sigma, \tau \in \Gamma$. We have

$$\alpha'_\sigma\, \sigma \cdot \alpha'_\tau = (a\alpha_\sigma\, \sigma \cdot a^{-1})\sigma \cdot (a\alpha_\tau\, \tau \cdot a^{-1}).$$

Since Γ acts on A by group automorphisms, we get

$$\alpha'_\sigma\, \sigma \cdot \alpha'_\tau = a\alpha_\sigma(\sigma \cdot \alpha_\tau)\sigma\tau \cdot a^{-1} = a\alpha_{\sigma\tau}\, \sigma\tau \cdot a^{-1} = \alpha'_{\sigma\tau}.$$

It remains to prove that α' is continuous. Let V be an open subset of A. We have to prove that $\alpha'^{-1}(V)$ is open in Γ. Since $\alpha'^{-1}(V) = \bigcup_{v \in V} \alpha'^{-1}(\{v\})$, it is enough to show that $\alpha'^{-1}(\{v\})$ is open for every $v \in V$. If v does not lie in the image of α', this is obvious, so we can assume that $\alpha'^{-1}(\{v\})$ is not empty. Let $\sigma \in \alpha'^{-1}(\{v\})$. We then have $\alpha'_\sigma = v$. By assumption, α is continuous. Since $\{1\}$ is open in A, $\alpha^{-1}(\{1\})$ is an open subgroup of Γ. Moreover, $\mathrm{Stab}_\Gamma(a)$ is open since Γ acts continuously on A. Therefore, $U = \alpha^{-1}(\{1\}) \cap \mathrm{Stab}_\Gamma(a)$ is open in Γ, and so is σU. Now for every $\tau \in U$, we have

$$\begin{aligned} \alpha'_{\sigma\tau} &= a\alpha_{\sigma\tau}\, \sigma\tau \cdot a^{-1} \\ &= a\alpha_\sigma\, \sigma \cdot \alpha_\tau\, \sigma\tau \cdot a^{-1} \\ &= a\alpha_\sigma\, \sigma\tau \cdot a^{-1} \\ &= a\alpha_\sigma\, \sigma \cdot a^{-1} \\ &= \alpha'_\sigma \\ &= v. \end{aligned}$$

Therefore for all $\sigma \in \alpha'^{-1}(\{v\})$, $\alpha'^{-1}(\{v\})$ contains an open neighbourhood of σ and thus $\alpha'^{-1}(\{v\})$ is open. This concludes the proof. $\square$

This leads to the following definition:

Definition II.3.10. Two 1-cocycles α, α' are said to be **cohomologous** if there exists $a \in A$ satisfying

$$\alpha'_\sigma = a\alpha_\sigma \, \sigma \cdot a^{-1} \text{ for all } \sigma \in \Gamma.$$

It is denoted by $\alpha \sim \alpha'$.

Remark II.3.11. The symbol '$\sigma \cdot a^{-1}$' may seem ambiguous at first sight, since it could denote $(\sigma \cdot a)^{-1}$ as well as $\sigma \cdot (a^{-1})$. However, these two elements are equal in our setting, since Γ acts on A by group automorphisms; we will keep this notation throughout.

Definition II.3.12. Let Γ be a profinite group, and let A be a Γ-group. The relation '$\sim$' is easily checked to be an equivalence relation on $Z^1(\Gamma, A)$. We denote by $H^1(\Gamma, A)$ the quotient set

$$H^1(\Gamma, A) = Z^1(\Gamma, A)/_\sim.$$

It is called **the first cohomology set of Γ with coefficients in A** .

The set $H^1(\Gamma, A)$ is not a group in general. However, it has a special element which is the class of the trivial cocycle. Therefore, $H^1(\Gamma, A)$ is a pointed set in the following sense:

Definition II.3.13. A **pointed set** is a pair (E, x), where E is a non-empty set and $x \in E$. The element x is called the **base point**. A **map of pointed sets** $f : (E, x) \longrightarrow (F, y)$ is a set-theoretic map such that $f(x) = y$. We will often forget to specify the base point when it is clear from the context.

Example II.3.14. The set $H^1(\Gamma, A)$ is a pointed set, and any abstract group G may be considered as a pointed set, whose base point is the neutral element.

Remark II.3.15. If A is a Γ-module, the set $Z^1(\Gamma, A)$ is an abelian group for the pointwise multiplication of functions. This operation is compatible with the equivalence relation, hence it induces an abelian group structure on $H^1(\Gamma, A)$.

We would like now to define higher cohomology groups.

Let A be a Γ-module (written additively here) and $n \geq 0$. We set $C^0(\Gamma, A) = A$, and if $n \geq 1$, we denote by $C^n(\Gamma, A)$ the set of all **continuous** maps from Γ^n to A.

We now define a map $d_n : C^n(\Gamma, A) \longrightarrow C^{n+1}(\Gamma, A)$ by the formulas:

$$d_0(a) = \sigma \cdot a - a$$

and for all $n \geq 1$

$$d_n(\alpha)_{\sigma_1,\ldots,\sigma_{n+1}} = \sigma_1 \cdot \alpha_{\sigma_2,\ldots,\sigma_{n+1}} + \sum_{i=1}^{n} (-1)^i \alpha_{\sigma_1,\ldots,\sigma_i\sigma_{i+1},\ldots,\sigma_{n+1}}$$

$$+(-1)^{n+1}\alpha_{\sigma_1,\ldots,\sigma_n}.$$

Definition II.3.16. An n-**cocycle** of Γ with values in A is a map $\alpha \in C^n(\Gamma, A)$ satisfying:

(1) $d_n(\alpha) = 0$

(2) $\alpha_{\sigma_1,\ldots,\sigma_n} = 0$ whenever $\sigma_i = 1$ for some i.

A map $\alpha \in C^n(\Gamma, A)$ is an n-**coboundary** of Γ with values in A if there exists $\beta \in C^{n-1}(\Gamma, A)$ such that:

(1) $\alpha = d_{n-1}(\beta)$

(2) $\beta_{\sigma_1,\ldots,\sigma_{n-1}} = 0$ whenever $\sigma_i = 1$ for some i.

Notice that for $n = 1$, condition (2) is empty. The set of n-cocycles and n-coboundaries are abelian subgroups of $C^n(\Gamma, A)$, denoted by $Z^n(\Gamma, A)$ and $B^n(\Gamma, A)$ respectively. One can check that $d_n d_{n-1} = 0$, so $B^n(\Gamma, A)$ is a subgroup of $Z^n(\Gamma, A)$, and we may define

$$H^n(\Gamma, A) = Z^n(\Gamma, A)/B^n(\Gamma, A).$$

The group $H^n(\Gamma, A)$ is called the nth **cohomology group of** Γ **with coefficients in** A. Two n-cocycles are said to be **cohomologous** if they have the same image in $H^n(\Gamma, A)$, i.e. if they differ by an n-coboundary. The constant map

$$\Gamma^n \longrightarrow A$$
$$\sigma \longmapsto 1$$

is an element of $Z^n(\Gamma, A)$, which is called the **trivial** n-cocycle.

Remarks II.3.17.

(1) In the existing literature, n-cocycles and n-coboundaries are often defined to be elements of $\ker(d_n)$ and $\mathrm{im}(d_{n-1})$ respectively, and cocycles and coboundaries satisfying the extra condition (2) are called normalized. However, one can show that the two quotient groups obtained with these two different definitions are canonically isomorphic.

(2) If Γ is a finite abstract group, we recover the classical definition of the cohomology groups associated to a finite group (modulo the previous remark).

(3) Assume that A is a Γ-module (denoted additively). Since a 1-cocycle α always satisfies $\alpha_1 = 0$, Definition II.3.12 is consistent with Definition II.3.16.

Now write A multiplicatively (it is more convenient for our purpose), and let us give explicitly the formulas defining a 2-cocycle. A 2-cocycle is a continuous map $\alpha : \Gamma \times \Gamma \longrightarrow A$ satisfying

$$\sigma \cdot \alpha_{\tau,\rho}\, \alpha_{\sigma,\tau\rho} = \alpha_{\sigma\tau,\rho}\, \alpha_{\sigma,\tau} \text{ for } \sigma, \tau, \rho \in \Gamma,$$

and

$$\sigma_{1,\tau} = \sigma_{\sigma,1} = 1 \text{ for all } \sigma, \tau \in \Gamma,$$

and two 2-cocycles α, α' are cohomologous if there exists a continuous map $\varphi : \Gamma \longrightarrow A$ satisfying $\varphi_1 = 1$ and

$$\alpha'_{\sigma,\tau} = (\sigma \cdot \varphi_\tau)\varphi_{\sigma\tau}^{-1}\varphi_\sigma \alpha_{\sigma,\tau} \text{ for all } \sigma, \tau \in \Gamma.$$

In the following, we will always assume implicitly that, when we write $H^n(\Gamma, A)$, the set A has the appropriate structure.

II.3.2 Functoriality

We now start to study the functorial properties of cohomology sets, which are their main interest. In the sequel, as well as in the other paragraphs, we will not check that the various maps we consider are continuous, and leave the verifications as an exercice for the reader.

Definition II.3.18. Let Γ, Γ' be two profinite groups. Let A be a Γ-set and A' be a Γ'-set. Moreover, let $\varphi : \Gamma' \longrightarrow \Gamma$ be a morphism of profinite groups (in particular, φ is continuous), and let $f : A \longrightarrow A'$ be a map. If A and A' are groups, we require that f is a group morphism.

We say that f and φ are **compatible** if

$$f(\varphi(\sigma')\cdot a) = \sigma'\cdot f(a) \text{ for } \sigma' \in \Gamma', a \in A.$$

Notice that it follows from the very definition that if a is fixed by Γ, then $f(a)$ is fixed by Γ'. Hence f induces by restriction a map of pointed sets

$$f_* : H^0(\Gamma, A) \longrightarrow H^0(\Gamma', A').$$

The following proposition shows that this is also true for higher cohomology sets.

Proposition II.3.19. *Let Γ, Γ', A, A' as above, and let $\varphi : \Gamma' \longrightarrow \Gamma$ and $f : A \longrightarrow A'$ be two compatible maps. For any n-cocycle $\alpha \in Z^n(\Gamma, A)$, the map*

$$f_*(\alpha)\colon \begin{array}{rcl} \Gamma'^n & \longrightarrow & A' \\ (\sigma'_1, \ldots, \sigma'_n) & \longmapsto & f(\alpha_{\varphi(\sigma'_1),\ldots,\varphi(\sigma'_n)}) \end{array}$$

is an n-cocycle, and the map

$$f_*\colon \begin{array}{rcl} H^n(\Gamma, A) & \longrightarrow & H^n(\Gamma', A') \\ {[\alpha]} & \longmapsto & [f_*(\alpha)] \end{array}$$

is a well-defined map of pointed sets (resp. a group morphism if A and A' are abelian).

Proof. We only prove the result for $n = 1$ in the case of Γ-groups. The remaining cases may be proved similarly.

Let $\alpha \in Z^1(\Gamma, A)$ and set $\beta = f_*(\alpha)$. By definition, we have $\beta_{\sigma'} = f(\alpha_{\varphi(\sigma')})$ for all $\sigma' \in \Gamma'$. Hence $\beta_{\sigma'\tau'} = f(\alpha_{\varphi(\sigma')\varphi(\tau')})$, since φ is a group morphism. Since α is a 1-cocycle, we get

$$\beta_{\sigma'\tau'} = f(\alpha_{\varphi(\sigma')}\varphi(\sigma')\cdot\alpha_{\varphi(\tau')}) = f(\alpha_{\varphi(\sigma')})f(\varphi(\sigma')\cdot\alpha_{\varphi(\tau')}).$$

By compatibility, we get that

$$\beta_{\sigma'\tau'} = f(\alpha_{\varphi(\sigma')})\sigma'\cdot f(\alpha_{\varphi(\tau')}) = \beta_{\sigma'}\,\sigma'\cdot\beta_{\tau'}.$$

Hence β is a 1-cocycle.

Now we have to show that if α and α' are cohomologous, then the corresponding β and β' are also cohomologous, so assume that

$$\alpha'_\sigma = a\alpha_\sigma\,\sigma\cdot a^{-1} \text{ for all } \sigma \in \Gamma,$$

for some $a \in A$. Applying this relation to $\sigma = \varphi(\sigma')$ and taking f on both sides gives

$$\beta'_{\sigma'} = f(a\alpha_{\varphi(\sigma')}\,\varphi(\sigma')\cdot a^{-1}).$$

Since f is a group morphism which is compatible with φ, we get

$$\beta'_{\sigma'} = f(a)f(\alpha_{\varphi(\sigma')})\,\sigma'\cdot f(a)^{-1} = f(a)\beta_{\sigma'}\,\sigma'\cdot f(a)^{-1}.$$

Hence β and β' are cohomologous. Finally, it is clear from the definition that f_* maps the trivial class onto the trivial class. □

Example II.3.20. Assume that $\Gamma = \Gamma'$ and $\varphi = \mathrm{Id}_\Gamma$. Then a compatible map $f : A \longrightarrow A'$ is just a morphism of Γ-sets (or Γ-groups, etc), and the map f_* just sends the cohomology class of α onto the cohomology class of $f \circ \alpha$. Moreover, if $g : A' \longrightarrow A''$ is a morphism of Γ-sets (or Γ-groups, etc), we have $(g \circ f)_* = g_* \circ f_*$.

Example II.3.21. Assume that $\Gamma = \Gamma'$, $\varphi = \mathrm{Int}(\rho)$, for a fixed element $\rho \in \Gamma$ and let

$$f : \begin{array}{l} A \longrightarrow A \\ a \longmapsto \rho^{-1}\cdot a. \end{array}$$

It is easy to see that f and φ are compatible.

Claim: For $n \geq 0$, the induced map $f_* : H^n(\Gamma, A) \longrightarrow H^n(\Gamma, A)$ is the identity.

For $n = 0$, this is clear. Again, we only prove the claim for $n = 1$ in the case of Γ-groups. We have to show that α and $\beta = f_*(\alpha)$ are cohomologous for every 1-cocycle $\alpha \in Z^1(\Gamma, A)$.

First notice that applying the cocyclicity relation to $\tau = \sigma^{-1}$ gives

$$\sigma^{-1}\cdot\alpha_\sigma = \alpha_{\sigma^{-1}}^{-1} \text{ for } \sigma \in \Gamma.$$

Now we have

$$\beta_\sigma = \rho^{-1}\cdot\alpha_{\rho\sigma\rho^{-1}} = \rho^{-1}\cdot(\alpha_\rho(\rho\cdot\alpha_{\sigma\rho^{-1}})) = (\rho^{-1}\cdot\alpha_\rho)\alpha_{\sigma\rho^{-1}},$$

since Γ acts on A by group automorphisms. Hence we get

$$\beta_\sigma = (\rho^{-1}\cdot\alpha_\rho)(\alpha_\sigma\,\sigma\cdot\alpha_{\rho^{-1}}) = \alpha_{\rho^{-1}}^{-1}\alpha_\sigma\,\sigma\cdot\alpha_{\rho^{-1}}.$$

Setting $a = \alpha_{\rho^{-1}}^{-1}$ then shows that α and β are cohomologous.

Example II.3.22. Let G, A be finite groups, where A is a G-group, and let H be a subgroup of G, acting on A by restricting the action of G. Now let $\varphi : H \longrightarrow G$ be the inclusion and let $f = \mathrm{Id}_A$. These two maps are compatible, and then we get a map

$$\mathrm{Res} : H^n(G, A) \longrightarrow H^n(H, A),$$

called the **restriction map from G to H.**

If $[\alpha] \in H^n(G, A)$, a cocycle representing $\mathrm{Res}([\alpha])$ is $\alpha_{|_{H^n}}$.

Example II.3.23. Let G, A be finite groups, and assume that G acts trivially on A. Let Γ be a profinite group acting continuously on G and acting trivially on A. Let $\varphi : \Gamma \longrightarrow G$ be a morphism of profinite groups, and let $f = \mathrm{Id}_A$. Then φ and f are compatible, and we get a map

$$\varphi^* : H^n(G, A) \longrightarrow H^n(\Gamma, A),$$

called **the inverse image** with respect to φ.

If $[\alpha] \in H^n(G, A)$, a cocycle β representing $\varphi^*([\alpha])$ is given by

$$\beta\colon \begin{array}{l} \Gamma^n \longrightarrow A \\ (\sigma_1, \ldots, \sigma_n) \longmapsto \alpha_{\varphi(\sigma_1), \ldots, \varphi(\sigma_n)}. \end{array}$$

We would like to observe now that the map φ^* depends on φ only up to conjugation. For, let $\rho \in G$ and set $\psi = \mathrm{Int}(\rho) \circ \varphi$. Then $\psi^*([\alpha])$ is represented by the cocycle γ defined by

$$\gamma\colon \begin{array}{l} \Gamma^n \longrightarrow A \\ (\sigma_1, \ldots, \sigma_n) \longmapsto \alpha_{\rho\varphi(\sigma_1)\rho^{-1}, \ldots, \rho\varphi(\sigma_n)\rho^{-1}}. \end{array}$$

Keeping in mind that G acts trivially on A, we see that the equality $\psi^*([\alpha]) = \varphi^*([\alpha])$ is a consequence of Example II.3.21.

Assume now that Γ also acts trivially on G, so that $H^1(\Gamma, G)$ is nothing but the set of conjugacy classes of continuous morphisms $\varphi : \Gamma \longrightarrow G$. The previous observations then imply that for each class $[\alpha] \in H^n(G, A)$, we have a well-defined map

$$\begin{aligned} H^1(\Gamma, G) &\longrightarrow H^n(\Gamma, A) \\ [\varphi] &\longmapsto \varphi^*([\alpha]). \end{aligned}$$

Example II.3.24. Let Γ be a profinite group acting continuously on A, and let $U, U' \in \mathcal{N}$ be two normal open subgroups of Γ, $U \supset U'$. Then Γ/U and Γ/U' acts continuously on A^U and $A^{U'}$ respectively. Let us denote by $\pi_U : \Gamma \longrightarrow \Gamma/U$ and $\pi_{U'} : \Gamma \longrightarrow \Gamma/U'$ the canonical projections. One can easily check that the maps $\varphi : \Gamma/U' \longrightarrow \Gamma/U$ and $f : A^U \longrightarrow A^{U'}$ are compatible, so we get an **inflation map**

$$\mathrm{inf}_{U,U'} : H^n(\Gamma/U, A^U) \longrightarrow H^n(\Gamma/U', A^{U'}).$$

If $[\alpha] \in H^n(\Gamma/U, A^U)$, the cohomology class $\mathrm{inf}_{U,U'}([\alpha])$ is represented by the cocycle

$$\beta : \begin{array}{rcl} (\Gamma/U)^n & \longrightarrow & A^U \\ (\pi_{U'}(\sigma_1), \ldots, \pi_{U'}(\sigma_n)) & \longmapsto & \alpha_{\pi_U(\sigma_1), \ldots, \pi_U(\sigma_n)}. \end{array}$$

Example II.3.25. Let Γ be a profinite group acting continuously on A, and let $U \in \mathcal{N}$ be a normal open subgroup of Γ. Then the maps $\pi_U : \Gamma \longrightarrow \Gamma/U$ and the inclusion $A^U \longrightarrow A$ are compatible, and give rise to a map

$$f_U : H^n(\Gamma/U, A^U) \longrightarrow H^n(\Gamma, A).$$

Proposition II.3.26. *For $i = 1, \ldots, 4$, let A_i be a Γ_i-set (Γ_i-group,etc). Assume that we have two commutative diagrams*

$$\begin{array}{ccc} A_1 & \xrightarrow{f_1} & A_2 \\ \downarrow{\scriptstyle f_3} & & \downarrow{\scriptstyle f_2} \\ A_4 & \xrightarrow{f_4} & A_3 \end{array}$$

and

$$\begin{array}{ccc} \Gamma_1 & \xleftarrow[\varphi_1]{} & \Gamma_2 \\ \uparrow{\scriptstyle \varphi_3} & & \uparrow{\scriptstyle \varphi_2} \\ \Gamma_4 & \xleftarrow[\varphi_4]{} & \Gamma_3 \end{array}$$

where φ_i is a morphism of profinite groups compatible with f_i, for $i = 1, \ldots, 4$. Then the diagram

$$\begin{array}{ccc} H^n(\Gamma_1, A_1) & \xrightarrow{f_{1*}} & H^n(\Gamma_2, A_2) \\ \downarrow{\scriptstyle f_{3*}} & & \downarrow{\scriptstyle f_{2*}} \\ H^n(\Gamma_4, A_4) & \xrightarrow{f_{4*}} & H^n(\Gamma_3, A_3) \end{array}$$

is commutative.

Proof. If $n = 0$, the result is clear, so we assume that $n \geq 1$. Let

$[\alpha] \in H^n(\Gamma_1, A_1)$. By definition, $f_{1*}([\alpha])$ is represented by the cocycle

$$\Gamma_2^n \longrightarrow A_2$$
$$(\sigma_1, \ldots, \sigma_n) \longmapsto f_1(\alpha_{\varphi_1(\sigma_1),\ldots,\varphi_1(\sigma_n)}).$$

Therefore, $(f_{2*} \circ f_{1*})([\alpha])$ is represented by the cocycle

$$\Gamma_3^n \longrightarrow A_3$$
$$(\sigma_1, \ldots, \sigma_n) \longmapsto f_2(f_1(\alpha_{\varphi_1(\varphi_2(\sigma_1)),\ldots,\varphi_1(\varphi_2(\sigma_n))})).$$

Similarly, $(f_{4*} \circ f_{3*})([\alpha])$ is represented by the cocycle

$$\Gamma_3^n \longrightarrow A_3$$
$$(\sigma_1, \ldots, \sigma_n) \longmapsto f_4(f_3(\alpha_{\varphi_3(\varphi_4(\sigma_1)),\ldots,\varphi_3(\varphi_4(\sigma_n))})).$$

Since $f_2 \circ f_1 = f_4 \circ f_3$ and $\varphi_1 \circ \varphi_2 = \varphi_3 \circ \varphi_4$ by assumption, we get the desired result. □

Convention: From now on, if $f : A \longrightarrow B$ is a morphism of Γ-sets (Γ-groups,etc), the symbol f_* will denote the map in cohomology obtained when taking $\varphi = \mathrm{Id}_\Gamma$ as in Example II.3.20, unless specified otherwise.

II.3.3 Cohomology sets as a direct limit

In this paragraph, we would like to relate the cohomology of profinite groups to the cohomology of its finite quotients. The key ingredient to do this is Proposition II.3.6, which says more or less that an n-cocycle $\alpha : \Gamma^n \longrightarrow A$ is locally defined by a family of n-cocycles

$$\alpha^{(U)} : (\Gamma/U)^n \longrightarrow A^U,$$

where U runs through the set of open normal subgroups of Γ. In view of the relation

$$\alpha_{\sigma_1,\ldots,\sigma_n} = \alpha^{(U)}_{\overline{\sigma}_1,\ldots,\overline{\sigma}_n} \text{ for all } \sigma_1, \ldots, \sigma_n \in \Gamma,$$

it easily implies that for all $U, U' \in \mathcal{N}, U \supset U'$, we have

$$\inf_{U,U'}([\alpha^{(U)}]) = [\alpha^{(U')}].$$

Conversely, we will see a family of cohomology classes satisfying this coherence condition may be patched together to define a cohomology class in $H^n(\Gamma, A)$.

We now go into details, and start with the definition of a direct limit.

Definition II.3.27. A **directed system** of sets (groups, rings, etc) is a family of sets (groups, rings, etc) $(X_i)_{i\in I}$, indexed by a directed set I, together with maps (resp. group morphisms, ring morphisms,etc) $\iota_{ij} : X_i \longrightarrow X_j$ for any $i, j \in I, i \leq j$, satisfying the following properties:

(1) $\iota_{ii} = \mathrm{Id}_{X_i}$ for all $i \in I$.

(2) For all $i, j, k \in I, i \leq j \leq k$, we have $\iota_{jk} \circ \iota_{ij} = \iota_{ik}$.

Example II.3.28. Let Γ be a profinite group, and let A be a Γ-set. If $U \in \mathcal{N}$, set $X_U = A^U$. For all $U, U' \in \mathcal{N}$, $U \supset U'$, we denote by $\iota_{U,U'}$ the inclusion $A^U \subset A^{U'}$. It is easy to check that we get a directed system of sets.

Definition II.3.29. Let $((X_i)_{i\in I}, (\iota_{ij}))$ be a directed system of sets (groups, rings, etc). We define an equivalence relation on the disjoint union $\coprod_{i\in I} X_i$ as follows: if $i, j \in I, i \leq j, x_i \in X_i, x_j \in X_j$, we say that $x_i \sim x_j$ if there exists $k \in I$ such that $k \geq i, j$ and $\iota_{jk}(x_j) = \iota_{ik}(x_i)$.

The **direct limit** of the sets (groups, rings, etc) $((X_i)_{i\in I}, (\iota_{ij}))$, denoted by $\varinjlim_{i\in I} X_i$, is the set (groups, ring, etc.) of equivalence classes $\coprod_{i\in I} X_i/\sim$.

The following lemma give a nicer description of direct limits in a particular case.

Lemma II.3.30. *Let $((X_i)_{i\in I}, (\iota_{ij}))$ be a directed system of sets (groups, rings, etc). Assume that we have injective maps (group morphisms, ring morphisms, etc) $f_i : X_i \longrightarrow X$ such that $f_i = f_j \circ \iota_{ij}$ for all $i \leq j$. Then*

$$\varinjlim_{i\in I} X_i \simeq \bigcup_{i\in I} f_i(X_i) \subset X.$$

In particular, if the X_i's are subsets (subgroups, subrings, etc) of a same set (group, ring, etc) X satisfying $X_i \subset X_j$ for all $i \leq j$, we have

$$\varinjlim_{i\in I} X_i \simeq \bigcup_{i\in I} X_i.$$

Proof. To see this, we define a map $f : \varinjlim_{i\in I} X_i \longrightarrow \bigcup_{i\in I} f_i(X_i)$ as follows: for $x_i \in X_i$, set

$$f(x_i/_\sim) = f_i(x_i).$$

If $x_i \sim x_j, i \leq j$, then there exists $k \in I, k \geq i, j$ such that $\iota_{jk}(x_j) = \iota_{ik}(x_i)$ and therefore $f_k(\iota_{jk}(x_j)) = f_k(\iota_{ik}(x_i))$, that is $f_j(x_j) = f_i(x_i)$

by assumption. Hence f is well-defined, and one can check easily that f is a group (ring, etc) morphism if the X_i's are.

We claim that f is bijective. Since surjectivity is clear by definition of f, we just need to check that f is injective. Assume that $f(x_i/_\sim) = f(x_j/_\sim)$ for some $i, j \in I, x_i \in X_i, x_j \in X_j$. Let $k \in I, k \geq i, j$. By assumption on the f_i's, we have $f(x_i/_\sim) = f_i(x_i) = f_k(\iota_{ik}(x_i))$, and similarly $f(x_j/_\sim) = f_k(\iota_{jk}(x_j))$. Now using the equality $f(x_i/_\sim) = f(x_j/_\sim)$ and the injectivity of f_k, we get $\iota_{ik}(x_i) = \iota_{jk}(x_j)$, and thus $x_i/_\sim = x_j/_\sim$. This concludes the proof. □

Example II.3.31. Let Γ be a profinite group, and let A be a Γ-set. It follows from Lemma II.3.3 and the previous result that $A \simeq \varinjlim_{U \in \mathcal{N}} A^U$.

Let Γ be a profinite group, and let A be a Γ-group (resp. a Γ-module if $n \geq 2$). Recall from Examples II.3.24 and II.3.25 that we have maps

$$\inf_{U,U'} : H^n(\Gamma/U, A^U) \longrightarrow H^n(\Gamma/U', A^{U'}),$$

for all $U, U' \in \mathcal{N}, U \supset U'$ and a map

$$f_U : H^n(\Gamma/U, A^U) \longrightarrow H^n(\Gamma, A).$$

The following lemma follows from direct computations.

Lemma II.3.32. *The sets $H^n(\Gamma/U, A^U)$ together with the maps $\inf_{U,U'}$ form a directed system of pointed sets (resp. of groups if A is abelian). Moreover, we have*

$$f_U = f_{U'} \circ \inf_{U,U'}.$$

We now come to the main result of this section.

Theorem II.3.33. *Let Γ be a profinite group, and let A be a Γ-group. Then we have an isomorphism of pointed sets (resp. an isomorphism of groups if A is abelian)*

$$\varinjlim_{U \in \mathcal{N}} H^n(\Gamma/U, A^U) \simeq H^n(\Gamma, A).$$

If $[\xi_U] \in H^n(\Gamma/U, A^U)$, this isomorphism maps $[\xi_U]/_\sim$ onto $f_U([\xi_U])$.

Proof. We first prove that there exists a well-defined map

$$f : \varinjlim_{U \in \mathcal{N}} H^n(\Gamma/U, A^U) \longrightarrow H^n(\Gamma, A),$$

which sends the equivalence class of $[\xi_U] \in H^n(\Gamma/U, A^U)$ onto $f_U([\xi_U])$.

Let $U, U' \in \mathcal{N}$, $[\xi_U] \in H^n(\Gamma/U, A^U)$ and $[\xi'_U] \in H^n(\Gamma/U', A^{U'})$ such that $[\xi_U]/_\sim = [\xi_{U'}]/_\sim$. By definition of the direct limit, there exists $V \in \mathcal{N}$ such that $U \supset V, U' \supset V$ and $\inf_{U,V}([\xi_U]) = \inf_{U',V}([\xi_{U'}])$. Applying f_V on both sides and using the previous lemma, we obtain that $f_U([\xi_U]) = f_{U'}([\xi_{U'}])$, proving that f is well-defined. Clearly, f maps the class of the trivial cocycle to the class of the trivial cocycle. Moreover if A is abelian, one can check easily that f is a group morphism.

Let us prove that f is bijective. For $U, U' \in \mathcal{N}, U \supset U'$, let us denote by

$$\pi_U : \Gamma \longrightarrow \Gamma/U \text{ and } \pi_{U,U'} : \Gamma/U' \longrightarrow \Gamma/U$$

the canonical projections. Let $[\alpha] \in H^n(\Gamma, A)$. By Proposition II.3.6, there exists $U \in \mathcal{N}$ and a map $\alpha^{(U)} : (\Gamma/U)^n \longrightarrow A^U$ such that

$$\alpha_{\sigma_1,\ldots,\sigma_n} = \alpha^{(U)}_{\pi_U(\sigma_1),\ldots,\pi_U(\sigma_n)}$$

for all $\sigma_1, \ldots, \sigma_n \in \Gamma$. Notice that by definition of the action of Γ/U on A^U, we have

$$\pi_U(\sigma) \cdot a = \sigma \cdot a \text{ for all } \sigma \in \Gamma, a \in A^U.$$

It follows easily that $\alpha^{(U)}$ is an n-cocycle. Moreover, by definition of f_U and $\alpha^{(U)}$, we have

$$f([\alpha^{(U)}]/_\sim) = f_U([\alpha^{(U)}]) = [\alpha].$$

Therefore, f is surjective.

It remains to prove the injectivity of f. Let us also denote by f_U and $\inf_{U,U'}$ the inflation maps at the level of cocycles. It is easy to check that we still have

$$f_U = f_{U'} \circ \inf_{U,U'} \text{ for all } U \supset U'.$$

Notice also that for all $U \in \mathcal{N}$, the map $f_U : Z^1(\Gamma/U, A^U) \to Z^1(\Gamma, A)$ is injective. Indeed, if $\xi, \xi' \in Z^1(\Gamma/U, A^U)$ are two cocycles such that $f_U(\xi) = f_U(\xi')$, then we have

$$\xi_{\pi_U(\sigma)} = \xi'_{\pi_U(\sigma)} \text{ for all } \sigma \in \Gamma.$$

Since $\pi_U : \Gamma \to \Gamma/U$ is surjective, we get $\xi = \xi'$.

We are now ready to prove that f is injective. Let $[\xi_U] \in H^n(\Gamma/U, A^U)$ and $[\xi'_{U'}] \in H^n(\Gamma/U', A^{U'})$ such that $f([\xi_U]/_\sim) = f([\xi'_{U'}]/_\sim)$. Let $U'' \in \mathcal{N}$ be a normal open subset of Γ contained in U and U'. Since

$[\inf_{U,U''}([\xi_U])]/_\sim = [\xi_U]/_\sim$ and $[\inf_{U',U''}([\xi'_{U'}])]/_\sim = [\xi'_{U'}]/_\sim$, we may assume without loss of generality that $U = U'$. The equality $f([\xi_U]/_\sim) = f([\xi'_U]/_\sim)$ then rewrites

$$f_U([\xi_U]) = f_U([\xi'_U]).$$

Therefore, there exists $a \in A$ such that

$$f_U(\xi'_U)_\sigma = a\, f_U(\xi_U)_\sigma\, \sigma{\cdot}a^{-1} \text{ for all } \sigma \in \Gamma.$$

By Lemma II.3.3, there exists $U_0 \in \mathcal{N}$ such that $a \in A^{U_0}$. Now let $V \in \mathcal{N}$ contained in U and U_0. In particular, we have $A^U \subset A^V$ and $A^{U_0} \subset A^V$. Let $\eta \in Z^1(\Gamma/V, A^V)$ be the cocycle defined by

$$\eta\colon \begin{array}{ccc} \Gamma/V & \longrightarrow & A^V \\ \pi_V(\sigma) & \longmapsto & a \inf_{U,V}(\xi_U)_{\pi_V(\sigma)}\, \pi_V(\sigma){\cdot}a. \end{array}$$

Taking into account that we have $\sigma{\cdot}a = \pi_V(\sigma){\cdot}a$ for all $\sigma \in \Gamma$, the previous equality yields

$$f_V(\inf_{U,V}(\xi'_U)) = f_V(\eta).$$

By injectivity of f_V, we get $\inf_{U,V}(\xi'_U) = \eta \in Z^1(\Gamma/V, A^V)$. Since η and $\inf_{U,V}(\xi_U)$ are cohomologous by construction, we have

$$\inf_{U,V}([\xi'_U)] = [\eta] = \inf_{U,V}([\xi_U]),$$

and thus $[\xi'_U]/_\sim = [\xi_U]/_\sim$. This concludes the proof. □

§II.4 Cohomology sequences

If G is a finite group and A is a finite G-module, the groups $H^n(G, A)$ are known to have interesting properties with respect to exact sequences of G-modules (see [11] or [26] for an account on cohomology of finite groups). We now proceed to show that similar properties hold in our more general setting.

Definition II.4.1. Let $f : A \longrightarrow B$ be a map of pointed sets. The **kernel** of f is the preimage by f of the base point of B.
A sequence of pointed sets

$$A \xrightarrow{\ f\ } B \xrightarrow{\ g\ } C$$

is called **exact** at B if $\operatorname{im} f = \ker g$.

A sequence of pointed sets

$$A_0 \longrightarrow A_1 \longrightarrow \cdots \longrightarrow A_{i-1} \longrightarrow A_i \longrightarrow A_{i+1} \longrightarrow \cdots$$

is called **exact** if it is exact at A_i for all $i \geq 1$.

An exact sequence of groups (resp. of Γ-groups, resp. of Γ-modules) is an exact sequence of pointed sets such that all the maps involved are group morphisms (resp. morphisms of Γ-groups, resp. of Γ-modules).

For example, the sequence

$$B \xrightarrow{g} C \longrightarrow 1$$

is exact if and only if g is surjective, and the sequence

$$1 \longrightarrow A \xrightarrow{f} B$$

is exact if and only if f has trivial kernel. This does **not** imply that f is injective, unless A and B are groups.

Assume that we have an exact sequence

$$1 \longrightarrow A \xrightarrow{f} B \xrightarrow{g} C \longrightarrow 1$$

of pointed Γ-sets. The goal of the next paragraphs is to derive some exact sequences in cohomology, under some reasonable conditions on A, B and C. We will keep this notation throughout.

II.4.1 The case of a subgroup

Assume that A and B are Γ-groups, that f is a group morphism (hence f is injective), and that g induces a bijection of Γ-sets $B/f(A) \simeq C$, where $B/f(A)$ is the set of left cosets modulo $f(A)$. In other words, g is surjective and for all $b, b' \in B$ we have

$$g(b) = g(b') \iff b' = bf(a) \text{ for some } a \in A.$$

For instance, these conditions are satisfied in the following cases:

(1) A is a Γ-subgroup of B, $C = B/A$, f is the inclusion and g is the natural projection.

(2) C is a Γ-group and g is a group morphism (this will be the case in the next subsection).

As pointed out previously in Example II.3.20, f and g induce maps on fixed points by restriction, namely $f_* : A^\Gamma \longrightarrow B^\Gamma$ and $g_* : B^\Gamma \longrightarrow C^\Gamma$. Our next goal is to define a map of pointed sets

$$\delta^0 : C^\Gamma \longrightarrow H^1(\Gamma, A).$$

Let $c \in C^\Gamma$, and let $b \in B$ any preimage of c under g, i.e. $g(b) = c$. By assumption, we have $c = \sigma \cdot c$ for all $\sigma \in \Gamma$. Therefore, we have

$$g(\sigma \cdot b) = \sigma \cdot g(b) = \sigma \cdot c = c = g(b).$$

By assumption on g, there exists a unique element $\alpha_\sigma \in A$ such that $f(\alpha_\sigma) = b^{-1}\sigma \cdot b$.

Lemma II.4.2. *The map $\alpha : \Gamma \longrightarrow A$ is a 1-cocycle, and its class in $H^1(\Gamma, A)$ does not depend on the choice of $b \in B$.*

Proof. Let us prove that α is a cocycle. By definition of α, for all $\sigma, \tau \in \Gamma$, we have

$$f(\alpha_{\sigma\tau}) = b^{-1}\,\sigma\tau \cdot b = b^{-1}\,\sigma \cdot (bb^{-1}\tau \cdot b) = (b^{-1}\,\sigma \cdot b)\,\sigma \cdot (b^{-1}\tau \cdot b).$$

Hence we have

$$f(\alpha_{\sigma\tau}) = f(\alpha_\sigma)\sigma \cdot f(\alpha_\tau) = f(\alpha_\sigma)f(\sigma \cdot \alpha_\tau) = f(\alpha_\sigma\sigma \cdot \alpha_\tau).$$

By injectivity of f, we get $\alpha_{\sigma\tau} = \alpha_\sigma\sigma \cdot \alpha_\tau$.

Let us prove now that the cohomology class of α does not depend on the choice of b. Let $b' \in B$ be another preimage of c under g. We then have $g(b') = c = g(b)$, so $b' = bf(a^{-1}) = bf(a)^{-1}$ for some $a \in A$ by assumption on g, and let α' be the corresponding 1-cocycle. We then have

$$f(\alpha'_\sigma) = f(a)b^{-1}\sigma \cdot (bf(a^{-1})) = f(a)f(\alpha_\sigma)\,\sigma \cdot f(a^{-1}) = f(a\alpha_\sigma\,\sigma \cdot a^{-1}),$$

so by injectivity of f, this implies that α and α' are cohomologous. This concludes the proof. □

Notice that a preimage under g of the base point of C^Γ is the neutral element $1 \in B$ (since g is a morphism of pointed sets). Since Γ acts by group automorphisms on B, we have $\sigma \cdot 1 = 1$ for all $\sigma \in \Gamma$, and therefore the base point of C^Γ is mapped onto the trivial cohomology class. We then have constructed a map of pointed sets

$$\delta^0 : \begin{aligned} C^\Gamma &\longrightarrow H^1(\Gamma, A) \\ c &\longmapsto [\alpha], \end{aligned}$$

where the cocycle α is defined by the relations

$$f(\alpha_\sigma) = b^{-1}\,\sigma \cdot b \text{ for all } \sigma \in \Gamma,$$

for an arbitrary preimage $b \in B$ of c.

Definition II.4.3. The map $\delta^0 : C^\Gamma \longrightarrow H^1(\Gamma, A)$ is called the 0th **connecting map**.

Proposition II.4.4. *The sequence of pointed sets*

$$1 \longrightarrow A^\Gamma \xrightarrow{f_*} B^\Gamma \xrightarrow{g_*} C^\Gamma \xrightarrow{\delta^0} H^1(\Gamma, A) \xrightarrow{f_*} H^1(\Gamma, B)$$

is exact.

Proof. The exactness of the sequence

$$1 \longrightarrow A^\Gamma \xrightarrow{f_*} B^\Gamma \xrightarrow{g_*} C^\Gamma$$

is left to the reader.

Exactness at C^Γ: We need to check that $\mathrm{im}(g_*) = \ker(\delta^0)$. Let $c \in C^G$, and let us denote by α the cocycle representing $\delta^0(c)$, as defined above. Assume first that $c \in \mathrm{im}(g_*)$, that is $c = g(b)$ for some $b \in B^\Gamma$. Then by definition α is the trivial cocycle, and c is mapped onto the trivial class. Therefore, $\mathrm{im}(g_*) \subset \ker(\delta^0)$. Conversely, assume that $\delta^0(c)$ is trivial, that is

$$\alpha_\sigma = a\,\sigma\cdot a^{-1} \text{ for all } \sigma \in \Gamma,$$

for some $a \in A$. Let $b \in B$ be a preimage of c under g. We then have

$$f(a\,\sigma\cdot a^{-1}) = b^{-1}\sigma\cdot b \text{ for all } \sigma \in \Gamma,$$

so $f(a)\sigma\cdot f(a)^{-1} = b^{-1}\sigma\cdot b$ for all $\sigma \in \Gamma$. Hence $bf(a) \in B^\Gamma$, and we have $c = g(b) = g(bf(a)) \in \mathrm{im}(g_*)$. Hence $\ker(\delta^0) = \mathrm{im}(g_*)$, which is what we wanted to prove.

Exactness at $H^1(\Gamma, A)$: We need to prove that $\mathrm{im}(\delta^0) = \ker(f_*)$. Let $c \in C^G$ and let $b \in B$ satisfying $c = g(b)$. Then by definition of f_* and $\delta^0(c)$, $f_*(\delta^0(c))$ is the class of the 1-cocycle

$$\begin{aligned} \Gamma &\longrightarrow B \\ \sigma &\longmapsto b^{-1}\,\sigma\cdot b, \end{aligned}$$

which is cohomologous to the trivial cocycle. Hence $\mathrm{im}(\delta^0) \subset \ker(f_*)$. Now if $[\alpha] \in H^1(\Gamma, A)$ satisfies $f_*([\alpha]) = 1$, then $f(\alpha_\sigma) = b^{-1}\,\sigma\cdot b$ for some b in B. Therefore, we have

$$\sigma\cdot g(b) = g(\sigma\cdot b) = g(bf(\alpha_\sigma)) = g(b) \text{ for all } \sigma \in \Gamma.$$

Hence $c = g(b)$ lies in C^Γ. Thus $b \in B$ is a preimage of $c \in C^\Gamma$ under g and $[\alpha] = \delta^0(c)$ by definition of δ^0. This concludes the proof. □

Before continuing, we need to define an action of B^Γ on C^Γ. Let $\beta \in B^\Gamma$ and $c \in C^\Gamma$. Let $b \in B$ be a preimage of c under g, and set

$$\beta\cdot c = g(\beta b) \in C.$$

Let us check that it does not depend on the choice of b. If $b' \in B$ is another preimage of c under g, then $b' = bf(a)$ for some $a \in A$, hence $g(\beta b') = g(\beta b f(a)) = g(\beta b)$ by assumption on g. Hence $\beta\cdot c$ does not depend the choice of b.

We now show that $\beta\cdot c \in C^\Gamma$. For $\sigma \in \Gamma$, we have

$$\sigma\cdot(\beta\cdot c) = \sigma\cdot g(\beta b) = g(\sigma\cdot(\beta b)) = g((\sigma\cdot\beta)(\sigma\cdot b)).$$

Since $\beta \in B^\Gamma$, we get $\sigma\cdot(\beta\cdot c) = g(\beta(\sigma\cdot b))$ for all $\sigma \in \Gamma$. Now $g(\sigma\cdot b) = \sigma\cdot g(b) = \sigma\cdot c = c$ since $c \in C^\Gamma$, so $\sigma\cdot b$ is also a preimage of c. Since $\beta\cdot c$ does not depend on the choice of a preimage of c, we conclude that $\sigma\cdot(\beta\cdot c) = \beta\cdot c$ for all $\sigma \in \Gamma$, so $\beta\cdot c \in C^\Gamma$.

It is then clear that the map

$$\begin{array}{ccc} B^\Gamma \times C^\Gamma & \longrightarrow & C^\Gamma \\ (\beta, c) & \longmapsto & \beta\cdot c \end{array}$$

gives rise to an action of B^Γ on C^Γ.
We will denote by C^Γ/B^Γ of the group B^Γ in C^Γ. Notice that this is a pointed set, whose base point is the orbit of 1. The next result identifies this orbit set.

Corollary II.4.5. *There is a natural bijection of pointed sets between the orbit set* C^Γ/B^Γ *and* $\ker(H^1(\Gamma, A) \longrightarrow H^1(\Gamma, B))$.

More precisely, the bijection sends the orbit of $c \in C^\Gamma$ *onto* $\delta^0(c)$.

Proof. By Proposition II.4.4, we have $\ker(H^1(\Gamma, A) \longrightarrow H^1(\Gamma, B)) = \operatorname{im}(\delta^0)$. Hence we have to construct a bijection between C^Γ/B^Γ and $\operatorname{im}(\delta^0)$.

Let $c, c' \in C^\Gamma$ lying in the same orbit, that is $c' = \beta\cdot c$ for some $\beta \in B^\Gamma$. Then $c' = g(\beta b)$, for some preimage $b \in B$ of c, and βb is a preimage of c'. Since we have $(\beta b)^{-1}\sigma\cdot(\beta b) = b^{-1}\beta^{-1}(\sigma\cdot\beta)(\sigma\cdot b) = b^{-1}\,\sigma\cdot b$, it turns out that $\delta^0(c') = \delta^0(c)$. Therefore, the map

$$\varphi\colon \begin{array}{c} C^\Gamma/B^\Gamma \longrightarrow \operatorname{im}(\delta^0) \\ B^\Gamma\cdot c \longmapsto \delta^0(c) \end{array}$$

is a well-defined surjective map. It remains to prove its injectivity. Let

$c, c' \in C^\Gamma$ such that $\delta^0(c') = \delta^0(c)$, and let α and α' be the cocycles representing $\delta^0(c)$ and $\delta^0(c')$ respectively. By assumption, there exists $a \in A$ such that $\alpha'_\sigma = a\,\alpha_\sigma\,\sigma a^{-1}$ for all $\sigma \in \Gamma$. If b (resp. b') is a preimage of c (resp. c') in B, applying f to this last equality implies that

$$b'^{-1}\sigma\cdot b' = f(a)b^{-1}(\sigma\cdot b)(\sigma\cdot f(a))^{-1}.$$

It easily turns out that $\beta = b' f(a)b^{-1} \in B^\Gamma$. Hence $c' = g(b') = g(b' f(a)) = g(\beta b) = \beta\cdot c$. Therefore, c and c' lie in the same orbit, showing that φ is injective. Finally, the orbit of 1 is mapped onto [1]. This concludes the proof. □

Proposition II.4.6. *Let A, B, C be Γ-sets, and let A', B', C' be Γ'-sets. Assume that we have a commutative diagram with exact rows*

$$\begin{array}{ccccccccc} 1 & \longrightarrow & A & \xrightarrow{f} & B & \xrightarrow{g} & C & \longrightarrow & 1 \\ & & \downarrow{\scriptstyle h_1} & & \downarrow{\scriptstyle h_2} & & \downarrow{\scriptstyle h_3} & & \\ 1 & \longrightarrow & A' & \xrightarrow{f'} & B' & \xrightarrow{g'} & C' & \longrightarrow & 1 \end{array}$$

satisfying the conditions explained at the beginning of the section. Let us denote by δ^0 and δ'^0 the respective connecting maps. If $\varphi : \Gamma' \longrightarrow \Gamma$ is compatible with h_1, h_2 and h_3, the diagram

$$\begin{array}{ccc} C^\Gamma & \xrightarrow{\delta^0} & H^1(\Gamma, A) \\ \downarrow{\scriptstyle h_{3*}} & & \downarrow{\scriptstyle h_{1*}} \\ C'^{\Gamma'} & \xrightarrow{\delta'^0} & H^1(\Gamma', A') \end{array}$$

is commutative.

Proof. Let $c \in C^\Gamma$, and let $b \in B$ be any preimage of c under g. The cohomology class $\delta^0(c)$ is represented by the cocycle α defined by the relations

$$f(\alpha_\sigma) = b^{-1}\sigma\cdot b \text{ for all } \sigma \in \Gamma.$$

Therefore, $h_{1*}(\delta^0(c))$ is represented by the cocycle

$$\beta\colon \begin{array}{l} \Gamma' \longrightarrow B' \\ \sigma' \longmapsto h_1(\alpha_{\varphi(\sigma')}). \end{array}$$

By commutativity of the diagram, we have

$$f'(\beta_{\sigma'}) = f'(h_1(\alpha_{\varphi(\sigma')})) = h_2(f(\alpha_{\varphi(\sigma')})) = h_2(b^{-1}\,\varphi(\sigma')\cdot b),$$

for all $\sigma' \in \Gamma'$. Since φ and h_2 are compatible (so h_2 is in particular a group morphism), we get

$$f'(\beta_{\sigma'}) = h_2(b)^{-1}\,\sigma'\cdot h_2(b) \text{ for all } \sigma' \in \Gamma'.$$

On the other hand, $h_{3*}(c) = h_3(c)$ by definition. By commutativity of the diagram, we have

$$h_3(c) = h_3(g(b)) = g'(h_2(b)).$$

Therefore, $h_2(b)$ is a preimage of $h_3(c)$ under g' and thus $\delta'^0(h_{3*}(c)) = \delta'^0(h_3(c))$ is represented by the cocycle β' defined by

$$f'(\beta'_{\sigma'}) = h_2(b)^{-1}\,\sigma'\cdot h_2(b) \text{ for all } \sigma' \in \Gamma'.$$

Hence we have

$$f'(\beta'_{\sigma'}) = f'(\beta_{\sigma'}) \text{ for all } \sigma' \in \Gamma'.$$

Since f' is injective, this yields $\beta' = \beta$. In other words, $h_{1*}(\delta^0(c))$ and $\delta'^0(h_{3*}(c))$ are represented by the same cocycle. This concludes the proof. □

II.4.2 The case of a normal subgroup

We now assume that we have an exact sequence of Γ-groups

$$1 \longrightarrow A \xrightarrow{f} B \xrightarrow{g} C \longrightarrow 1$$

so A can be identified with a normal subgroup of B.

Proposition II.4.7. *The sequence of pointed sets*

$$1 \longrightarrow A^\Gamma \longrightarrow B^\Gamma \longrightarrow C^\Gamma \longrightarrow H^1(\Gamma, A) \longrightarrow H^1(\Gamma, B) \longrightarrow H^1(\Gamma, C)$$

is exact.

Proof. By Proposition II.4.4, only the exactness at $H^1(\Gamma, B)$ needs a proof. If $[\beta] = f_*([\alpha])$ for some $[\alpha] \in H^1(\Gamma, A)$, then we have

$$g_*([\beta]) = g_*(f_*([\alpha])) = (g \circ f)_*([\alpha]) = 1,$$

since $g \circ f$ is trivial by assumption. Hence $\mathrm{im}(f_*) \subset \ker(g_*)$.
Conversely, let $[\beta] \in H^1(\Gamma, B)$ such that $g_*([\beta]) = 1$. Then there exists $c \in C$ such that

$$g(\beta_\sigma) = c^{-1}\sigma\cdot c \text{ for all } \sigma \in \Gamma.$$

Write $c = g(b)$. We then have $g(\beta_\sigma) = g(b^{-1}\sigma\cdot b)$, so $\beta_\sigma = b^{-1}(\sigma\cdot b)f(a_\sigma)$, for some $a_\sigma \in A$. Since $f(A)$ is normal in B, $(\sigma\cdot b)f(a_\sigma)(\sigma\cdot b)^{-1} \in f(A)$, so $\beta_\sigma = b^{-1}f(\alpha_\sigma)\,\sigma\cdot b$ for some $\alpha_\sigma \in A$, and thus

$$b\beta_\sigma\,\sigma\cdot b^{-1} = f(\alpha_\sigma) \text{ for all } \sigma \in \Gamma.$$

The fact that the map

$$\begin{aligned} \Gamma &\longrightarrow B \\ \sigma &\longmapsto b\beta_\sigma\,\sigma\cdot b^{-1} \end{aligned}$$

is a 1-cocycle and the injectivity of f imply easily that α is a cocycle. Moreover, by construction of α, we have $[\beta] = f_*([\alpha]) \in \mathrm{im}(g_*)$. This concludes the proof. □

II.4.3 The case of a central subgroup

Assume now that A is abelian and that $f(A)$ is a central subgroup of B.

Our next goal is to define a map of pointed sets

$$\delta^1 : H^1(\Gamma, C) \longrightarrow H^2(\Gamma, A).$$

Given a 1-cocycle $\gamma \in Z^1(\Gamma, C)$, we will denote by $\beta_\sigma \in B$ a preimage of γ_σ under g. We have

$$g(\beta_\sigma(\sigma\cdot\beta_\tau)\beta_{\sigma\tau}^{-1}) = \gamma_\sigma\,\sigma\cdot\gamma_\tau\gamma_{\sigma\tau}^{-1} = 1,$$

so $\beta_\sigma(\sigma\cdot\beta_\tau)\beta_{\sigma\tau}^{-1} = f(\alpha_{\sigma,\tau})$ for some (unique) $\alpha_{\sigma,\tau} \in A$.

Lemma II.4.8. *The map $\alpha : \Gamma \times \Gamma \longrightarrow A$ is a 2-cocycle, whose class in $H^2(\Gamma, A)$ only depends on $[\gamma]$.*

Proof. Let us check that α is a 2-cocycle. First of all, we have

$$f(\alpha_{\sigma,1}) = \beta_\sigma(\sigma\cdot\beta_1)\beta_\sigma^{-1} = \beta_\sigma\beta_\sigma^{-1} = 1,$$

since $\beta_1 = 1$. By injectivity of f, we get $\alpha_{\sigma,1} = 1$ for all $\sigma \in \Gamma$. Similar arguments show that we have $\alpha_{1,\tau} = 1$ for all $\tau \in \Gamma$ as well. Moreover, we have

$$f((\sigma\cdot\alpha_{\tau,\rho})\alpha_{\sigma,\tau\rho}) = f(\sigma\cdot\alpha_{\tau,\rho})f(\alpha_{\sigma,\tau\rho}) = f(\sigma\cdot\alpha_{\tau,\rho})\beta_\sigma(\sigma\cdot\beta_{\tau\rho})\beta_{\sigma\tau\rho}^{-1}.$$

Since $f(A)$ is central in B, we get

$$f((\sigma\cdot\alpha_{\tau,\rho})\alpha_{\sigma,\tau\rho}) = \beta_\sigma f(\sigma\cdot\alpha_{\tau,\rho})(\sigma\cdot\beta_{\tau\rho})\beta_{\sigma\tau\rho}^{-1} = \beta_\sigma(\sigma\cdot f(\alpha_{\tau,\rho}))(\sigma\cdot\beta_{\tau\rho})\beta_{\sigma\tau\rho}^{-1}.$$

Since σ acts by group automorphisms on B and $f(\alpha_{\tau,\rho}) = \beta_\tau(\tau\cdot\beta_\rho)\beta_{\tau\rho}^{-1}$, this yields

$$f((\sigma\cdot\alpha_{\tau,\rho})\alpha_{\sigma,\tau\rho}) = \beta_\sigma(\sigma\cdot\beta_\tau)(\sigma\tau\cdot\beta_\rho)(\sigma\cdot\beta_{\tau\rho}^{-1})(\sigma\cdot\beta_{\tau\rho})\beta_{\sigma\tau\rho}^{-1}.$$

Thus we get

$$f((\sigma\cdot\alpha_{\tau,\rho})\alpha_{\sigma,\tau\rho}) = \beta_\sigma(\sigma\cdot\beta_\tau)(\sigma\tau\cdot\beta_\rho)\beta_{\sigma\tau\rho}^{-1}.$$

On the other hand, we have

$$f(\alpha_{\sigma\tau,\rho}\alpha_{\sigma,\tau}) = f(\alpha_{\sigma,\tau}\alpha_{\sigma\tau,\rho}) = f(\alpha_{\sigma,\tau})f(\alpha_{\sigma\tau,\rho}) = \beta_\sigma(\sigma\cdot\beta_\tau)(\sigma\tau\cdot\beta_\rho)\beta_{\sigma\tau\rho}^{-1}.$$

Hence $f((\sigma\cdot\alpha_{\tau,\rho})\alpha_{\sigma,\tau\rho}) = f(\alpha_{\sigma\tau,\rho}\alpha_{\sigma,\tau})$, and the injectivity of f shows that α is a 2-cocycle.

Now take another preimage β'_σ of γ_σ under g for each σ, and denote by α' the corresponding 2-cocycle. Since $g(\beta'_\sigma) = g(\beta_\sigma)$, we have $\beta'_\sigma = \beta_\sigma f(\varphi_\sigma)$ for some $\varphi_\sigma \in A$. Since $\beta'_1 = \beta_1 = 1$ and f is injective, we get that $\varphi_1 = 1$. One can easily see, using the fact that $f(A)$ is central in B, that

$$f(\alpha'_{\sigma,\tau}) = f((\sigma\cdot\varphi_\tau)\varphi_{\sigma\tau}^{-1}\varphi_\sigma\alpha_{\sigma,\tau}) = f(d_1(\varphi)_{\sigma,\tau}\alpha_{\sigma,\tau}).$$

Using the injectivity of f, we get that α and α' are cohomologous.

Finally, let γ' be a 1-cocycle cohomologous to γ, so $\gamma'_\sigma = c\gamma_\sigma\,\sigma\cdot c^{-1}$ for some $c \in C$. Writing $c = g(b)$, one sees that $\beta'_\sigma = b\beta_\sigma\,\sigma\cdot b^{-1}$ is a preimage of γ'_σ. Then we have

$$\begin{aligned} f(\alpha'_{\sigma,\tau}) &= (b\beta_\sigma\,\sigma\cdot b^{-1})(\sigma\cdot b\,\sigma\cdot\beta_\tau\,\sigma\tau\cdot b^{-1})(\sigma\tau\cdot b\,\beta_{\sigma\tau}^{-1}b^{-1}) \\ &= b\beta_\sigma\,\sigma\cdot\beta_\tau\,\beta_{\sigma\tau}^{-1}b^{-1} \\ &= bf(\alpha_{\sigma,\tau})b^{-1} \\ &= f(\alpha_{\sigma,\tau}), \end{aligned}$$

since $f(A)$ is central. Hence we get $\alpha' = \alpha$ in this case. This concludes the proof. □

If $\gamma = 1$, we may take $\beta_\sigma = 1$ for all $\sigma \in \Gamma$. In this case, α is the trivial 2-cocycle, and thus $\delta^1([1]) = [1]$. Therefore, we get a map of pointed sets

$$\delta^1 : H^1(\Gamma, C) \longrightarrow H^2(\Gamma, A)$$

which sends $[\gamma] \in H^1(\Gamma, C)$ to the class $[\alpha] \in H^2(\Gamma, A)$, where α is defined by the relations

$$f(\alpha_{\sigma,\tau}) = \beta_\sigma(\sigma\cdot\beta_\tau)\beta_{\sigma\tau}^{-1} \text{ for all } \sigma, \tau \in \Gamma,$$

where $\beta_\sigma \in B$ is any preimage of γ_σ under g.

Definition II.4.9. The map $\delta^1 : H^1(\Gamma, C) \longrightarrow H^2(\Gamma, A)$ is called the **first connecting map**.

Proposition II.4.10. *The sequence of pointed sets*

$$1 \to A^\Gamma \to B^\Gamma \to C^\Gamma \to H^1(\Gamma, A) \to H^1(\Gamma, B) \to H^1(\Gamma, C) \to H^2(\Gamma, A)$$

is exact.

Proof. We keep the notation above. By Proposition II.4.7, we only have to prove the exactness at $H^1(\Gamma, C)$. Assume that $[\gamma] = g_*([\beta])$ for some $[\beta] \in H^1(\Gamma, B)$. Then there exists $c \in C$ such that

$$\gamma_\sigma = cg(\beta_\sigma)\,\sigma\cdot c^{-1} \text{ for all } \sigma \in \Gamma.$$

Let $b \in B$ be a preimage of c under g. We then have

$$\gamma_\sigma = g(b)g(\beta_\sigma)\,\sigma\cdot g(b)^{-1} = g(b\beta_\sigma\,\sigma\cdot b^{-1}) \text{ for all } \sigma \in \Gamma.$$

Therefore, replacing β by a cohomologous cocycle if necessary, we may assume without loss of generality that $\gamma_\sigma = g(\beta_\sigma)$ for all $\sigma \in \Gamma$. In this case, β_σ is a preimage of γ_σ under g, and $\delta^1([\gamma])$ is represented by the 2-cocycle $\alpha : \Gamma \times \Gamma \longrightarrow A$ defined by the relations

$$f(\alpha_{\sigma,\tau}) = \beta_\sigma\,\sigma\cdot\beta_\tau\,\beta_{\sigma\tau}^{-1} \text{ for all } \sigma, \tau \in \Gamma.$$

Since β is a 1-cocycle, we get

$$f(\alpha_{\sigma,\tau}) = 1 \text{ for all } \sigma, \tau \in \Gamma,$$

and the injectivity of f implies that $\alpha = 1$. Therefore, $\delta^1([\gamma]) = 1$ and we get $\mathrm{im}(g_*) \subset \ker(\delta^1)$.

Conversely, assume that $\delta^1([\gamma]) = 1$ is the trivial class, so that $\alpha_{\sigma,\tau} = \varphi_\sigma(\sigma\cdot\varphi_\tau)\varphi_{\sigma\tau}^{-1}$, for some continuous map $\varphi : \Gamma \longrightarrow A$ satisfying $\varphi_1 = 1$. We then have

$$f(\alpha_{\sigma,\tau}) = \beta_\sigma(\sigma\cdot\beta_\tau)\beta_{\sigma\tau}^{-1} = f(\varphi_\sigma)(\sigma\cdot f(\varphi_\tau))f(\varphi_{\sigma\tau})^{-1}.$$

It easily follows from the fact that $f(A)$ is central that the map

$$\beta' : \begin{array}{l} \Gamma \longrightarrow B \\ \sigma \longmapsto \beta_\sigma f(\varphi_\sigma)^{-1} \end{array}$$

is a cocycle. Moreover, the cohomology class $g_*([\beta'])$ is represented by the cocycle

$$\begin{array}{l} \Gamma \longrightarrow C \\ \sigma \longmapsto g(\beta'_\sigma). \end{array}$$

Now we have

$$g(\beta'_\sigma) = g(\beta_\sigma)g(f(\varphi_\sigma))^{-1} = g(\beta_\sigma) = \gamma_\sigma,$$

since $g \circ f = 1$. Therefore $[\gamma] = g_*([\beta']) \in \mathrm{im}(g_*)$ and we are done. □

Proposition II.4.11. *Let A, B, C be Γ-groups, and let A', B', C' be Γ'-groups. Assume that we have a commutative diagram with exact rows*

$$\begin{array}{ccccccccc} 1 & \longrightarrow & A & \xrightarrow{f} & B & \xrightarrow{g} & C & \longrightarrow & 1 \\ & & \downarrow{\scriptstyle h_1} & & \downarrow{\scriptstyle h_2} & & \downarrow{\scriptstyle h_3} & & \\ 1 & \longrightarrow & A' & \xrightarrow{f'} & B' & \xrightarrow{g'} & C' & \longrightarrow & 1 \end{array}$$

Assume that $f(A)$ and $f'(A')$ are central subgroups of B and B' respectively, and let us denote by δ^1 and δ'^1 the respective first connecting maps. If $\varphi : \Gamma' \longrightarrow \Gamma$ is compatible with h_1, h_2 and h_3, the diagram

$$\begin{array}{ccc} H^1(\Gamma, C) & \xrightarrow{\delta^1} & H^2(\Gamma, A) \\ \downarrow{\scriptstyle h_{3*}} & & \downarrow{\scriptstyle h_{1*}} \\ H^1(\Gamma', C') & \xrightarrow{\delta'^1} & H^2(\Gamma', A') \end{array}$$

is commutative.

Proof. Let $[\gamma] \in H^1(\Gamma, C)$. For all $\sigma \in \Gamma$, let $\beta_\sigma \in B$ be a preimage of γ_σ under g. The cohomology class $\delta^1([\gamma])$ is represented by the cocycle $\alpha : \Gamma \times \Gamma \longrightarrow A$ defined by the relations

$$f(\alpha_{\sigma,\tau}) = \beta_\sigma(\sigma\cdot\beta_\tau)\beta_{\sigma\tau}^{-1} \text{ for all } \sigma, \tau \in \Gamma.$$

Hence $h_{1*}(\delta^1([\gamma]))$ is represented by the cocycle

$$\eta\colon \begin{array}{l} \Gamma' \times \Gamma' \longrightarrow B' \\ (\sigma', \tau') \longmapsto h_1(\alpha_{\varphi(\sigma'),\varphi(\tau')}). \end{array}$$

By commutativity of the diagram, we have

$$f'(\eta_{\sigma',\tau'}) = h_2(f(\alpha_{\varphi(\sigma'),\varphi(\tau')})) = h_2(\beta_{\varphi(\sigma')}(\varphi(\sigma')\cdot\beta_{\varphi(\tau')})\beta_{\varphi(\sigma')\varphi(\tau')}^{-1}),$$

for all $\sigma', \tau' \in \Gamma'$. Since φ is a group morphism, we get

$$f'(\eta_{\sigma',\tau'}) = h_2(\beta_{\varphi(\sigma')}(\varphi(\sigma')\cdot\beta_{\varphi(\tau')})\beta_{\varphi(\sigma'\tau')}^{-1}) \text{ for all } \sigma', \tau' \in \Gamma'.$$

Now since φ and h_2 are compatible (so h_2 is in particular a group morphism), we get

$$f'(\eta_{\sigma',\tau'}) = h_2(\beta_{\varphi(\sigma')})(\sigma'\cdot h_2(\beta_{\varphi(\tau')}))h_2(\beta_{\varphi(\sigma'\tau')}^{-1}) \text{ for all } \sigma', \tau' \in \Gamma'.$$

On the other hand, $h_{3*}([\gamma])$ is represented by the cocycle

$$\begin{aligned} \Gamma' &\longrightarrow A' \\ \sigma' &\longmapsto h_3(\gamma_{\varphi(\sigma')}). \end{aligned}$$

By commutativity of the diagram, we have

$$h_3(\gamma_{\varphi(\sigma')}) = h_3(g(\beta_{\varphi(\sigma')})) = g'(h_2(\beta_{\varphi(\sigma')})).$$

Therefore, $h_2(\beta_{\varphi(\sigma')})$ is a preimage of $h_3(\gamma_{\varphi(\sigma')})$ under g' for all $\sigma' \in \Gamma'$, and thus the cohomology class $\delta'^1(h_{3*}([\gamma]))$ is represented by the cocycle

$$\eta' : \begin{aligned} \Gamma' \times \Gamma' &\longrightarrow B' \\ (\sigma', \tau') &\longmapsto \eta'_{\sigma',\tau'}, \end{aligned}$$

where $\eta'_{\sigma',\tau'}$ satisfies

$$f'(\eta'_{\sigma',\tau'}) = h_2(\beta_{\varphi(\sigma')})(\sigma' \cdot h_2(\beta_{\varphi(\tau')}))h_2(\beta^{-1}_{\varphi(\sigma'\tau')}) \text{ for all } \sigma', \tau' \in \Gamma'.$$

Therefore,

$$f'(\eta'_{\sigma',\tau'}) = f'(\eta_{\sigma',\tau'}) \text{ for all } \sigma', \tau' \in \Gamma'.$$

By injectivity of f', we get $\eta' = \eta$. Hence $h_{1*}(\delta^1([\gamma]))$ and $\delta'^1(h_{3*}([\gamma]))$ are represented by the same cocycle. This concludes the proof. □

§II.5 Twisting

Let $f : A \longrightarrow B$ be a morphism of Γ-groups. Even if the map of pointed sets $f_* : H^1(\Gamma, A) \longrightarrow H^1(\Gamma, B)$ may have trivial kernel in some cases, it does not mean that it will be injective. To study injectivity of f_*, the main idea is to identify the fiber $f_*^{-1}([\beta])$ of an element $[\beta] \in H^1(\Gamma, B)$ under f_* to the kernel of some appropriate induced map in cohomology. For this, we need the method of twisting.

Lemma II.5.1. *Let A be a Γ-group, and let $\alpha \in Z^1(\Gamma, A)$ be a cocycle. Then the map*

$$\begin{aligned} \Gamma \times A &\longrightarrow A \\ (\sigma, a) &\longmapsto \sigma * a = \alpha_\sigma \, \sigma \cdot a \, \alpha_\sigma^{-1} \end{aligned}$$

endows A with the structure of a Γ-group.

Proof. We first prove that $*$ is indeed an action of Γ on A. Let $a \in A$.

Since $\alpha_1 = 1$, it is clear that $1 \in \Gamma$ acts trivially on a. Now if $\sigma, \tau \in \Gamma$, we have

$$\begin{aligned}
\sigma\tau * a &= \alpha_{\sigma\tau}\, \sigma\tau \cdot a\, \alpha_{\sigma\tau}^{-1} \\
&= \alpha_\sigma\, \sigma \cdot \alpha_\tau\, \sigma\tau \cdot a\, \sigma \cdot \alpha_\tau^{-1} \alpha_\sigma^{-1} \\
&= \alpha_\sigma (\sigma \cdot (\alpha_\tau\, \tau \cdot a \alpha_\tau^{-1})) \alpha_\sigma^{-1} \\
&= \alpha_\sigma\, \sigma \cdot (\tau * a)\, \alpha_\sigma^{-1} \\
&= \sigma * (\tau * a).
\end{aligned}$$

Moreover, the action $*$ is an action by group automorphisms. Indeed, if $a, a' \in A$ and $\sigma \in \Gamma$, we have

$$\begin{aligned}
\sigma * (aa') &= \alpha_\sigma\, \sigma \cdot aa'\, \alpha_\sigma^{-1} \\
&= \alpha_\sigma\, (\sigma \cdot a)(\sigma \cdot a')\, \alpha_\sigma^{-1} \\
&= \alpha_\sigma\, \sigma \cdot a\, \alpha_\sigma^{-1} \alpha_\sigma\, (\sigma \cdot a')\, \alpha_\sigma^{-1} \\
&= (\sigma * a)(\sigma * a').
\end{aligned}$$

It remains to prove that this action is continuous. For, let $\sigma \in \Gamma$ such that $\sigma * a = a$. We need to prove that the stabilizer of a with respect to the action $*$ contains an open neighbourhood of σ. Let $U = \mathrm{Stab}_\Gamma(a) \cap \alpha^{-1}(\{1\})$, where $\mathrm{Stab}_\Gamma(a)$ is the stabilizer of a under the untwisted action of Γ. Then U is an open subgroup of Γ since α is continuous and A is a Γ-group. For all $\tau \in U$, we then have

$$\sigma\tau \cdot a = \sigma \cdot (\tau \cdot a) = \sigma \cdot a,$$

and

$$\alpha_{\sigma\tau} = \alpha_\sigma\, \sigma \cdot \alpha_\tau = \alpha_\sigma\, \sigma \cdot 1 = \alpha_\sigma.$$

Putting things together, for all $\tau \in U$, we have

$$\sigma\tau * a = \sigma * a = a.$$

Hence the stabilizer of a under the action $*$ contains σU. This concludes the proof. □

Definition II.5.2. The Γ-group A described in the previous lemma will be denoted by A_α in the sequel. We will say that A_α is obtained by **twisting** A by the cocycle α.

We then have the following proposition:

Proposition II.5.3. *Let A be a Γ-group and $\alpha \in Z^1(\Gamma, A)$. Then the map*

$$\theta_\alpha : \begin{array}{ccc} H^1(\Gamma, A_\alpha) & \longrightarrow & H^1(\Gamma, A) \\ {[\gamma]} & \longmapsto & [\gamma\alpha] \end{array}$$

is a well-defined bijection which maps the class of the trivial cocycle of $H^1(\Gamma, A_\alpha)$ *onto* $[\alpha]$.

Proof. Let $\gamma \in Z^1(\Gamma, A_\alpha)$. We first have to check that $\gamma\alpha \in Z^1(\Gamma, A)$. For all $\sigma, \tau \in \Gamma$, we have $\gamma_{\sigma\tau} = \gamma_\sigma\, \sigma{*}\gamma_\tau$, so by definition of the action of Γ on A_α, we have $\gamma_{\sigma\tau} = \gamma_\sigma\alpha_\sigma(\sigma\cdot\gamma_\tau)\alpha_\sigma^{-1}$. Thus

$$\gamma_{\sigma\tau}\alpha_{\sigma\tau} = \gamma_\sigma\alpha_\sigma(\sigma\cdot\gamma_\tau)(\sigma\cdot\alpha_\tau) = \gamma_\sigma\alpha_\sigma\,\sigma\cdot(\gamma_\tau\alpha_\tau),$$

and therefore $\gamma\alpha$ is a cocycle with values in A. Now assume that $\gamma' \in Z^1(\Gamma, A_\alpha)$ is cohomologous to γ, so that $\gamma'_\sigma = a\gamma_\sigma\, \sigma{*}a^{-1}$ for some $a \in A_\alpha$. We then have $\gamma'_\sigma = a\gamma_\sigma\,\alpha_\sigma(\sigma\cdot a^{-1})\alpha_\sigma^{-1}$, and so $\gamma'_\sigma\alpha_\sigma = a\gamma_\sigma\alpha_\sigma\,\sigma\cdot a^{-1}$. Therefore, $\gamma'\alpha$ and $\gamma\alpha$ are cohomologous, proving that θ_α is well-defined.

To check that θ_α is a bijection, observe that α^{-1} is a 1-cocycle with values in A_α. Indeed, for all $\sigma, \tau \in \Gamma$, we have

$$\begin{aligned}
\alpha_{\sigma\tau}^{-1} &= (\alpha_\sigma\,\sigma\cdot\alpha_\tau)^{-1} \\
&= \sigma\cdot\alpha_\tau^{-1}\alpha_\sigma^{-1} \\
&= \alpha_\sigma^{-1}\alpha_\sigma\,\sigma\cdot\alpha_\tau^{-1}\alpha_\sigma^{-1} \\
&= \alpha_\sigma^{-1}\,\sigma{*}\alpha_\tau^{-1}.
\end{aligned}$$

We may then twist A_α by α^{-1}. By definition, we have $(A_\alpha)_{\alpha^{-1}} = A$, so we get a map

$$\theta_{\alpha^{-1}} : \begin{array}{c} H^1(\Gamma, A) \longrightarrow H^1(\Gamma, A_\alpha) \\ [\xi] \longmapsto [\xi\alpha^{-1}] \end{array}.$$

It is then easy to check that θ_α and $\theta_{\alpha^{-1}}$ are mutually inverse. □

Remark II.5.4. If A is commutative, then $A_\alpha = A$ as Γ-modules and θ_α is just the translation by $[\alpha]$.

We now describe the functorial properties of twisting.

First, assume that $f : A \longrightarrow B$ is a morphism of Γ-groups, and let $\beta = f_*(\alpha) \in Z^1(\Gamma, B)$. Then the map f, considered as a map $f_\alpha : A_\alpha \longrightarrow B_\beta$, is a morphism of Γ-groups and the diagram

$$\begin{array}{ccc}
H^1(\Gamma, A_\alpha) & \xrightarrow{\theta_\alpha} & H^1(\Gamma, A) \\
\Big\downarrow{\scriptstyle (f_\alpha)_*} & & \Big\downarrow{\scriptstyle f_*} \\
H^1(\Gamma, B_\beta) & \xrightarrow{\theta_\beta} & H^1(\Gamma, B)
\end{array}$$

is commutative. In particular, we get immediatly that θ_α induces a bijection between $\ker((f_\alpha)_*)$ and the fiber $f_*^{-1}([\beta])$. Hence we get the following lemma:

Lemma II.5.5. *Let $f : A \longrightarrow B$ be a map of Γ-groups. The induced map $f_* : H^1(\Gamma, A) \longrightarrow H^1(\Gamma, B)$ is injective if and only if $(f_\alpha)_*$ has trivial kernel for every $\alpha \in Z^1(\Gamma, A)$.*

Before continuing, we would like to generalize the definition of a twisted Γ-group. Let A be a Γ-group. We let Γ act on $\mathrm{Aut}(A)$ as follows: if $f \in \mathrm{Aut}(A)$ and $\sigma \in \Gamma$, we set

$$(\sigma \cdot f)(a) = \sigma \cdot (f(\sigma^{-1} \cdot a)) \text{ for all } a \in A.$$

This is an action by group automorphisms. We will assume in the sequel that this action is continuous. Let us point out that this is not necessarily the case (see the exercises for a counterexample).
Let $\xi \in Z^1(\Gamma, \mathrm{Aut}(A))$. As above, we may define a new continuous action $*$ of Γ on A by

$$\sigma * a = \xi_\sigma(\sigma \cdot a) \text{ for all } \sigma \in \Gamma, a \in A.$$

This new Γ-group is denoted by A_ξ. Now if $\alpha \in Z^1(\Gamma, A)$, we denote by $\overline{\alpha} \in Z^1(\Gamma, \mathrm{Aut}(A))$ the image of α under the map $\mathrm{Int} : A \longrightarrow \mathrm{Aut}(A)$. Hence the Γ-group $A_{\overline{\alpha}}$ is nothing but the Γ-group A_α defined above. However, we will keep the notation A_α in the sequel for this particular twisted Γ-group.

Now assume that we have an exact sequence of Γ-groups

$$1 \longrightarrow A \xrightarrow{f} B \xrightarrow{g} C \longrightarrow 1$$

where A is abelian and $f(A)$ is a central subgroup of B.

Let $\gamma \in Z^1(\Gamma, C)$. If $c \in C$, let $b \in B$ such that $c = g(b)$. Then the inner automorphism $\mathrm{Int}(b) \in \mathrm{Aut}(B)$ does not depend on the choice of b, since $f(A)$ is a central subgroup of B.

We then get a well-defined map $\psi : C \longrightarrow \mathrm{Aut}(B)$. Let $\beta = \psi_*(\gamma) \in Z^1(\Gamma, \mathrm{Aut}(B))$. Finally, set $[\varepsilon] = \delta^1([\gamma])$, where δ^1 is the connecting map $\delta^1 : H^1(\Gamma, B) \longrightarrow H^2(\Gamma, A)$.

Proposition II.5.6. *With the previous notation, the sequence of Γ-groups*

$$1 \longrightarrow A \longrightarrow B_\beta \longrightarrow C_\gamma \longrightarrow 1$$

is exact and the diagram

$$\begin{array}{ccc} H^1(\Gamma, C_\gamma) & \xrightarrow{\theta_\gamma} & H^1(\Gamma, C) \\ \downarrow{\scriptstyle \delta^1_\gamma} & & \downarrow{\scriptstyle \delta^1} \\ H^2(\Gamma, A) & \xrightarrow{\mu} & H^2(\Gamma, A) \end{array}$$

commutes, where δ^1_γ is the connecting map with respect to the exact sequence above, and μ is multiplication by $[\varepsilon]$.

Proof. We let the reader check that the morphisms involved in the exact sequence respect the new actions of Γ. Let $\alpha \in Z^1(\Gamma, C_\gamma)$. Let x_σ be a preimage of α_σ in B_β under $g_\beta = g$ and let $y_\sigma \in B$ be a preimage of γ_σ under g.

By definition of ε, we have $f(\varepsilon_{\sigma,\tau}) = y_\sigma(\sigma \cdot y_\tau)y_{\sigma\tau}^{-1}$. Since $x_\sigma y_\sigma$ is a preimage of $\alpha_\sigma\gamma_\sigma$ for all $\sigma \in \Gamma$, $\delta^1(\theta_\gamma(\alpha)) = \delta^1([\alpha\gamma])$ is represented by the cocycle $\eta : \Gamma \times \Gamma \longrightarrow A$ satisfying

$$f(\eta_{\sigma,\tau}) = x_\sigma y_\sigma\,(\sigma\cdot(x_\tau y_\tau))y_{\sigma\tau}^{-1}x_{\sigma\tau}^{-1},$$

for all $\sigma, \tau \in \Gamma$. Hence we get

$$f(\eta_{\sigma,\tau}) = x_\sigma y_\sigma(\sigma\cdot x_\tau)y_\sigma^{-1}f(\varepsilon_{\sigma,\tau})x_{\sigma\tau}^{-1}.$$

By definition of β and B_β, we have $\sigma * a = y_\sigma\,\sigma\cdot a\,y_\sigma^{-1}$, and therefore,

$$f(\eta_{\sigma,\tau}) = x_\sigma(\sigma * x_\tau)x_{\sigma\tau}^{-1}f(\varepsilon_{\sigma,\tau}) = x_\sigma(\sigma * x_\tau)f(\varepsilon_{\sigma,\tau})x_{\sigma\tau}^{-1}.$$

Now $\delta^1_\gamma([\alpha])$ is represented by the cocycle $\eta' : \Gamma \times \Gamma \longrightarrow A$ satisfying

$$f(\eta'_{\sigma,\tau}) = x_\sigma(\sigma * x_\tau)x_{\sigma\tau}^{-1} \text{ for all } \sigma, \tau \in \Gamma.$$

Therefore, we get

$$f(\eta_{\sigma,\tau}) = f(\eta'_{\sigma,\tau})f(\varepsilon_{\sigma,\tau}) = f(\eta'_{\sigma,\tau}\varepsilon_{\sigma,\tau}),$$

for all $\sigma, \tau \in \Gamma$. By injectivity of f, this means that $\eta = \eta'\varepsilon$. In other words,

$$\delta^1(\theta_\gamma([\alpha])) = \mu(\delta^1_\gamma([\alpha])).$$

This completes the proof of the proposition. □

§II.6 Cup-products

In this section, we introduce briefly the cup-product, which is a useful tool to construct higher cohomology classes.

Let A and B be two Γ-modules. Since A and B are abelian groups, that is $\mathbb{Z}$-modules, we may consider $A \otimes_{\mathbb{Z}} B$. Then Γ acts continuously on $A \otimes_{\mathbb{Z}} B$ as follows:

$$\sigma\cdot(a \otimes b) = (\sigma\cdot a) \otimes (\sigma\cdot b) \text{ for all } \sigma \in \Gamma, a \in A, b \in B.$$

This endows $A \otimes_{\mathbb{Z}} B$ with a structure of a Γ-module. We will keep this notation until the end of the section.

Proposition II.6.1. *Let $p, q \geq 1$ be two integers. Let $\alpha \in Z^p(\Gamma, A)$ and $\beta \in Z^q(\Gamma, B)$ be two cocycles. The map*

$$\alpha \cup \beta : \begin{array}{l} \Gamma^{p+q} \longrightarrow A \otimes_{\mathbb{Z}} B \\ (\sigma_1, \ldots, \sigma_{p+q}) \longmapsto \alpha_{\sigma_1,\ldots,\sigma_p} \otimes \sigma_1 \cdots \sigma_p \cdot \beta_{\sigma_{p+1},\ldots,\sigma_{p+q}} \end{array}$$

is a $(p+q)$-cocycle, whose cohomology class only depends on the cohomology classes $[\alpha]$ and $[\beta]$, and the map

$$\cup : \begin{array}{l} H^p(\Gamma, A) \times H^q(\Gamma, B) \longrightarrow H^{p+q}(\Gamma, A \otimes_{\mathbb{Z}} B) \\ ([\alpha], [\beta]) \longmapsto [\alpha \cup \beta] \end{array}$$

is $\mathbb{Z}$-bilinear. Moreover, after identifying canonically the $\mathbb{Z}$-modules $B \otimes_{\mathbb{Z}} A$ and $A \otimes_{\mathbb{Z}} B$, we have

$$[\alpha] \cup [\beta] = (-1)^{pq} [\beta] \cup [\alpha] \text{ for all } [\alpha] \in H^p(\Gamma, A), [\beta] \in H^q(\Gamma, B).$$

Proof. The Γ-modules A and B will be denoted additively. We will only sketch a proof of the first part, and refer to [11] or [26] for a detailed proof. Tedious but straightforward computations show that we have the formula

$$d_{p+q}(\alpha \cup \beta) = d_p(\alpha) \cup \beta + (-1)^p \alpha \cup d_q(\beta),$$

for all continuous maps $\alpha : \Gamma^p \longrightarrow A$ and $\beta : \Gamma^q \longrightarrow B$. In particular, if α and β are cocycles, so is $\alpha \cup \beta$.

To prove that the cohomology class of $\alpha \cup \beta$ only depends on $[\alpha]$ and $[\beta]$, we have to prove that if α or β is a coboundary, so is $\alpha \cup \beta$.

Assume for example that α is a coboundary, so $\alpha = d_{p-1}(\gamma)$ for some continuous map $\gamma : \Gamma^{p-1} \longrightarrow A$. Then we have

$$d_{p-1+q}(\gamma \cup \beta) = \alpha \cup \beta + (-1)^{p-1} \gamma \cup d_q(\beta) = \alpha \cup \beta,$$

since β is a cocycle. Hence $\alpha \cup \beta$ lies in the image of d_{p+q-1}, and is therefore a coboundary. The other case is proved in a similar way.

□

Remark II.6.2. The last part of the proposition is a property which is difficult to establish at the level of cocycles, and its proof needs shifting techniques in cohomology (see [26] for example for a detailed proof). However, we will use it only in the case where Γ acts trivially on A and B and $2A = 2B = 0$. In this particular case, it is immediate.

Definition II.6.3. The map $\cup$ is called the **cup-product**.

Let C be a third Γ-module and let $\theta : A \times B \longrightarrow C$ be a $\mathbb{Z}$-bilinear map satisfying

$$\theta(\sigma\cdot a, \sigma\cdot b) = \sigma\cdot\theta(a, b) \text{ for all } \sigma \in \Gamma, a \in A, b \in B.$$

Such a map will be called a bilinear map of Γ-modules. It induces a map of Γ-modules $A \otimes_{\mathbb{Z}} B \longrightarrow C$, still denoted by θ. We therefore get an induced map

$$\theta_* : H^n(\Gamma, A \otimes_{\mathbb{Z}} B) \longrightarrow H^n(\Gamma, C).$$

Definition II.6.4. The map $\cup_\theta : H^p(\Gamma, A) \times H^q(\Gamma, B) \longrightarrow H^{p+q}(\Gamma, C)$ defined by

$$[\alpha] \cup_\theta [\beta] = \theta_*([\alpha] \cup [\beta]) = [\theta \circ (\alpha \cup \beta)]$$

is called the **cup-product relative to** θ.

Remark II.6.5. It follows from the properties of the cup-product that $\cup_\theta$ is also $\mathbb{Z}$-bilinear and satisfies

$$[\alpha] \cup_\theta [\beta] = (-1)^{pq}[\beta] \cup_\theta [\alpha] \text{ for all } [\alpha] \in H^p(\Gamma, A), [\beta] \in H^q(\Gamma, B).$$

We end this chapter by studying the functorial properties of the cup-product.

Proposition II.6.6. *Let Γ and Γ' be two profinite groups, and let $\theta : A_1 \times A_2 \longrightarrow A_3$ and $\theta' : A_1' \times A_2' \longrightarrow A_3'$ be two bilinear maps of Γ-modules and Γ'-modules respectively.*

Assume that $\varphi : \Gamma' \longrightarrow \Gamma$ is compatible with $f_i : A_i \longrightarrow A_i'$ for $i = 1, 2, 3$, and that the diagram

$$\begin{array}{ccc} A_1 \times A_2 & \xrightarrow{\theta} & A_3 \\ \downarrow{\scriptstyle (f_1,f_2)} & & \downarrow{\scriptstyle f_3} \\ A_1' \times A_2' & \xrightarrow{\theta'} & A_3' \end{array}$$

is commutative. Then for all $p, q \geq 0$, *the diagram*

$$\begin{array}{ccc} H^p(\Gamma, A_1) \times H^q(\Gamma, A_2) & \xrightarrow{\cup_\theta} & H^{p+q}(\Gamma, A_3) \\ \downarrow{\scriptstyle ((f_1)_*,(f_2)_*)} & & \downarrow{\scriptstyle (f_3)_*} \\ H^p(\Gamma', A_1') \times H^q(\Gamma', A_2') & \xrightarrow{\cup_{\theta'}} & H^{p+q}(\Gamma', A_3') \end{array}$$

is commutative. In other words, for all $[\alpha] \in H^p(\Gamma, A_1)$ *and all* $[\beta] \in H^q(\Gamma, A_2)$, *we have*

$$(f_3)_*([\alpha] \cup_\theta [\beta]) = (f_1)_*([\alpha]) \cup_{\theta'} (f_2)_*([\beta]).$$

Proof. Notice first that the diagram

$$\begin{array}{ccc} A_1 \otimes_{\mathbb{Z}} A_2 & \xrightarrow{\theta} & A_3 \\ \downarrow{\scriptstyle f_1 \otimes f_2} & & \downarrow{\scriptstyle f_3} \\ A_1' \otimes_{\mathbb{Z}} A_2' & \xrightarrow{\theta'} & A_3' \end{array}$$

is commutative. Let $\alpha \in Z^p(\Gamma, A_1)$ and let $\beta \in Z^q(\Gamma, A_2)$. Then $[\alpha]\cup_\theta[\beta]$ is represented by the cocycle

$$\begin{aligned} \Gamma^{p+q} &\longrightarrow A_1 \otimes_{\mathbb{Z}} A_2 \\ (\sigma_1, \ldots, \sigma_{p+q}) &\longmapsto \theta(\alpha_{\sigma_1,\ldots,\sigma_p} \otimes \sigma_1 \cdots \sigma_p \cdot \beta_{\sigma_{p+1},\ldots,\sigma_{p+q}}), \end{aligned}$$

and $(f_3)_*([\alpha] \cup_\theta [\beta])$ is thus represented by the cocycle $\xi : \Gamma'^{p+q} \longrightarrow A_3$ defined for all $\sigma_1', \ldots, \sigma_{p+q}' \in \Gamma'$ by

$$\xi_{\sigma_1',\ldots,\sigma_{p+q}'} = f_3(\theta(\alpha_{\varphi(\sigma_1'),\ldots,\varphi(\sigma_p')} \otimes \varphi(\sigma_1') \cdots \varphi(\sigma_p') \cdot \beta_{\varphi(\sigma_{p+1}'),\ldots,\varphi(\sigma_{p+q}')})).$$

On the other hand, $(f_1)_*([\alpha]) \cup_{\theta'} (f_2)_*([\beta])$ is represented by the cocycle $\xi' : \Gamma'^{p+q} \longrightarrow A_3'$ defined for all $\sigma_1', \ldots, \sigma_{p+q}' \in \Gamma'$ by

$$\xi'_{\sigma_1',\ldots,\sigma_{p+q}'} = \theta'(f_1(\alpha_{\varphi(\sigma_1'),\ldots,\varphi(\sigma_p')}) \otimes \sigma_1' \cdots \sigma_p' \cdot f_2(\beta_{\varphi(\sigma_{p+1}'),\ldots,\varphi(\sigma_{p+q}')})).$$

To prove the proposition, it is enough to check that $\xi' = \xi$. Let $\sigma'_1, \ldots, \sigma'_{p+q} \in \Gamma'$. Since φ is compatible with f_2, we have

$$\xi'_{\sigma'_1,\ldots,\sigma'_{p+q}} = \theta'(f_1(\alpha_{\varphi(\sigma'_1),\ldots,\varphi(\sigma'_p)}) \otimes f_2(\varphi(\sigma'_1\cdots\sigma'_p)\cdot\beta_{\varphi(\sigma'_{p+1}),\ldots,\varphi(\sigma'_{p+q})})).$$

Using the fact that φ is a group morphism and the commutativity of the diagram above, we get the desired equality.

□

Example II.6.7. Let Γ and Γ' be two profinite groups, acting trivially on three abelian groups A, B, C. Let $\theta : A \times B \longrightarrow C$ be a bilinear map, and let $\varphi : \Gamma' \longrightarrow \Gamma$ be a morphism of profinite groups.

If $[\alpha] \in H^n(\Gamma, A)$, we will denote by $[\alpha]_{\Gamma'} \in H^n(\Gamma', A)$ its image under the map $H^n(\Gamma, A) \longrightarrow H^n(\Gamma', A)$ induced by φ and Id_A (and similarly for the other modules). For all $[\alpha] \in H^p(\Gamma, A), [\beta] \in H^q(\Gamma, B)$, we then have

$$([\alpha] \cup_\theta [\beta])_{\Gamma'} = [\alpha]_{\Gamma'} \cup_\theta [\beta]_{\Gamma'} \in H^{p+q}(\Gamma', C).$$

Notes

Our treatment of cohomology of profinite groups follows [58], which is the standard reference on this topic, and [30], from which we borrowed notation. The reader will find more results on twisting in [58], as well as an account on cohomology of groups in [57] or [26].

Exercises

1. Let Γ be a profinite group, and let A be a discrete set on which Γ acts on the left. Show that Γ acts continuously on A if and only if the map

$$\begin{aligned} \Gamma \times A &\longrightarrow A \\ (\sigma, a) &\longmapsto \sigma\cdot a \end{aligned}$$

is continuous.

2. Let Γ be a profinite group, let $n \geq 0$ be an integer, and let A, A', B be three Γ-groups (resp. three Γ-modules if $n \geq 2$).

(a) Show that the action of Γ on $A \times B$ defined by

$$\sigma \cdot (a,b) = (\sigma \cdot a, \sigma \cdot b) \text{ for all } \sigma \in \Gamma, a \in A, b \in B$$

endows $A \times B$ with the structure of a Γ-group, and that we have a bijection of pointed sets (resp. a group isomorphism if $n \geq 2$)

$$H^n(\Gamma, A \times B) \simeq H^n(\Gamma, A) \times H^n(\Gamma, B).$$

(b) Show that if the two Γ-groups A and A' are isomorphic, then we have a bijection of pointed sets (resp. a group isomorphism if $n \geq 2$)

$$H^n(\Gamma, A) \simeq H^n(\Gamma, A').$$

3. Let $\alpha \in Z^1(\Gamma, A)$ and let $\beta \in Z^1(\Gamma, B)$ be the image of α under the map induced by the inclusion $A \subset B$. Show that there is a natural bijection between the fiber of $H^1(\Gamma, A) \longrightarrow H^1(\Gamma, B)$ over $[\beta]$ and the orbit set of the group B_β^Γ in $(B_\beta/A_\alpha)^\Gamma$.

4. Let E be the additive group of complex sequences $u = (u_n)_{n\in\mathbb{Z}}$ with finite support, considered as a topological discrete group, and let p be a prime integer. For $x \in \mathbb{Z}_p$ and $u \in E$, denote by $x \cdot u$ the element of E defined by

$$(x \cdot u)_n = \begin{cases} u_n & \text{if } n \leq 0 \\ e^{\frac{2i\pi x}{p^n}} u_n & \text{if } n \geq 1. \end{cases}$$

Here $e^{\frac{2i\pi x}{p^n}}$ has to be understood as $e^{\frac{2i\pi x_n}{p^n}}$, where $x_n \in \mathbb{Z}$ is any representative of the class of x modulo $p^{n+1}\mathbb{Z}_p$.

(a) Show that the map

$$\mathbb{Z}_p \times E \longrightarrow E$$
$$u \longmapsto x \cdot u$$

endows E with a structure of a $\mathbb{Z}_p$-module.

(b) Let $f \in \mathrm{Aut}(E)$ be the automorphism of E defined by

$$f \colon \begin{array}{l} E \longrightarrow E \\ (u_n)_{n\in\mathbb{Z}} \longmapsto (u_{n-1})_{n\in\mathbb{Z}}. \end{array}$$

Show that the stabilizer of f for the action of $\mathbb{Z}_p$ induced by its action on E is $\{0\}$.

(c) Deduce that the action of $\mathbb{Z}_p$ on $\mathrm{Aut}(E)$ is not continuous.

5. Let B a Γ-group, let A be a Γ-subgroup and let $C = B/A$. Assume that A is normal in B. In particular, C is a Γ-group.

   (a) If $c = bA \in C^\Gamma$, and $\alpha \in Z^1(\Gamma, A)$, show that
   $$\beta_\sigma = b\alpha_\sigma \sigma \cdot b^{-1} \in A \text{ for all } \sigma \in \Gamma,$$
   and that the map $\beta : \Gamma \longrightarrow A$ is a 1-cocycle, whose cohomology class only depends on c and $[\alpha]$.

   (b) Show that the map
   $$\begin{aligned} C^\Gamma \times H^1(\Gamma, A) &\longrightarrow H^1(\Gamma, A) \\ (c, [\alpha]) &\longmapsto [\beta] \end{aligned}$$
   endows $H^1(\Gamma, A)$ with a left action of C^Γ, and that there is a natural bijection between $\ker(H^1(\Gamma, B) \longrightarrow H^1(\Gamma, C))$ and the orbit set of the group C^Γ in $H^1(\Gamma, A)$.

   (c) Let $\beta \in Z^1(\Gamma, B)$. Let $\gamma \in Z^1(\Gamma, C)$ be the image of β under the map induced by the canonical projection $B \longrightarrow C$, and let $\alpha \in Z^1(\Gamma, \mathrm{Aut}(A))$ be the image of β under the map induced by the conjugation map $B \longrightarrow \mathrm{Aut}(A)$.
   Show that there is a natural bijection between the fiber of $H^1(\Gamma, B) \longrightarrow H^1(\Gamma, C)$ over $[\gamma]$ and the orbit set of the group C_γ^Γ in $H^1(\Gamma, A_\alpha)$.

6. Let B a Γ-group, let A be a Γ-subgroup and let $C = B/A$. Assume that A is central in B.

   (a) Show that $H^1(\Gamma, A)$ acts naturally on $H^1(\Gamma, B)$ by
   $$[\alpha] \cdot [\beta] = [\alpha\beta].$$

   (b) Show that there is a natural bijection between the kernel of the connecting map $\delta^1 : H^1(\Gamma, C) \longrightarrow H^2(\Gamma, A)$ and the orbit set of the group $H^1(\Gamma, A)$ in $H^1(\Gamma, B)$.

   (c) Let $\gamma \in Z^1(\Gamma, C)$. The conjugation map $C \longrightarrow \mathrm{Aut}(B)$ induces a 1-cocycle $\beta \in Z^1(\Gamma, B)$. Let $[\varepsilon] = \delta^1([\gamma])$.
   Show that there is a natural bijection between the fiber of $\delta^1 : H^1(\Gamma, C) \longrightarrow H^2(\Gamma, A)$ over $[\varepsilon]$ and the orbit set of the group $H^1(\Gamma, A)$ in $H^1(\Gamma, B_\beta)$.

7. Let A be a Γ-group. A **principal homogeneous space** over A is a non-empty Γ-set P endowed with a **simply transitive** right action $*$ of A which is compatible with the left action of Γ, that is

$$\sigma\cdot(x*a) = (\sigma\cdot x)*(\sigma\cdot a), \text{ for all } \sigma \in \Gamma, a \in A, x \in P.$$

A morphism of principal homogeneous spaces is a map which is A-equivariant and Γ-equivariant. We denote by $\mathbf{Tors}(\Gamma, A)$ the set of isomorphism classes of principal homogeneous spaces over A.

(a) Check that a morphism of principal homogeneous spaces is an isomorphism.

(b) Let $\alpha \in Z^1(\Gamma, A)$ be a 1-cocycle, and let P_α be the set A, endowed with the following actions of Γ and A:

$$\begin{aligned} \Gamma \times P_\alpha &\longrightarrow P_\alpha \\ (\sigma, x) &\longmapsto \alpha_\sigma(\sigma\cdot x) \end{aligned}$$

$$\begin{aligned} P_\alpha \times A &\longrightarrow P_\alpha \\ (x, a) &\longmapsto xa. \end{aligned}$$

Show that P_α is a principal homogeneous space, and that for all $\alpha, \alpha' \in Z^1(\Gamma, A)$, we have

$$\alpha \sim \alpha' \Rightarrow P_\alpha \simeq P_{\alpha'}.$$

(c) Let P be a principal homogeneous space, and let $x \in P$. For every $\sigma \in \Gamma$, justify the existence of a unique $\alpha_\sigma \in A$ such that $\sigma\cdot x = x * \alpha_\sigma$. Show that the map $\alpha : \Gamma \longrightarrow A$ is a 1-cocycle, whose cohomology class does not depend on the choice of x.

(d) Conclude that $H^1(\Gamma, A) \simeq \mathbf{Tors}(\Gamma, A)$, that is $H^1(\Gamma, A)$ classifies principal homogeneous spaces up to isomorphism.

8. Let $f : A \longrightarrow B$ be a morphism of Γ-groups.

(a) If P is a principal homogeneous space over A, consider $P \times B$ endowed with the diagonal action of Γ. Show that the groups A and B acts on P by

$$\begin{aligned} (P \times B) \times A &\longrightarrow P \times B \\ ((x, b), a) &\longmapsto (x*a, f(a^{-1})b) \end{aligned}$$

$$
\begin{aligned}
(P \times B) \times B &\longrightarrow P \times B \\
((x, b), b') &\longmapsto (x, bb'),
\end{aligned}
$$

and that these two actions commute.

(b) Show that the induced action of B on the set of A-orbits $f_*(P) = (P \times B)/A$ endows $f_*(P)$ with a structure of a principal homogeneous space over B and that the isomorphism class of $f_*(P)$ only depends on the isomorphism class of P. We then get a map

$$f_* : \mathbf{Tors}(\Gamma, A) \longrightarrow \mathbf{Tors}(\Gamma, B).$$

(c) Show that diagram

$$
\begin{array}{ccc}
H^1(\Gamma, A) & \longrightarrow & H^1(\Gamma, B) \\
\downarrow & & \downarrow \\
\mathbf{Tors}(\Gamma, A) & \longrightarrow & \mathbf{Tors}(\Gamma, B)
\end{array}
$$

is commutative.

III
Galois cohomology

In this chapter, we study the specific properties of the cohomology sets associated to the Galois group of a Galois field extension. Our main goal is to express the obstruction of a Galois descent problem associated to algebraic objects in terms of Galois cohomology. Of course, this kind of question only makes sense if the algebraic objects may be 'defined over an arbitrary field', and if we have a notion of 'scalar extension' of our objects. For example, to each field extension K/k of a field k, we can consider the set $\mathrm{M}_n(K)$ of matrices with coefficients in K, and if $\iota : K \longrightarrow L$ is a morphism of extensions of k, one can associate a map

$$\begin{aligned} \mathrm{M}_n(K) &\longrightarrow \mathrm{M}_n(L) \\ (m_{ij}) &\longmapsto (\iota(m_{ij})). \end{aligned}$$

Another example may be obtained by considering the set $\mathbf{Alg}_n(K)$ of K-algebras of dimension n over K. In this case, the map $\mathbf{Alg}_n(K) \longrightarrow \mathbf{Alg}_n(L)$ associated to a morphism $K \longrightarrow L$ is given by the tensor product $\otimes_K L$. In both cases, scalar extension maps satisfy some natural properties: the scalar extension map induced by the trivial extension is the identity map, and extending scalars from K to L, then from L to M is the same as extending scalars from K to M.

We therefore start this chapter by formalizing this situation and introducing the concept of a functor between two categories.

§III.7 Warm-up

III.7.1 Digression: categories and functors

In this section, we recall some basic facts on categories and covariant functors. The reader is referred to [35] for details. Roughly speaking, the notion of a category is here to palliate to the (mostly inconvenient) fact that there is no 'set of all sets'. More seriously, when studying various objects such as sets, groups, rings, algebras, varieties, a common setting appears constantly. We define special kinds of maps between them, namely morphisms, sharing some similar properties: the composition of two morphisms (when defined) is a morphism, the identity map is a morphism, the composition of morphisms is associative.

The notion of a category allows us to axiomatize this common setting. As we will see, the axioms of a category are essentially related to the morphisms between the objects.

Definition III.7.1. A **category** $\mathcal{C}$ consists of:

(1) A collection of **objects** $\mathrm{Ob}(\mathcal{C})$.

(2) For two objects $A, B \in \mathrm{Ob}(\mathcal{C})$ a set $\mathrm{Mor}_{\mathcal{C}}(A,B)$ (possibly empty), called the set of **morphisms** $f : A \longrightarrow B$ from A to B.

(3) For three objects $A, B, C \in \mathrm{Ob}(\mathcal{C})$ a **law of composition**

$$\begin{aligned} \mathrm{Mor}_{\mathcal{C}}(B,C) \times \mathrm{Mor}_{\mathcal{C}}(A,B) &\longrightarrow \mathrm{Mor}_{\mathcal{C}}(A,C) \\ (g,f) &\longmapsto g \circ f \end{aligned}$$

satisfying the following axioms:

(a) Two sets $\mathrm{Mor}_{\mathcal{C}}(A,B)$ and $\mathrm{Mor}_{\mathcal{C}}(A',B')$ are disjoint unless $A = A'$ and $B = B'$, in which case they are equal.

(b) For all $A \in \mathrm{Ob}(\mathcal{C})$, there is a morphism $\mathrm{Id}_A \in \mathrm{Mor}_{\mathcal{C}}(A,A)$, called the identity morphism, acting as left and right identity on the elements of $\mathrm{Mor}_{\mathcal{C}}(A,B)$ and $\mathrm{Mor}_{\mathcal{C}}(B,A)$ respectively.

(c) The law of composition is associative (when defined).

We may then define the notion of isomorphism, endomorphism and automorphism in an obvious way.

Remark III.7.2. If $\mathcal{C}$ is a category, then for all $A \in \mathrm{Ob}(\mathcal{C})$, the identity morphism Id_A is unique. Indeed, assume that Id'_A is another identity morphism. Then we have

$$\mathrm{Id}_A = \mathrm{Id}_A \circ \mathrm{Id}'_A = \mathrm{Id}'_A,$$

by definition of the identity morphisms.

Examples III.7.3. The following categories will be used frequently in the sequel:

(1) The category **Sets** is the category whose objects are sets, and whose morphisms are set-theoretic maps.

(2) The category of pointed sets **Sets*** is the category whose objects are pointed sets and whose morphisms are maps of pointed sets.

(3) The category of groups **Grps** is the category whose objects are groups and whose morphisms are group morphisms. The category of abelian groups **AbGrps** is the category whose objects are abelian groups and whose morphisms are group morphisms.

(4) Let k be a field. The category $\mathbf{Alg}_k$ is the category whose objects are (associative unital) commutative k-algebras and whose morphisms are k-algebra morphisms.

(5) Let k be a field. The category $\mathfrak{C}_k$ is the category whose objects are the field extensions (K, ι) of k and whose morphisms are morphisms of extensions of k.

(6) If Γ is a profinite group, then Γ-sets, Γ-groups and Γ-modules, together with the appropriate morphisms, form three categories that we denote by $\mathbf{Sets}_\Gamma$, $\mathbf{Grps}_\Gamma$ and $\mathbf{Mod}_\Gamma$ respectively.

We now introduce the notion of a subcategory.

Definition III.7.4. Let $\mathcal{C}$ be a category. A **subcategory** of $\mathcal{C}$ is a category $\mathcal{C}'$ such that

(1) $\mathrm{Ob}(\mathcal{C}')$ is a subcollection of $\mathrm{Ob}(\mathcal{C})$.

(2) For every pair of objects $A, B \subset \mathrm{Ob}(\mathcal{C})$, we have $\mathrm{Mor}_{\mathcal{C}'}(A, B) \subset \mathrm{Mor}_{\mathcal{C}}(A, B)$.

(3) For all $A \in \mathrm{Ob}(\mathcal{C}')$, $\mathrm{Id}_A \in \mathrm{Mor}_{\mathcal{C}'}(A, A)$.

(4) The composition law in $\mathcal{C}'$ is the restriction of the composition law of $\mathcal{C}$.

Example III.7.5. The categories **Grps** and $\mathbf{Alg}_k$ are subcategories of **Sets**.

Definition III.7.6. Let $\mathcal{C}, \mathcal{C}'$ be two categories. A **covariant functor** is a rule $\mathbf{F} : \mathcal{C} \longrightarrow \mathcal{C}'$ which associates to each object $A \in \mathrm{Ob}(\mathcal{C})$ an object $\mathbf{F}(A)$ in $\mathrm{Ob}(\mathcal{C}')$ and to each morphism $f : A \longrightarrow B$ of $\mathcal{C}$ associates a morphism $\mathbf{F}(f) : \mathbf{F}(A) \longrightarrow \mathbf{F}(B)$ of $\mathcal{C}'$ such that:

(1) For all A in $\mathrm{Ob}(\mathcal{C})$, we have $\mathbf{F}(\mathrm{Id}_A) = \mathrm{Id}_{\mathbf{F}(A)}$.

(2) If $f : A \longrightarrow B$ and $g : B \longrightarrow C$ are two morphisms of $\mathcal{C}$, then

$$\mathbf{F}(g \circ f) = \mathbf{F}(g) \circ \mathbf{F}(f).$$

The definition is perhaps a bit more enlightening when $\mathcal{C} = \mathfrak{C}_k$ and $\mathcal{C}' = \mathbf{Sets}$. In this case, a functor may be viewed (roughly speaking) as a family of sets associated to field extensions, together with 'scalar extension maps'.

Examples III.7.7. Here are some classical examples of functors:

(1) The forgetful functor $\mathbf{F} : \mathbf{Grps} \longrightarrow \mathbf{Sets}$, which maps a group G to itself (viewed as a set), and a morphism of groups to itself (viewed as a map between sets).

(2) Let k be a field. For any field extension L/k, set $\mathbf{F}(L/k) = \mathrm{M}_n(L)$. If $\varphi : L \longrightarrow L'$ is a morphism of field extensions of k, we define $\mathbf{F}(\varphi)$ to be the map

$$\mathbf{F}(\varphi) \colon \begin{array}{l} \mathrm{M}_n(L) \longrightarrow \mathrm{M}_n(L') \\ (m_{ij}) \longmapsto (\varphi(m_{ij})). \end{array}$$

Clearly, axioms (1) and (2) are satisfied, so we obtain a functor from the category $\mathfrak{C}_k$ to the category $\mathbf{Sets}$, that we denote by $\mathbf{M}_n$.

(3) If $\mathcal{C}'$ is a subcategory of a category $\mathcal{C}$, the obvious rule $\mathbf{F} : \mathcal{C}' \longrightarrow \mathcal{C}$ is a functor.

(4) Let $n \geq 1$, and let k be a field. Let $\mathcal{C} = \mathbf{Alg}_k$ and $\mathcal{C}' = \mathbf{Grps}$. For any commutative k-algebra R, set $\mathbf{F}(R) = \mathrm{GL}_n(R)$. If $\varphi : R \longrightarrow R'$ is a morphism of k-algebras, we define a map $\mathbf{F}(\varphi)$ by

$$\mathbf{F}(\varphi) \colon \begin{array}{l} \mathbf{F}(R) \longrightarrow \mathbf{F}(R') \\ (m_{ij}) \longmapsto (\varphi(m_{ij})). \end{array}$$

It is easy to check that this map is a group morphism, and that axioms (1) and (2) are once again satisfied. Hence we obtain a functor $\mathbf{GL}_n : \mathbf{Alg}_k \longrightarrow \mathbf{Grps}$. If $n = 1$, it is denoted by $\mathbb{G}_m$.

(5) Let $\mathcal{C}$ be a category and let $A \in \mathrm{Ob}(\mathcal{C})$ be an object of $\mathcal{C}$. For every $B, C \in \mathrm{Ob}(\mathcal{C})$, and every $f : B \longrightarrow C \in \mathrm{Mor}_{\mathcal{C}}(B, C)$, let $h_A(B) = \mathrm{Mor}_{\mathcal{C}}(A, B)$ and

$$h_A(f) \colon \begin{array}{l} \mathrm{Mor}_{\mathcal{C}}(A, B) \longrightarrow \mathrm{Mor}_{\mathcal{C}}(A, C) \\ \varphi \longmapsto f \circ \varphi. \end{array}$$

We then obtain a covariant functor $h_A : \mathcal{C} \longrightarrow \mathbf{Sets}$.

(6) From Example II.3.20, we see that we have covariant functors

$$H^1(\Gamma, \_\,) : \mathbf{Grps}_\Gamma \longrightarrow \mathbf{Sets}^*$$

and

$$H^n(\Gamma, \_\,) : \mathbf{Mod}_\Gamma \longrightarrow \mathbf{AbGrps}.$$

Definition III.7.8. Let $\mathcal{C}, \mathcal{C}'$ be two categories, and let $\mathbf{F}_1 : \mathcal{C} \longrightarrow \mathcal{C}'$ and $\mathbf{F}_2 : \mathcal{C} \longrightarrow \mathcal{C}'$ be two covariant functors. A **morphism** (or **natural transformation**) of functors $\Theta : \mathbf{F}_1 \longrightarrow \mathbf{F}_2$ is a rule assigning to each object $A \in \mathrm{Ob}(\mathcal{C})$ an element $\Theta_A \in \mathrm{Mor}_{\mathcal{C}'}(\mathbf{F}_1(A), \mathbf{F}_2(A))$, $\Theta_A : \mathbf{F}_1(A) \longrightarrow \mathbf{F}_2(A)$, such that for every morphism $f : A \longrightarrow B$ of $\mathcal{C}$ the diagram

$$\begin{array}{ccc} \mathbf{F}_1(A) & \xrightarrow{\Theta_A} & \mathbf{F}_2(A) \\ \Big\downarrow{\scriptstyle \mathbf{F}_1(f)} & & \Big\downarrow{\scriptstyle \mathbf{F}_2(f)} \\ \mathbf{F}_1(B) & \xrightarrow{\Theta_B} & \mathbf{F}_2(B) \end{array}$$

is commutative.

Example III.7.9. Let k be a field. For every commutative k-algebra R, let $\det_R$ be the group morphism

$$\det_R : \begin{array}{l} \mathrm{GL}_n(R) \longrightarrow R^\times \\ \quad M \longmapsto \det(M). \end{array}$$

The properties of the determinant of matrices imply immediately that the rule $\det : \mathbf{GL}_n \longrightarrow \mathbb{G}_m$ obtained in this way is a natural transformation of functors.

Clearly, the composition of two morphisms is a morphism, and this composition is associative. Moreover, for any functor $\mathbf{F} : \mathcal{C} \longrightarrow \mathcal{C}'$, there is an identity element $\mathbf{1}_\mathbf{F}$ for the composition, which assigns to each $A \in \mathrm{Ob}(\mathcal{C})$ the morphism $\mathrm{Id}_{\mathbf{F}(A)}$. Therefore, with this notion of morphism, the covariant functors from $\mathcal{C} \longrightarrow \mathcal{C}'$ form a category. In particular, we get the notion of isomorphism of functors.

We now define the notion of a subfunctor of a functor.

Definition III.7.10. Let $\mathcal{C}$ be a category, let $\mathcal{C}'$ be a subcategory of **Sets**, and let $\mathbf{F} : \mathcal{C} \longrightarrow \mathcal{C}'$ be a functor. A functor $\mathbf{F}' : \mathcal{C} \longrightarrow \mathcal{C}'$ is a **subfunctor** of $\mathbf{F}$ if the following properties hold:

(1) For all $A \in \mathrm{Ob}(\mathcal{C})$, we have $\mathbf{F}'(A) \subset \mathbf{F}(A)$.

(2) For all $A, B \in \mathrm{Ob}(\mathcal{C})$, and every map $f \in \mathrm{Mor}_{\mathcal{C}}(A, B)$, the induced morphism $\mathbf{F}'(f) : \mathbf{F}'(A) \longrightarrow \mathbf{F}'(B)$ is the restriction of $\mathbf{F}(f) : \mathbf{F}(A) \longrightarrow \mathbf{F}(B)$. In other words, the diagram

$$\begin{array}{ccc} \mathbf{F}'(A) & \xrightarrow{\mathbf{F}'(f)} & \mathbf{F}'(B) \\ \downarrow & & \downarrow \\ \mathbf{F}(A) & \xrightarrow{\mathbf{F}(f)} & \mathbf{F}(B) \end{array}$$

commutes.

Example III.7.11. The functor of $n \times n$ of matrices of trace zero is a subfunctor of $\mathbf{M}_n$.

Definition III.7.12. Let $\mathbf{F} : \mathcal{C} \longrightarrow$ **Sets** be a covariant functor. We say that $\mathbf{F}$ is **representable** if $\mathbf{F} \simeq h_A$ for some object $A \in \mathrm{Ob}(\mathcal{C})$, where h_A is the functor defined in Example III.7.7(5). In this case, we will say that **F is represented by** A.

The reader may wonder if the object A above is uniquely defined up to isomorphism. The answer is affirmative, and is given by Yoneda's Lemma.

Lemma III.7.13 (Yoneda's Lemma)**.** *Let $\mathcal{C}$ be a category. For every pair of objects $A, B \in \mathrm{Ob}(\mathcal{C})$, there is a one-to-one correspondence between the set of morphisms $\varphi : B \longrightarrow A$ and the set of natural transformations $\Theta : h_A \longrightarrow h_B$. More precisely, let $\Psi_{A,B}$ and $\Psi'_{A,B}$ be the maps defined as follows:*

if $\varphi : B \longrightarrow A$ is a morphism of $\mathcal{C}$, we define $\Psi_{A,B}(\varphi) : h_A \longrightarrow h_B$ by

$$\Psi_{A,B}(\varphi)_C : \begin{array}{rcl} \mathrm{Mor}_{\mathcal{C}}(A, C) & \longrightarrow & \mathrm{Mor}_{\mathcal{C}}(B, C) \\ f & \longmapsto & f \circ \varphi \end{array}$$

for all $C \in \mathrm{Ob}(\mathcal{C})$.

If $\Theta : h_A \longrightarrow h_B$ is a natural transformation of functors, we define $\Psi'_{A,B}(\Theta) : B \longrightarrow A$ to be the morphism $\Theta_A(\mathrm{Id}_A) \in h_B(A) = \mathrm{Mor}_{\mathcal{C}}(B, A)$.

Then $\Psi_{A,B}$ and $\Psi'_{A,B}$ are mutually inverse. Moreover, for all $A, B, C \in \mathrm{Ob}(\mathcal{C})$, all $\varphi \in \mathrm{Mor}_{\mathcal{C}}(B, A)$ and all $\varphi' \in \mathrm{Mor}_{\mathcal{C}}(C, B)$ we have:

(1) $\Psi_{A,A}(\mathrm{Id}_A) = \mathbf{1}_{h_A}$.

(2) $\Psi_{A,C}(\varphi \circ \varphi') = \Psi_{B,C}(\varphi') \circ \Psi_{A,B}(\varphi)$.

In particular, for any pair of objects $A, B \in \mathrm{Ob}(\mathcal{C})$, we have

$$h_A \simeq h_B \iff A \simeq B.$$

Proof. The fact that $\Psi_{A,B}(\varphi)$ is actually a natural transformation is left as an exercise for the reader. Let us prove that $\Psi_{A,B}$ and $\Psi'_{A,B}$ are mutually inverse. For any morphism $\varphi : B \longrightarrow A$, we have

$$\Psi'_{A,B}(\Psi_{A,B}(\varphi)) = \Psi_{A,B}(\varphi)_A(\mathrm{Id}_A) = \mathrm{Id}_A \circ \varphi = \varphi,$$

so $\Psi'_{A,B} \circ \Psi_{A,B}$ is the identity map. Now if $\Theta : h_A \longrightarrow h_B$ is a natural transformation, for any object $C \in \mathrm{Ob}(\mathcal{C})$ and any morphism $f \in \mathrm{Mor}_{\mathcal{C}}(A, C)$, we have

$$\Psi_{A,B}(\Psi'_{A,B}(\Theta))_C(f) = f \circ \Psi'_{A,B}(\Theta) = f \circ \Theta_A(\mathrm{Id}_A).$$

Since Θ is a natural transformation, the diagram

$$\begin{array}{ccc} h_A(A) & \xrightarrow{\Theta_A} & h_B(A) \\ \Big\downarrow{\scriptstyle h_A(f)} & & \Big\downarrow{\scriptstyle h_B(f)} \\ h_A(C) & \xrightarrow{\Theta_C} & h_B(C) \end{array}$$

is commutative. In particular, we have

$$h_B(f)(\Theta_A(\mathrm{Id}_A)) = \Theta_C(h_A(f)(\mathrm{Id}_A)).$$

By definition of $h_A(f)$ and $h_B(f)$, this reads

$$f \circ \Theta_A(\mathrm{Id}_A) = \Theta_C(f).$$

Hence we get $\Psi_{A,B}(\Psi'_{A,B}(\Theta))_C(f) = \Theta_C(f)$ for all $C \in \mathrm{Ob}(\mathcal{C})$ and all $f \in \mathrm{Mor}_{\mathcal{C}}(A, C)$, that is

$$\Psi_{A,B}(\Psi'_{A,B}(\Theta)) = \Theta.$$

Thus $\Psi_{A,B} \circ \Psi'_{A,B}$ is the identity map as well; this concludes the proof of the first part. Relations (1) and (2) are clear, in view of the definition of $\Psi_{A,B}$.

It remains to prove the last statement. If $\varphi : B \longrightarrow A$ is an isomorphism, relations (1) and (2) readily imply that $\Psi_{A,B}(\varphi) : h_A \longrightarrow h_B$ is an isomorphism of functors, whose inverse is the natural transformation $\Psi_{B,A}(\varphi^{-1}) : h_B \longrightarrow h_A$. Conversely, if $\Theta : h_A \longrightarrow h_B$ is an isomorphism of functors, write $\Theta = \Psi_{A,B}(\varphi)$ for some $\varphi \in \mathrm{Mor}_{\mathcal{C}}(B, A)$, and $\Theta^{-1} = \Psi_{B,A}(\varphi')$ for some $\varphi' \in \mathrm{Mor}_{\mathcal{C}}(A, B)$. Now we have

$$\Psi_{B,B}(\mathrm{Id}_B) = \mathbf{1}_{h_B} = \Theta \circ \Theta^{-1} = \Psi_{A,B}(\varphi) \circ \Psi_{B,A}(\varphi') = \Psi_{B,B}(\varphi' \circ \varphi).$$

Since $\Psi_{B,B}$ is a bijection, we get $\varphi' \circ \varphi = \mathrm{Id}_B$. Similarly, $\varphi \circ \varphi' = \mathrm{Id}_A$, so $\varphi : B \longrightarrow A$ is an isomorphism. □

We now give a fundamental example of a representable functor:

Lemma III.7.14. *Let k be a field, let $A = k[X_1, \dots, X_n]/I$ be a finitely generated k-algebra. Then the functor $V(I) : \mathbf{Alg}_k \longrightarrow \mathbf{Sets}$ defined by*

$$V(I)(R) = \{(r_1, \dots, r_n) \in R^n \mid f(r_1, \dots, r_n) = 0 \text{ for all } f \in I\}$$

is isomorphic to h_A.

Proof. Let R be a commutative unital k-algebra. For $\mathbf{r} = (r_1, \dots, r_n) \in V(I)(R)$, let

$$ev_{\mathbf{r}} : \begin{array}{r} k[X_1, \dots, X_n] \longrightarrow R \\ X_i \longmapsto r_i \end{array}$$

be the evaluation at $\mathbf{r}$. By definition of $V(I)$, the kernel of $ev_{\mathbf{r}}$ contains I, so it induces a morphism

$$\varphi_R(\mathbf{r}) : \begin{array}{r} k[X_1, \dots, X_n]/I \longrightarrow R \\ \overline{X}_i \longmapsto r_i. \end{array}$$

We then get a map $\varphi_R : V(I)(R) \longrightarrow h_{k[X_1,\dots,X_n]/I}(R)$. It is easy to check that the corresponding rule $\varphi : V(I) \longrightarrow h_{k[X_1,\dots,X_n]/I}$ is a natural transformation of functors.

Conversely, let $u : k[X_1, \dots, X_n]/I \longrightarrow R$ be a k-algebra morphism. Then $\psi_R(u) = (u(\overline{X}_1), \dots, u(\overline{X}_n)) \in V(I)(R)$ since we have

$$f(u(\overline{X}_1), \dots, u(\overline{X}_n)) = u(f(\overline{X}_1, \dots, \overline{X}_n)) = u(\overline{f}) = u(\overline{0}) = 0,$$

for all $f \in I$. We then get a map $\psi_R : h_{k[X_1,\dots,X_n]/I}(R) \longrightarrow V(I)(R)$. Once again, it is easy to check that the rule $\psi : h_{k[X_1,\dots,X_n]/I} \longrightarrow V(I)$ is a natural transformation of functors. Clearly, φ and ψ are mutually inverse. This concludes the proof. □

III.7.2 Algebraic group-schemes

Let K/k be a field extension, and let Ω/K be a Galois extension of K with Galois group $\mathcal{G}_\Omega$. In order to define cohomology sets of the profinite group $\mathcal{G}_\Omega$, we need $\mathcal{G}_\Omega$-groups. The first step is to obtain groups on which $\mathcal{G}_\Omega$ acts by group automorphisms. We proceed to show that a natural way to achieve this is to consider Ω-points of group-valued functors.

Let $\mathbf{F} : \mathfrak{C}_k \longrightarrow \mathbf{Sets}$ be a covariant functor. If K/k is a field extension, we will denote $\mathbf{F}(K/k)$ simply by $\mathbf{F}(K)$. If $K \longrightarrow K'$ is a morphism of extensions of k, for every $x \in \mathbf{F}(K)$, we will denote by $x_{K'} \in \mathbf{F}(K')$ the image of x under the map $\mathbf{F}(K) \longrightarrow \mathbf{F}(K')$ if there is no ambiguity on the choice of the map $K \longrightarrow K'$.

Notice that since $\mathbf{F}$ is a functor, every $\sigma \in \mathcal{G}_\Omega$ induces a map

$$\mathbf{F}(\sigma) : \mathbf{F}(\Omega) \longrightarrow \mathbf{F}(\Omega).$$

Lemma III.7.15. *The map*

$$\begin{aligned} \mathcal{G}_\Omega \times \mathbf{F}(\Omega) &\longrightarrow \mathbf{F}(\Omega) \\ (\sigma, x) &\longmapsto \sigma \cdot x = \mathbf{F}(\sigma)(x) \end{aligned}$$

gives rise to an action of $\mathcal{G}_\Omega$ on $\mathbf{F}(\Omega)$.

If Ω/K and Ω'/K are two Galois extensions such that $\Omega \subset \Omega'$, we have

$$\sigma' \cdot x_{\Omega'} = (\sigma'_{|\Omega} \cdot x)_{\Omega'} \text{ for all } x \in \mathbf{F}(\Omega), \sigma' \in \mathcal{G}_{\Omega'}.$$

Moreover, if $\mathbf{F} : \mathfrak{C}_k \longrightarrow \mathbf{Grps}$ is a group-valued functor, the action above is an action by group automorphisms, that is

$$\sigma \cdot (xy) = (\sigma \cdot x)(\sigma \cdot y) \text{ for all } \sigma \in \mathcal{G}_\Omega, x, y \in \mathbf{F}(\Omega).$$

Proof. Since $\mathbf{F}$ is a functor, we have $\mathbf{F}(\mathrm{Id}_\Omega) = \mathrm{Id}_{\mathbf{F}(\Omega)}$. Therefore,

$$\mathrm{Id}_\Omega \cdot x = x \text{ for all } x \in \mathbf{F}(\Omega).$$

Now let $\sigma, \tau \in \mathcal{G}_\Omega$. Since $\mathbf{F}$ is a covariant functor, we have

$$\mathbf{F}(\sigma \circ \tau) = \mathbf{F}(\sigma) \circ \mathbf{F}(\tau),$$

and thus $(\sigma \circ \tau) \cdot x = \sigma \cdot (\tau \cdot x)$ for all $x \in \mathbf{F}(\Omega)$. This proves the first part of the lemma.

Let Ω/K and Ω'/K be two Galois extensions such that $\Omega \subset \Omega'$, and let

$\mathbf{F}(\iota) : \mathbf{F}(\Omega) \longrightarrow \mathbf{F}(\Omega')$ be the map induced by the inclusion $\iota : \Omega \subset \Omega'$. For all $\sigma' \in \mathcal{G}_{\Omega'}$, the diagram

$$\begin{array}{ccc} \Omega & \xrightarrow{\sigma'_{|\Omega}} & \Omega \\ \downarrow \iota & & \downarrow \iota \\ \Omega' & \xrightarrow{\sigma'} & \Omega' \end{array}$$

is commutative. Therefore, the induced diagram

$$\begin{array}{ccc} \mathbf{F}(\Omega) & \xrightarrow{\mathbf{F}(\sigma'_{|\Omega})} & \mathbf{F}(\Omega) \\ \downarrow \mathbf{F}(\iota) & & \downarrow \mathbf{F}(\iota) \\ \mathbf{F}(\Omega') & \xrightarrow{\mathbf{F}(\sigma')} & \mathbf{F}(\Omega') \end{array}$$

commutes. In other words, we have $\mathbf{F}(\iota) \circ \mathbf{F}(\sigma'_{|\Omega})(x) = \mathbf{F}(\sigma') \circ \mathbf{F}(\iota)(x)$ for all $x \in \mathbf{F}(\Omega)$, that is

$$(\sigma'_{|\Omega} \cdot x)_{\Omega'} = \sigma' \cdot x_{\Omega'} \text{ for all } x \in \mathbf{F}(\Omega).$$

Finally, if $\mathbf{F}$ is a group-valued functor, $\mathbf{F}(\sigma)$ is a group morphism and the last part follows. This concludes the proof. □

Unfortunately, there is no reason for this action to be continuous. However this is the case for a representable functor $\mathbf{F} : \mathbf{Alg}_k \longrightarrow \mathbf{Sets}$ under some mild assumption, as we proceed to show now.

Lemma III.7.16. *Let $\mathbf{F} : \mathbf{Alg}_k \longrightarrow \mathbf{Sets}$ be a representable functor, and let A be a commutative k-algebra such that $\mathbf{F} \simeq h_A$. Then the following properties hold:*

(1) *For every Galois extension Ω/K, the map $\mathbf{F}(K) \longrightarrow \mathbf{F}(\Omega)$ is injective and induces a bijection (resp. a group isomorphism if $\mathbf{F}$ is a group-valued functor)*

$$\mathbf{F}(K) \simeq \mathbf{F}(\Omega)^{\mathcal{G}_\Omega}.$$

(2) *Assume that A is finitely generated over k, and let Ω/K be a Galois extension. For every finite Galois extension L/K contained in Ω, denote by $\iota_L : \mathbf{F}(L) \longrightarrow \mathbf{F}(\Omega)$ the map induced by the inclusion $L \subset \Omega$. Then $\mathcal{G}_\Omega$ acts continuously on $\mathbf{F}(\Omega)$, and we have*

$$\mathbf{F}(\Omega) = \bigcup_{L \subset \Omega} \iota_L(\mathbf{F}(L)).$$

Proof. Let $\Psi : \mathbf{F} \xrightarrow{\sim} h_A$ be an isomorphism of functors. Notice first that since Ψ is a natural transformation of functors, then for every morphism of k-algebras $\varphi : R \longrightarrow S$, we have a commutative diagram

$$\begin{array}{ccc} \mathbf{F}(R) & \xrightarrow{\Psi_R} & \mathrm{Hom}_{k-alg}(A, R) \\ \downarrow{\scriptstyle \mathbf{F}(\varphi)} & & \downarrow \\ \mathbf{F}(S) & \xrightarrow{\Psi_S} & \mathrm{Hom}_{k-alg}(A, S) \end{array}$$

where the second vertical map is left composition by φ. In other words, if $a \in \mathbf{F}(R)$ and $f = \Psi_R(a) \in \mathrm{Hom}_{k-alg}(A, R)$ is the corresponding morphism, then $\mathbf{F}(\varphi)(a)$ corresponds to the morphism $\varphi \circ f \in \mathrm{Hom}_{k-alg}(A, S)$.

We may now start the proof of the lemma. Let Ω/K be a Galois extension and let $\varepsilon : K \longrightarrow \Omega$ be the corresponding injective morphism. The previous considerations show that the map $\mathbf{F}(\varepsilon) : \mathbf{F}(K) \longrightarrow \mathbf{F}(\Omega)$ identifies to the map

$$\begin{aligned} \mathrm{Hom}_{k-alg}(A, K) &\longrightarrow \mathrm{Hom}_{k-alg}(A, \Omega) \\ f &\longmapsto \varepsilon \circ f. \end{aligned}$$

The injectivity of $\mathbf{F}(\varepsilon)$ then comes from the injectivity of ε.

We now prove that the image of $\mathbf{F}(\varepsilon)$ is $\mathbf{F}(\Omega)^{\mathcal{G}_\Omega}$. For, let $\sigma \in \mathcal{G}_\Omega$, let $a \in \mathbf{F}(\Omega)$ and let $f \in \mathrm{Hom}_{k-alg}(A, \Omega)$ be the corresponding morphism, that is $f = \Psi_\Omega(a)$. If a lies in the image of $\mathbf{F}(\varepsilon)$, then there exists $f' \in \mathrm{Hom}_{k-alg}(A, K)$ such that $f = \varepsilon \circ f'$. Now the action of σ on $\mathbf{F}(\Omega)$ corresponds to left composition by σ, so $\sigma \cdot a$ corresponds to the morphism

$$\sigma \circ f = (\sigma \circ \varepsilon) \circ f' = \varepsilon \circ f' = f,$$

the second equality coming from the K-linearity of σ. Hence $\sigma \cdot a = a$ for all $\sigma \in \mathcal{G}_\Omega$. Conversely, assume that $a \in \mathbf{F}(\Omega)$ satisfies

$$\sigma \cdot a = a \text{ for all } \sigma \in \mathcal{G}_\Omega.$$

We then have

$$\sigma \circ f = f \text{ for all } \sigma \in \mathcal{G}_\Omega.$$

It implies that for all $x \in A$, $f(x)$ is fixed by $\mathcal{G}_\Omega$, that is $f(x) \in K$ since

Ω/K is a Galois extension. Consider now the k-algebra morphism

$$f' : \begin{array}{l} A \longrightarrow K \\ x \longmapsto f(x), \end{array}$$

and let a' be the corresponding element of $\mathbf{F}(K)$. Since $f = \varepsilon \circ f'$ by definition, a is the image of a' under the map $\mathbf{F}(\varepsilon)$. This proves (1).

Let us prove (2). Let $a \in \mathbf{F}(\Omega)$, let $f \in \mathrm{Hom}_{k-alg}(A, \Omega)$ be the corresponding morphism and let $\sigma \in \mathcal{G}_\Omega$. The equality $\sigma \cdot a = a$ is equivalent to $\sigma \circ f = f$. Let $\alpha_1, \ldots, \alpha_n$ be a set of generators of A, and let $K' = K(f(\alpha_1), \ldots, f(\alpha_n))$. This is a finite extension of K contained in Ω. Now $\sigma \circ f = f$ if and only if σ restricts to the identity on K', that is $\sigma \in \mathrm{Gal}(\Omega/K')$. Hence the stabilizer of a is $\mathrm{Gal}(\Omega/K')$, which is an open subgroup of $\mathcal{G}_\Omega$ since K'/K is finite (this follows from the definition of the Krull topology). Therefore, the action of $\mathcal{G}_\Omega$ on $\mathbf{F}(\Omega)$ is continuous. Using Lemma II.3.3 and the first point, we get immediately the equality

$$\mathbf{F}(\Omega) = \bigcup_{L \subset \Omega} \iota_L(\mathbf{F}(L)).$$

This concludes the proof. □

Remark III.7.17. The two previous lemmas show in particular that if $G : \mathbf{Alg}_k \longrightarrow \mathbf{Grps}$ is represented by a finite-dimensional k-algebra, then for every Galois extension Ω/K, the group $G(\Omega)$ is a $\mathcal{G}_\Omega$-group. Such a functor deserves a special name.

Definition III.7.18. Let k be a field. A **group-scheme** defined over k is a covariant functor $G : \mathbf{Alg}_k \longrightarrow \mathbf{Grps}$. An **affine group-scheme** defined over k is a group-scheme $G : \mathbf{Alg}_k \longrightarrow \mathbf{Grps}$ which is representable as a functor $\mathbf{Alg}_k \longrightarrow \mathbf{Sets}$. An **algebraic group-scheme** defined over k is an affine group-scheme G which is represented by a finitely generated k-algebra A. If furthermore $A_{k_{alg}}$ is reduced, we say that G is an **algebraic group** defined over k.

Examples III.7.19.

(1) Let V be a finite dimensional k-vector space. For any commutative unital k-algebra R, set

$$\mathbf{GL}(V)(R) = \mathbf{GL}(V_R),$$

where $V_R = V \otimes_k R$. If $V = k^n$, we just denote it by $\mathbf{GL}_n$ (this is consistent with the definition given in the first paragraph). We obtain an algebraic group-scheme $\mathbf{GL}(V)$. Indeed, if we choose a basis of V, then we can see that $\mathbf{GL}(V)$ is isomorphic to $\mathbf{GL}_n$ as a group-scheme. Hence, this is enough to check that $\mathbf{GL}_n$ is an algebraic group-scheme. Let G be the group-scheme defined by

$$G(R) = \{(x_{11}, x_{12}, \dots, x_{nn}, y) \in R^{n^2+1} \mid \det((x_{ij}))y = 1\}$$

for any commutative k-algebra R. Then the natural transformation $\varphi : \mathbf{GL}_n \longrightarrow G$ defined by

$$\varphi_R : \begin{array}{rcl} \mathbf{GL}_n(R) & \longrightarrow & G(R) \\ (m_{ij}) & \longmapsto & (m_{11}, m_{12}, \dots, m_{nn}, \det((m_{ij}))^{-1}) \end{array}$$

is an isomorphism. Now by Lemma III.7.14, the group-scheme G (and therefore $\mathbf{GL}_n$) is isomorphic as a functor $\mathbf{Alg}_k \longrightarrow \mathbf{Sets}$ to the representable functor h_A, where

$$A = k[X_{ij}, Y]/(\det((X_{ij}))Y - 1).$$

It follows that $\mathbf{GL}_n$ is an algebraic group-scheme. The reader may show as an exercice that $\mathbf{GL}_n$ is in fact an algebraic group.

(2) If A is a finite dimensional k-algebra, the functor $\mathbf{GL}_1(A)$ defined by

$$\mathbf{GL}_1(A)(R) = A_R^\times$$

is an algebraic group-scheme. To see this, for all $a \in A_R$, let us denote by ℓ_a the endomorphism of right R-modules given by left multiplication by a. Then $a \in A_R^\times$ if and only if ℓ_a is an isomorphism. Since A_R is a free R-module, this is equivalent to $\det(\ell_a) \in R^\times$. Now if $e_1, \dots, e_n$ is a k-basis of A, let us denote by $P_A \in R[X_1, \dots, X_n]$ the polynomial

$$P_A = \det(\ell_{e_1 \otimes 1} X_1 + \ldots + \ell_{e_n \otimes 1} X_n).$$

It is easy to check that $P_A \in k[X_1, \dots, X_n]$. Moreover, $a = \sum_{i=1}^{n} e_i \otimes \lambda_i \in A_R^\times$ if and only if $P_A(\lambda_1, \dots, \lambda_n) \in R^\times$. Arguing as in the previous example, we see that $\mathbf{GL}_1(A)$ is represented by the k-algebra

$$k[X_1, \dots, X_n, Y]/(P_A Y - 1).$$

In particular, if $A = \mathrm{End}(V)$, we recover the result of (1).

(3) The additive group-scheme $\mathbb{G}_a$ is defined by

$$\mathbb{G}_a(R) = R.$$

By Lemma III.7.14, it is represented by $k[X]$. Clearly, this is an algebraic group.

(4) The multiplicative group-scheme $\mathbb{G}_m$ is defined by

$$\mathbb{G}_m(R) = R^\times.$$

By (1), this is an affine group-scheme represented by

$$k[X,Y]/(XY-1),$$

and therefore an algebraic group.

(5) The group-scheme of nth-roots of unity μ_n is defined by

$$\mu_n(R) = \{r \in R \mid r^n = 1\}.$$

By Lemma III.7.14, it is an affine group-scheme, represented by $k[X]/(X^n-1)$. This is an algebraic group if and only if n is prime to the characteristic of k.

(6) Let G be an abstract finite group of order n. If R is a commutative k-algebra, we will index the coordinates of an element $\mathbf{r}$ of R^n by the elements of G. Set

$$G(R) = \left\{\mathbf{r} \in R^n \Big| \sum_g r_g = 1, r_g r_h = \delta_{g,h} r_g \text{ for all } g,h \in G\right\}.$$

We define the product of $r = (r_g)_{g\in G}$ and $r' = (r'_g)_{g\in G}$ by

$$(r.r')_g = \sum_{h\in G} r_h r'_{h^{-1}g}.$$

One can check that $G(R)$ is a group for this multiplication law. Notice that if R has no non-trivial idempotents, then $G(R)$ is isomorphic to the abstract group G via the isomorphism

$$\begin{aligned} G &\xrightarrow{\sim} G(R) \\ g &\longmapsto (\delta_{g,h})_{h\in G}. \end{aligned}$$

One can show that the corresponding functor is an affine group scheme, still denoted by G, called **the constant group scheme** (associated to) G. The reader will show that the k-algebra representing G is $\mathrm{Map}(G,k)$. In particular, G is an algebraic group.

(7) If A is a finite dimensional k-algebra (not necessarily commutative), we define the group-scheme $\mathbf{Aut}_{alg}(A)$ by

$$\mathbf{Aut}_{alg}(A)(R) = \mathrm{Aut}_{R-alg}(A_R),$$

where $A_R = A \otimes_k R$. The reader will check as an exercise that this group-scheme is affine.

If $G : \mathfrak{C}_k \longrightarrow \mathbf{Grps}$ is a group-scheme, the action $\mathcal{G}_\Omega$ on $G(\Omega)$ may seem a bit mysterious. We would like to make it more explicit in the case of algebraic group-schemes. We start with an example.

Example III.7.20. Let $G = \mathbf{GL}_n$. In this case, G is represented by the finite dimensional k-algebra $k[X_{ij}, Y]/(\det(X_{ij})Y - 1)$ by Example III.7.19 (1). An invertible matrix $M \in \mathrm{GL}_n(\Omega)$ corresponds to the morphism $f : A \longrightarrow \Omega$, which sends $\overline{X}_{ij}$ to m_{ij} and $\overline{Y}$ to $1/\det(m_{ij})$. Then $\sigma \cdot M$ corresponds to the morphism $f' = \sigma \circ f$, as pointed out at the beginning of the proof of Lemma III.7.16. By definition of f, this morphism sends $\overline{X}_{ij}$ to $\sigma(m_{ij})$ and $\overline{Y}$ to $\sigma(1/\det(m_{ij}))$. Since we have $\sigma(1/\det(m_{ij})) = 1/\det(\sigma(m_{ij}))$, it follows that f' corresponds to the matrix $(\sigma(m_{ij}))$. Therefore, the action of $\mathcal{G}_\Omega$ on $\mathrm{GL}_n(\Omega)$ is just the usual action coefficient by coefficient.

To describe this action in the general case, we need first to introduce the concept of a closed subgroup.

Definition III.7.21. Let G and H be two affine group-schemes defined over k, represented by A and B respectively. We say that H is a **closed subgroup** of G if there exists a natural transformation $\Theta : H \longrightarrow G$ such that the morphism of k-algebras $\varphi : A \longrightarrow B$ corresponding to Θ via Yoneda's lemma is surjective. In this case Θ is called a **closed embedding**.

Remark III.7.22. In this setting, Θ is injective, that is $\Theta_R : H(R) \longrightarrow G(R)$ is injective for all R. In particular, $H(R)$ identifies to a subgroup of $G(R)$ for all R.

Indeed, if $f_1, f_2 \in \mathrm{Hom}_{k-alg}(B, R)$ satisfy $f_1 \circ \varphi = f_2 \circ \varphi$, then we have $f_1 = f_2$ by surjectivity of φ. But by Yoneda's Lemma the map

$$\begin{aligned} \mathrm{Hom}_{k-alg}(B, R) &\longrightarrow \mathrm{Hom}_{k-alg}(A, R) \\ f &\longmapsto f \circ \varphi \end{aligned}$$

is Θ_R, so we are done.

This explains the term 'subgroup' in the previous definition. We now would like to explain in which sense H is 'closed' in G when G is algebraic. In this case, since A is finitely generated, we may write $A = k[X_1, \dots, X_n]/I$. Since φ is surjective, B is isomorphic to a quotient of A, so $B \simeq k[X_1, \dots, X_n]/J$ for some ideal J of $k[X_1, \dots, X_n]$ containing I. Recall now that $G \simeq V(I)$ and $H \simeq V(J)$ by Lemma III.7.14, and notice that $V(J)$ is a subfunctor of $V(I)$. Now if we identify $H(k_{alg})$ and $G(k_{alg})$ to $V(J)(k_{alg})$ and $V(I)(k_{alg})$ respectively, then $H(k_{alg})$ is a closed subset of $G(k_{alg})$ for the Zariski topology.

As a final remark, let us mention that if G and H are both algebraic group-schemes, then a natural transformation $\Theta : H \to G$ is a closed embedding if and only if Θ_R is injective for all R (see [30, Proposition 22.2] for example).

Before going further, we need the following result, which is proved in [69]:

Proposition III.7.23. *Any algebraic group-scheme G defined over k is a closed subgroup of $\mathbf{GL}_n$, for some $n \geq 1$.*

We are now ready to describe the action of $\mathcal{G}_\Omega$ on $G(\Omega)$ in more explicit terms.

Lemma III.7.24. *Let G be an algebraic group-scheme defined over k represented by B, and let $\Theta : G \longrightarrow \mathbf{GL}_n$ be a closed embedding. Let K/k be a field extension, let Ω/K be a Galois extension, and assume that $\mathcal{G}_\Omega$ acts naturally on $\mathrm{GL}_n(\Omega)$. Then for every $\sigma \in \mathcal{G}_\Omega$ and every $g \in G(\Omega)$, $\sigma \cdot g$ is the unique element $g' \in G(\Omega)$ satisfying*

$$\Theta_\Omega(g') = \sigma \cdot \Theta_\Omega(g).$$

Proof. Notice that there is at most one element $g' \in G(\Omega)$ satisfying the equality above, since Θ_Ω is injective by Remark III.7.22. Now if $g \in G(\Omega)$ and $\sigma \in \mathcal{G}_\Omega$, the naturality of Θ shows that we have

$$\Theta_\Omega(G(\sigma)(g)) = \Theta_\Omega(\sigma)(g),$$

that is

$$\Theta_\Omega(\sigma \cdot g) = \sigma \cdot \Theta_\Omega(g).$$

The results follows. □

Remark III.7.25. The previous lemma and Example III.7.20 make the natural Galois action on an algebraic group-scheme G totally explicit.

In particular, it shows that if $G \subset \mathbf{GL}_n$, the action of $\mathcal{G}_\Omega$ on $G(\Omega)$ is simply the restriction of the natural action of $\mathcal{G}_\Omega$ on matrices.

Example III.7.26. Let us describe the action of $\mathcal{G}_\Omega$ on $\mathrm{M}_n(\Omega)$. The map

$$\Theta \colon \begin{array}{l} \mathbf{M}_n \longrightarrow \mathbf{GL}_{2n} \\ M \longmapsto \begin{pmatrix} I_n & M \\ 0 & I_n \end{pmatrix} \end{array}$$

is easily seen to be a closed embedding. It follows from the previous lemma that the action of $\mathcal{G}_\Omega$ on $\mathrm{M}_n(\Omega)$ is nothing but the action of $\mathcal{G}_\Omega$ entrywise.

III.7.3 The Galois cohomology functor

We now introduce the general setting in which we are going to work, in order to get some functorial properties of cohomology of Galois groups of field extensions. Notice first that if G is an algebraic group-scheme defined over a field k, then by Remark III.7.17 the group $G(\Omega)$ is a $\mathcal{G}_\Omega$-group for every Galois extension Ω/k. Therefore, one may consider the set $H^n(\mathcal{G}_\Omega, G(\Omega))$. Moreover, for every finite Galois subextension L/k of Ω/k, $G(\Omega)^{\mathrm{Gal}(\Omega/L)}$ identifies to $G(L)$ by Lemma III.7.16 (1). Using this property and the fact that $\mathcal{G}_\Omega/\mathrm{Gal}(\Omega/L)$ is isomorphic to $\mathcal{G}_L$, Theorem II.3.33 shows that $H^n(\mathcal{G}_\Omega, G(\Omega))$ is the direct limit of the sets $H^n(\mathcal{G}_L, G(L))$, a property we were looking for in order to generalize our approach of the conjugacy problem of matrices to infinite Galois extensions.

However, limiting ourselves to consider cohomology sets of algebraic-group schemes is a bit too restrictive, since the group-schemes involved in other descent problems may not be representable (even if it will be true in most of the cases) and may not even be defined on the category $\mathbf{Alg}_k$. The idea is then to consider group-schemes G having the properties listed in Lemma III.7.16, which are really the only ones used to obtain cohomology sets behaving well.

Definition III.7.27. A group-scheme $G : \mathfrak{C}_k \longrightarrow \mathbf{Grps}$ is a **Galois functor** if for every field extension K/k and every Galois extension Ω/K, the following conditions are satisfied:

(1) The map $G(K) \longrightarrow G(\Omega)$ is injective, and induces a group iso-

morphism

$$G(K) \simeq G(\Omega)^{\mathcal{G}_\Omega}.$$

(2) $G(\Omega) = \bigcup_{L \subset \Omega} \iota_{L,\Omega}(G(L))$, where L/K runs over the set of finite Galois subextensions of Ω and $\iota_{L,\Omega} : G(L) \longrightarrow G(\Omega)$ is the map induced by the inclusion $L \subset \Omega$.

Example III.7.28. An algebraic group-scheme, viewed as a functor from $\mathfrak{C}_k$ to **Grps**, is a Galois functor by Lemma III.7.16.

We will see other examples of Galois functors in the next paragraph.

Remark III.7.29. Let $G : \mathfrak{C}_k \longrightarrow$ **Grps** be a Galois functor. By Lemma II.3.3, conditions (1) and (2) imply that for every field extension K/k and every Galois extension Ω/K, the profinite group $\mathcal{G}_\Omega$ acts continuously on $G(\Omega)$ via

$$\begin{aligned} \mathcal{G}_\Omega \times G(\Omega) &\longrightarrow G(\Omega) \\ (\sigma, g) &\longmapsto \sigma \cdot g = G(\sigma)(g). \end{aligned}$$

Lemma III.7.15 ensures that this action is an action by group automorphisms, so $G(\Omega)$ is a $\mathcal{G}_\Omega$-group. We may then consider the pointed set $H^n(\mathcal{G}_\Omega, G(\Omega))$.

Let $G : \mathfrak{C}_k \longrightarrow$ **Grps** be a Galois functor. We would like now to use Theorem II.3.33 to relate the Galois cohomology of $\mathcal{G}_\Omega$ to the Galois cohomology of its finite Galois subextensions. For every (not necessarily) finite Galois subextensions L/K and L'/K of Ω/K such that $L \subset L'$, by Lemma III.7.15, the maps

$$r_{L,L'} : \begin{aligned} \mathcal{G}_{L'} &\longrightarrow \mathcal{G}_L \\ \sigma &\longmapsto \sigma_{|L} \end{aligned}$$

and

$$\iota_{L,L'} : G(L) \longrightarrow G(L')$$

are compatible. We then get a map

$$\rho_{L,L'} : H^n(\mathcal{G}_L, G(L)) \longrightarrow H^n(\mathcal{G}_{L'}, G(L')).$$

It is easy to check that we obtain this way a directed system of pointed sets. When $L' = \Omega$, we will simply denote these three maps by r_L, ι_L and ρ_L respectively.

We then have the following theorem.

Theorem III.7.30. *Let $G : \mathfrak{C}_k \longrightarrow$ **Grps** be a Galois functor. For every field extension K/k and every Galois extension Ω/K, we have an isomorphism of pointed sets (resp. an isomorphism of groups if G takes values in* **AbGrps** *)*

$$\varinjlim_{L\subset\Omega} H^n(\mathcal{G}_L, G(L)) \simeq H^n(\mathcal{G}_\Omega, G(\Omega)),$$

where L/K runs through the finite Galois subextensions of Ω/K. If $[\alpha] \in H^n(\mathcal{G}_L, G(L))$, this isomorphism maps $[\alpha]/_\sim$ onto $\rho_L([\alpha])$.

Proof. We start with some preliminary remarks. Since G is a Galois functor, the map ι_L induces a group isomorphism

$$G(L) \xrightarrow{\sim} G(\Omega)^{\mathrm{Gal}(\Omega/L)},$$

that we still denote by ι_L. Moreover, we have an isomorphism

$$\theta_L : \begin{array}{l} \mathcal{G}_\Omega/\mathrm{Gal}(\Omega/L) \xrightarrow{\sim} \mathcal{G}_L \\ \qquad\qquad \overline{\sigma} \longmapsto \sigma_{|L} \end{array}$$

induced by r_L. The maps θ_L and ι_L are compatible. Indeed, for every $\overline{\sigma} \in \mathcal{G}_\Omega/\mathrm{Gal}(\Omega/L) \xrightarrow{\sim} \mathcal{G}_L$ and $g \in G(L)$ we have

$$\iota_L(\theta_L(\overline{\sigma})\cdot g) = \iota_L(\sigma_{|L}\cdot g) = (\sigma_{|L}\cdot g)_\Omega.$$

By Lemma III.7.15, we get $\iota_L(\theta_L(\overline{\sigma})\cdot g) = \sigma\cdot g_\Omega = \overline{\sigma}\cdot g_\Omega$, the last equality coming from the definition of the action of $\overline{\sigma}$ on $G(\Omega)^{\mathrm{Gal}(\Omega/L)}$. This reads

$$\iota_L(\theta_L(\overline{\sigma})\cdot g) = \overline{\sigma}\cdot\iota_L(g),$$

which is what we wanted to prove. Therefore, we get a bijection

$$\eta_L : H^n(\mathcal{G}_L, G(L)) \xrightarrow{\sim} H^n(\mathcal{G}_\Omega/\mathrm{Gal}(\Omega/L), G(\Omega)^{\mathrm{Gal}(\Omega/L)})$$

for all L/K induced by the two previous compatible maps. Let us denote by $f_L : H^n(\mathcal{G}_\Omega/\mathrm{Gal}(\Omega/L), G(\Omega)^{\mathrm{Gal}(\Omega/L)}) \longrightarrow H^n(\mathcal{G}_\Omega, G(\Omega))$ the map induced in cohomology by the two compatible maps

$$\pi_L : \mathcal{G}_\Omega \longrightarrow \mathcal{G}_\Omega/\mathrm{Gal}(\Omega/L) \text{ and the inclusion } G(\Omega)^{\mathrm{Gal}(\Omega/L)} \longrightarrow G(\Omega).$$

Notice that the two diagrams

$$\begin{array}{ccc} G(L) & \xrightarrow{\iota_L} & G(\Omega) \\ \downarrow{\scriptstyle \iota_L} & & \| \\ G(\Omega)^{\mathrm{Gal}(\Omega/L)} & \longrightarrow & G(\Omega) \end{array}$$

and

$$\begin{array}{ccc} \mathcal{G}_L & \xleftarrow{\ r_L\ } & \mathcal{G}_\Omega \\ \uparrow{\scriptstyle \theta_L} & & \| \\ \mathcal{G}_\Omega/\mathrm{Gal}(\Omega/L) & \xleftarrow{\ \pi_L\ } & \mathcal{G}_\Omega \end{array}$$

are commutative. Applying Proposition II.3.26, we get the equality $f_L \circ \eta_L = \rho_L$.

We are now ready to prove the theorem. First of all, Theorem II.3.33 shows that we have

$$\varinjlim_{L\subset\Omega} H^n(\mathcal{G}_\Omega/\mathrm{Gal}(\Omega/L), G(\Omega)^{\mathrm{Gal}(\Omega/L)}) \simeq H^n(\mathcal{G}_\Omega, G(\Omega)),$$

since it follows from the Galois correspondence that open normal subgroups of $\mathcal{G}_\Omega$ have the form $\mathrm{Gal}(\Omega/L)$, where L/K is a finite Galois subextension of Ω/K. If $[\xi] \in H^n(\mathcal{G}_\Omega/\mathrm{Gal}(\Omega/L), G(\Omega)^{\mathrm{Gal}(\Omega/L)})$, this isomorphism maps $[\xi]/_\sim$ onto $f_L([\xi])$. Now for every finite Galois extensions L/K and L'/K such that $L \subset L'$, the corresponding inflation map will be denoted by $\inf_{L,L'}$. Using again Proposition II.3.26, one can check that we have a commutative diagram

$$\begin{array}{ccc} H^n(\mathcal{G}_L, G(L)) & \longrightarrow & H^n(\mathcal{G}_\Omega/\mathrm{Gal}(\Omega/L), G(\Omega)^{\mathrm{Gal}(\Omega/L)}) \\ \downarrow{\scriptstyle \rho_{L,L'}} & & \downarrow{\scriptstyle \inf_{L,L'}} \\ H^n(\mathcal{G}_{L'}, G(L')) & \longrightarrow & H^n(\mathcal{G}_\Omega/\mathrm{Gal}(\Omega/L'), G(\Omega)^{\mathrm{Gal}(\Omega/L')}) \end{array}$$

where the horizontal maps are the bijections defined above. It follows easily that we have a well-defined bijection

$$u : \varinjlim_{L\subset\Omega} H^n(\mathcal{G}_L, G(L)) \xrightarrow{\ \sim\ } \varinjlim_{L\subset\Omega} H^n(\mathcal{G}_\Omega/\mathrm{Gal}(\Omega/L), G(\Omega)^{\mathrm{Gal}(\Omega/L)}).$$

If $[\alpha] \in H^n(\mathcal{G}_L, G(L))$, then u maps $[\alpha]/_\sim$ onto $\eta_L([\alpha])/_\sim$. Details are left for the reader as an exercise. Composing u with the previous map, we get a bijection

$$\varinjlim_{L\subset\Omega} H^n(\mathcal{G}_L, G(L)) \simeq H^n(\mathcal{G}_\Omega, G(\Omega)),$$

mapping the equivalence class of $[\alpha]$ onto $f_L(\eta_L([\alpha]))$. Since $f_L\circ\eta_L = \rho_L$, we are done. □

We now give an application of this result to the computation of the Galois cohomology of the algebraic group-scheme $\mathbb{G}_a$.

Proposition III.7.31. *Let k be a field. For every $n \geq 1$ and every Galois extension Ω/k, we have $H^n(\mathcal{G}_\Omega, \Omega) = 0$.*

Proof. By Theorem III.7.30, it is enough to prove $H^n(\mathcal{G}_L, L) = 0$ for every finite Galois extension L/k. Let L/k be such an extension. By Dedekind's lemma, the elements of $\mathcal{G}_L$ are linearly independent over L (see [42] for a proof of this fact, for example). In particular, $\sum\limits_{\sigma \in \mathcal{G}_L} \sigma \neq 0$. Let $y \in L$ such that $\sum\limits_{\sigma \in \mathcal{G}_L} \sigma(y) \neq 0$. Notice that $\sum\limits_{\sigma \in \mathcal{G}_L} \sigma(y) \in k$ since it is fixed by every element of $\mathcal{G}_L$. Then the element $x = (\sum\limits_{\sigma \in \mathcal{G}_L} \sigma(y))^{-1} y \in L$ satisfies

$$\sum_{\sigma \in \mathcal{G}_L} \sigma \cdot x = \sum_{\sigma \in \mathcal{G}_L} \sigma(x) = 1.$$

Let $\alpha \in Z^n(\mathcal{G}_L, L)$, and let

$$a\colon \begin{array}{rcl} \mathcal{G}_L^{n-1} & \longrightarrow & L \\ (\sigma_1, \ldots, \sigma_{n-1}) & \longmapsto & \sum\limits_{\rho \in \mathcal{G}_L} \alpha_{\sigma_1, \ldots, \sigma_{n-1}, \rho} \, \sigma_1 \cdots \sigma_{n-1} \rho \cdot x. \end{array}$$

We are going to prove that $\alpha = d_{n-1}((-1)^n a)$, which will imply that α is cohomologous to the trivial cocycle. By definition, the element $d_{n-1}(a)_{\sigma_1, \ldots, \sigma_n}$ is equal to

$$\sum_{\rho \in \mathcal{G}_L} \left(\sigma_1 \cdot \alpha_{\sigma_2, \ldots, \sigma_n, \rho} + \sum_{i=1}^{n-1} (-1)^i \alpha_{\sigma_1, \ldots, \sigma_i \sigma_{i+1}, \ldots, \sigma_n, \rho} \right) \sigma_1 \cdots \sigma_n \rho \cdot x$$
$$+ (-1)^n \sum_{\rho \in \mathcal{G}_L} \alpha_{\sigma_1, \ldots, \sigma_{n-1}, \rho} \, \sigma_1 \cdots \sigma_{n-1} \rho \cdot x.$$

Performing the change of indices $\sigma_n \rho \leftrightarrow \rho$ in the second sum, we see that we have

$$d_{n-1}(a)_{\sigma_1, \ldots, \sigma_n} = \sum_{\rho \in \mathcal{G}_L} \left(d_n(\alpha)_{\sigma_1, \ldots, \sigma_n, \rho} - (-1)^{n+1} \alpha_{\sigma_1, \ldots, \sigma_n} \right) \sigma_1 \cdots \sigma_n \rho \cdot x.$$

Since α is a n-cocycle, we have $d_n(\alpha) = 0$ and we get

$$\begin{aligned} d_{n-1}(a)_{\sigma_1,\ldots,\sigma_n} &= (-1)^n \alpha_{\sigma_1,\ldots,\sigma_n} \sum_{\rho\in\mathcal{G}_L} \sigma_1\cdots\sigma_n\rho\cdot x \\ &= (-1)^n \alpha_{\sigma_1,\ldots,\sigma_n} \sum_{\rho\in\mathcal{G}_L} \rho\cdot x \\ &= (-1)^n \alpha_{\sigma_1,\ldots,\sigma_n}. \end{aligned}$$

This concludes the proof. □

We now establish some functoriality properties of Galois cohomology. In the sequel, G will denote a Galois functor.

Let $\iota : K \longrightarrow K'$ be a morphism of field extensions of k. Let Ω/K and Ω'/K' be two Galois extensions, and assume that we have a morphism $\varphi : \Omega \longrightarrow \Omega'$ of field extensions of k which extends ι. Let $\overline{\varphi} : \mathcal{G}_{\Omega'} \longrightarrow \mathcal{G}_{\Omega}$ the continuous group morphism associated to φ by Corollary I.2.10.

Lemma III.7.32. *The maps $\overline{\varphi} : \mathcal{G}_{\Omega'} \longrightarrow \mathcal{G}_{\Omega}$ and $G(\varphi) : G(\Omega) \longrightarrow G(\Omega')$ are compatible.*

Proof. We need to prove that for all $\sigma' \in \mathcal{G}_{\Omega'}$ and all $g \in G(\Omega)$, we have

$$G(\varphi)(\overline{\varphi}(\sigma')\cdot g) = \sigma'\cdot G(\varphi)(g).$$

By definition of the action of $\mathcal{G}_{\Omega}$ on $G(\Omega)$, we have

$$G(\varphi)(\overline{\varphi}(\sigma')\cdot g) = (G(\varphi)\circ G(\overline{\varphi}(\sigma')))(g).$$

Since G is a functor, this yields

$$G(\varphi)(\overline{\varphi}(\sigma')\cdot g) = G(\varphi\circ\overline{\varphi}(\sigma'))(g).$$

By definition of $\overline{\varphi}$, we have $\varphi\circ\overline{\varphi}(\sigma') = \sigma'\circ\varphi$. We then get

$$G(\varphi)(\overline{\varphi}(\sigma')\cdot g) = G(\sigma'\circ\varphi)(g) = G(\sigma')(G(\varphi)(g)) = \sigma'\cdot G(\varphi)(g).$$

This proves the lemma. □

In view of the previous lemma and Proposition II.3.19, we have an induced map

$$R_\varphi : H^n(\mathcal{G}_\Omega, G(\Omega)) \longrightarrow H^n(\mathcal{G}_{\Omega'}, G(\Omega')).$$

Proposition III.7.33. *Let $\varphi : \Omega \longrightarrow \Omega$" be an extension of ι. Then the map*

$$R_\varphi : H^n(\mathcal{G}_\Omega, G(\Omega)) \longrightarrow H^n(\mathcal{G}_{\Omega'}, G(\Omega'))$$

only depends on ι.

Proof. Let $\varphi' : \Omega \longrightarrow \Omega'$ be another extension of ι. By Corollary I.2.10, there exists $\rho \in \mathcal{G}_\Omega$ such that $\varphi = \varphi' \circ \rho$, and we have

$$\overline{\varphi}' = \mathrm{Int}(\rho) \circ \overline{\varphi}.$$

In particular, we have

$$G(\varphi')(g) = G(\varphi \circ \rho^{-1})(g) = (G(\varphi) \circ G(\rho^{-1}))(g) = G(\varphi)(\rho^{-1} \cdot g),$$

for all $g \in G(\Omega)$. We then have two commutative diagrams

$$\begin{array}{ccc} G(\Omega) & \xrightarrow{\rho^{-1}\cdot} & G(\Omega) \\ \downarrow{\scriptstyle G(\varphi')} & & \downarrow{\scriptstyle G(\varphi)} \\ G(\Omega') & = & G(\Omega') \end{array}$$

$$\begin{array}{ccc} \mathcal{G}_\Omega & \xleftarrow[\mathrm{Int}(\rho)]{} & \mathcal{G}_\Omega \\ \uparrow{\scriptstyle \overline{\varphi}'} & & \uparrow{\scriptstyle \overline{\varphi}} \\ \mathcal{G}_{\Omega'} & = & \mathcal{G}_{\Omega'} \end{array}$$

for wich all the corresponding maps are compatible, by the previous lemma and Example II.3.21. By Proposition II.3.26 and the same example, we get

$$\begin{array}{ccc} H^n(\mathcal{G}_\Omega, G(\Omega)) & = & H^n(\mathcal{G}_\Omega, G(\Omega)) \\ \downarrow{\scriptstyle R_{\varphi'}} & & \downarrow{\scriptstyle R_\varphi} \\ H^n(\mathcal{G}_{\Omega'}, G(\Omega')) & = & H^n(\mathcal{G}_{\Omega'}, G(\Omega')) \end{array}$$

and this completes the proof of the proposition. □

We are ready to apply these results to define a Galois cohomology functor

$$H^n(\_, G) : \mathfrak{C}_k \longrightarrow \mathbf{Sets}^*.$$

Let K/k and L/k be two field extensions, and let $\iota : K \longrightarrow L$ be a morphism of extensions. Now let K_{alg} and L_{alg} be algebraic closures of K and L respectively. Let K_s and L_s be the corresponding separable closures. Finally, let $\varphi : K_s \longrightarrow L_s$ be any extension of ι (such an extension exists by Corollary I.1.20), and let $\overline{\varphi} : \mathcal{G}_{L_s} \longrightarrow \mathcal{G}_{K_s}$ be the continuous group morphism associated to φ by Corollary I.2.10. By Proposition III.7.33, we get a map

$$R_\varphi : H^n(\mathcal{G}_{K_s}, G(K_s)) \longrightarrow H^n(\mathcal{G}_{L_s}, G(L_s)),$$

which only depends on ι.

Remark III.7.34. If $K = L$ and $\iota = \mathrm{Id}_K$, we obtain the identity map, since we may take $\varphi = \mathrm{Id}_{K_s}$.

Lemma III.7.35. *Let $K/k, L/k$ and M/k be three field extensions of k, and let two morphisms of extensions $\iota : K \longrightarrow L$ and $\eta : L \longrightarrow M$. Let $\varphi : K_s \longrightarrow L_s$ and $\psi : L_s \longrightarrow M_s$ be extensions of ι and η respectively. Then we have*

$$R_{\psi\circ\varphi} = R_\psi \circ R_\varphi.$$

Proof. The map $\psi \circ \varphi : K_s \longrightarrow M_s$ is an extension of $K \longrightarrow M$. Moreover, it is easy to check that we have $\overline{\psi \circ \varphi} = \overline{\psi} \circ \overline{\varphi}$. Now, consider the commutative diagrams

$$\begin{array}{ccc} G(K_s) & \xrightarrow{G(\varphi)} & G(L_s) \\ \Big\downarrow{\scriptstyle G(\psi\circ\varphi)} & & \Big\downarrow{\scriptstyle G(\psi)} \\ G(M_s) & = & G(M_s) \end{array}$$

and

$$\begin{array}{ccc} \mathcal{G}_{K_s} & \xleftarrow[\overline{\varphi}]{} & \mathcal{G}_{L_s} \\ {\scriptstyle \overline{\psi\circ\varphi}}\Big\uparrow & & \Big\uparrow{\scriptstyle \overline{\psi}} \\ \mathcal{G}_{M_s} & = & \mathcal{G}_{M_s} \end{array}$$

and apply Proposition II.3.26. □

Corollary III.7.36. *Let K be a field, let K_{alg} be an algebraic closure of K and let K_s be the corresponding separable closure. Then the set $H^n(\mathcal{G}_{K_s}, G(K_s))$ does not depend on the choice of K_{alg}, up to canonical bijection.*

Proof. Let $K_{alg}^{(1)}$ and $K_{alg}^{(2)}$ be two algebraic closures of K. We denote by $K_s^{(1)}$ and $K_s^{(2)}$ the respective separable closures. Let $\varphi : K_{alg}^{(1)} \longrightarrow K_{alg}^{(2)}$ be a K-isomorphism. Since φ maps separable elements to separable elements (we have already seen this during the proof of Corollary I.1.20), we obtain a K-embedding

$$\varphi_s : K_s^{(1)} \longrightarrow K_s^{(2)}.$$

Now the inverse map

$$\psi_s : K_{alg}^{(2)} \longrightarrow K_{alg}^{(1)}$$

induces a K-embedding $\psi_s : K_s^{(2)} \longrightarrow K_s^{(1)}$. By definition, we have

$$\begin{aligned} \varphi_s \circ \psi_s &= \mathrm{Id}_{K_s^{(2)}}, \\ \psi_s \circ \varphi_s &= \mathrm{Id}_{K_s^{(1)}}. \end{aligned}$$

Applying Lemma III.7.35 and Remark III.7.34, we get that that $\overline{\varphi}_{s*}$ is bijective with inverse $\overline{\psi}_{s*}$. This bijection is canonical, since it only depends on the restriction of φ to K, which is the identity. □

Definition III.7.37. Let $G : \mathbf{Alg}_k \longrightarrow \mathbf{Grps}$ be a Galois functor, and let K/k be a field extension. We define the nth **Galois cohomology set** of G by

$$H^n(K, G) = H^n(\mathcal{G}_{K_s}, G(K_s)).$$

If G is abelian (i.e. $G(R)$ is an abelian group for all R), it is a commutative group. If $\iota : K \longrightarrow L$ is a morphism of field extensions of k, the corresponding map $H^n(K, G) \longrightarrow H^n(L, G)$ is called **the restriction map**, and is denoted by $\mathrm{Res}_{L/K}$. It follows from Remark III.7.34 and Lemma III.7.35 that if $\iota = \mathrm{Id}_K$, $\mathrm{Res}_{K/K}$ is the identity map, and that for any tower of field extensions $K \longrightarrow L \longrightarrow M$, we have

$$\mathrm{Res}_{M/K} = \mathrm{Res}_{M/L} \circ \mathrm{Res}_{L/K}.$$

Therefore, we get a functor

$$H^1(\_, G) : \mathfrak{C}_k \longrightarrow \mathbf{Sets}^*.$$

If G is abelian, the restriction map is a group morphism, and for $n \geq 1$, we obtain a functor

$$H^n(\_, G) : \mathfrak{C}_k \longrightarrow \mathbf{AbGrps}.$$

Remark III.7.38. The restriction map is easier to describe if L/K is separable: indeed, in this case $K_s = L_s$ and $\mathcal{G}_{L_s}$ is a closed subgroup of $\mathcal{G}_{K_s}$; therefore applying the restriction map is just restricting the cocycles.

We would like to continue this section by a translation of some results of the previous chapter in our situation. Until the end, G, G', H, H', N and N' will denote Galois functors.

Theorem III.7.39. *Let K/k and L/k be two field extensions of k, let $K \longrightarrow L$ be a morphism of field extensions, and let $K_s \longrightarrow L_s$ be any extension to the separable closures.*

(1) *Assume that we have an exact sequence of $\mathcal{G}_{K_s}$-groups*

$$1 \longrightarrow N(K_s) \longrightarrow G(K_s) \longrightarrow H(K_s) \longrightarrow 1.$$

Then the exact sequence of pointed sets

$$1 \longrightarrow N(K) \longrightarrow G(K) \longrightarrow H(K)$$

may be extended to

$$\longrightarrow H(K) \xrightarrow{\delta_K^0} H^1(K,N) \longrightarrow H^1(K,G) \longrightarrow H^1(K,H).$$

If moreover $N(K_s)$ identifies to a central subgroup of $G(K_s)$, then the exact sequence above may be extended further to

$$\longrightarrow H^1(K,H) \xrightarrow{\delta_K^1} H^2(K,N).$$

(2) *Assume that the diagram*

$$\begin{array}{ccc} G(K_s) & \longrightarrow & H(K_s) \\ \downarrow & & \downarrow \\ G'(K_s) & \longrightarrow & H'(K_s) \end{array}$$

commutes. Then the diagram

$$\begin{array}{ccc} H^n(K,G) & \longrightarrow & H^n(K,H) \\ \downarrow & & \downarrow \\ H^n(K,G') & \longrightarrow & H^n(L,H') \end{array}$$

commutes.

(3) *Assume that the diagram*

$$\begin{array}{ccc} G(K_s) & \longrightarrow & H(K_s) \\ \downarrow & & \downarrow \\ G(L_s) & \longrightarrow & H(L_s) \end{array}$$

commutes. Then the diagram

$$\begin{array}{ccc} H^n(K,G) & \longrightarrow & H^n(K,H) \\ \downarrow & & \downarrow \\ H^n(L,G) & \longrightarrow & H^n(L,H) \end{array}$$

commutes, where the vertical maps are the restriction maps.

(4) *Assume that we have a commutative diagram with exact rows*

$$\begin{array}{ccccccccc} 1 & \longrightarrow & N(K_s) & \longrightarrow & G(K_s) & \longrightarrow & H(K_s) & \longrightarrow & 1 \\ & & \downarrow & & \downarrow & & \downarrow & & \\ 1 & \longrightarrow & N'(K_s) & \longrightarrow & G'(K_s) & \longrightarrow & H'(K_s) & \longrightarrow & 1 \end{array}$$

and let us denote by δ^0_K and δ'^0_K the respective 0th connecting maps. Then the diagram

$$\begin{array}{ccc} H(K) & \xrightarrow{\delta^0_K} & H^1(K,N) \\ \downarrow & & \downarrow \\ H'(K) & \xrightarrow{\delta'^0_K} & H^1(K,N') \end{array}$$

is commutative.

If moreover $N(K_s)$ and $N'(K_s)$ identify to central subgroups of $G(K_s)$ and $G'(K_s)$ respectively, then the diagram

$$\begin{array}{ccc} H^1(K,H) & \xrightarrow{\delta^1_K} & H^2(K,N) \\ \downarrow & & \downarrow \\ H^1(K,H') & \xrightarrow{\delta'^1_K} & H^2(K,N') \end{array}$$

is commutative, where δ^1_K and δ'^1_K denote the respective first connecting maps.

(5) *Assume that we have a commutative diagram with exact rows*

$$\begin{array}{ccccccccc} 1 & \longrightarrow & N(K_s) & \longrightarrow & G(K_s) & \longrightarrow & H(K_s) & \longrightarrow & 1 \\ & & \downarrow & & \downarrow & & \downarrow & & \\ 1 & \longrightarrow & N(L_s) & \longrightarrow & G(L_s) & \longrightarrow & H(L_s) & \longrightarrow & 1 \end{array}$$

and let us denote by δ^0_K *and* δ^0_L *the respective* 0th *connecting maps. Then the diagram*

$$\begin{array}{ccc} H(K) & \xrightarrow{\delta^0_K} & H^1(K,N) \\ \downarrow & & \downarrow \\ H(L) & \xrightarrow{\delta^0_L} & H^1(L,N) \end{array}$$

is commutative.

If moreover $N(K_s)$ *and* $N(L_s)$ *identify to central subgroups of* $G(K_s)$ *and* $G(L_s)$ *respectively, then the diagram*

$$\begin{array}{ccc} H^1(K,H) & \xrightarrow{\delta^1_K} & H^2(K,N) \\ \downarrow & & \downarrow \\ H^1(L,H) & \xrightarrow{\delta^1_L} & H^2(L,N) \end{array}$$

is commutative, where δ^1_K *and* δ^1_L *denote the respective first connecting maps.*

(6) *For all* $n \geq 1$*, we have a bijection of pointed sets (resp. a group isomorphism if* G *is abelian)*

$$H^n(K,G) \simeq \varinjlim_L H^n(\mathcal{G}_L, G(L)),$$

where L/K *runs through the finite Galois extensions of* K*.*

Proof. (1) is a direct application of Propositions II.4.7 and II.4.10, together with Lemma III.7.16 (2). To prove points (2) – (5), apply Propositions II.3.26, II.4.6 and II.4.11, as well as Lemma III.7.32 for points (3) and (5). Finally, (6) is just an application of Theorem III.7.30 to $\Omega = K_s$. □

§III.8 Abstract Galois descent

Now that the decor is set and the actors are in place, we are ready to expose the theory of Galois descent. As explained in the introduction, Galois cohomology measures in which extent two objects defined over k are isomorphic, provided they are isomorphic over a field extension of k. We would like to generalize the approach used to solve the Galois descent problem for conjugacy of matrices to arbitrary algebraic objects. We start this section by having a closer look at this case.

III.8.1 Matrices reloaded

In this paragraph, we would like to extract the essential arguments of our solution to the conjugacy problem, and rewrite them in a more concise and formal way, in order to find a method to attack the general Galois descent problem. Let us first reformulate the result we obtained. In fact, we have proved in the introduction that the set of $G(k)$-conjugacy classes of matrices $M \in \mathrm{M}_n(\Omega)$ which are $G(\Omega)$-conjugate to a given matrix $M_0 \in \mathrm{M}_n(k)$ is in one-to-one correspondence with the set of cohomology classes $[\alpha] \in H^1(\mathcal{G}_\Omega, Z_G(M_0)(\Omega))$, which may be written $[\alpha] = [\alpha^C]$ for some $C \in G(\Omega)$, where α^C is the cocycle

$$\alpha^C : \begin{array}{l} \mathcal{G}_\Omega \longrightarrow Z_G(M_0)(\Omega) \\ \sigma \longmapsto C(\sigma{\cdot}C)^{-1}. \end{array}$$

This set of cohomology classes is nothing but the kernel of the map

$$H^1(\mathcal{G}_\Omega, Z_G(M_0)(\Omega)) \longrightarrow H^1(\mathcal{G}_\Omega, G(\Omega))$$

induced by the inclusion $Z_G(M_0)(\Omega) \subset G(\Omega)$.

This observation will allow us to give a more conceptual (and less miraculous) explanation of our result. Notice first that the congugacy class of a matrix may be reinterpreted as an orbit under the action of $G \subset \mathbf{GL}_n$ by conjugation. This action will be denoted by $*$ in the sequel. The next crucial observation is then the following one: if $M_0 \in \mathrm{M}_n(k)$, we may rewrite $Z_G(M_0)(\Omega)$ as

$$Z_G(M_0)(\Omega) = \{C \in G(\Omega) \mid C * M_0 = M_0\}.$$

In other words, $Z_G(M_0)(\Omega)$ is nothing but the stabilizer of M_0 (viewed as an element of $\mathrm{M}_n(\Omega)$) with respect to the action of $G(\Omega)$ on $\mathrm{M}_n(\Omega)$. The second important point is that the action of G_Ω on $G(\Omega)$ restricts to an action on $Z_G(M_0)(\Omega)$. To see this, recall that the action of $\mathcal{G}_\Omega$ on a matrix $S = (s_{ij}) \in \mathrm{M}_n(\Omega)$ is given by

$$\sigma{\cdot}S = (\sigma(s_{ij})).$$

In particular, the following properties hold:

(i) $\mathrm{M}_n(\Omega)^{\mathcal{G}_\Omega} = \mathrm{M}_n(k)$

(ii) For all $S \in \mathrm{M}_n(\Omega), C \in G(\Omega)$ and $\sigma \in \mathcal{G}_\Omega$, we have

$$\sigma{\cdot}(C * S) = (\sigma{\cdot}C) * (\sigma{\cdot}S).$$

We have in fact an even more general property. If $\iota : K \longrightarrow L$ is a morphism of field extensions of k and $S \in \mathrm{M}_n(K)$, set

$$\iota\cdot S = (\iota(s_{ij})).$$

We then have

(*ii'*) For all morphisms of extensions $\iota : K \longrightarrow L, S \in \mathrm{M}_n(K)$ and all $C \in G(K)$, we have

$$\iota\cdot(C * S) = (\iota\cdot C) * (\iota\cdot S).$$

If now $C \in Z_G(M_0)(\Omega)$ and $\sigma \in G(\Omega)$, we have

$$(\sigma\cdot C) * M_0 = (\sigma\cdot C) * (\sigma\cdot M_0)$$

by (*i*), since $M_0 \in \mathrm{M}_n(k)$. Using (*ii*), we then get

$$(\sigma\cdot C) * M_0 = \sigma\cdot(C * M_0) = \sigma\cdot M_0 = M_0,$$

the second equality coming from the fact that $C \in Z_G(M_0)(\Omega)$. Hence the action of $\mathcal{G}_\Omega$ on $G(\Omega)$ restricts to an action on $Z_G(M_0)(\Omega)$ as claimed. Now it follows from elementary group theory that we have a bijection

$$G(\Omega)/Z_G(M_0)(\Omega) \simeq G(\Omega) * M_0.$$

Equivalently, we have an exact sequence

$$1 \longrightarrow Z_G(M_0)(\Omega) \longrightarrow G(\Omega) \longrightarrow G(\Omega) * M_0 \longrightarrow 1,$$

which may be easily seen to be an exact sequence of G_Ω-sets satisfying the conditions explained in § II.4.1.

The apparition of $\ker[H^1(\mathcal{G}_\Omega, Z_G(M_0)(\Omega)) \longrightarrow H^1(\mathcal{G}_\Omega, G(\Omega))]$ is not a real surprise then, in view of Corollary II.4.5. The same corollary says that this kernel is in one-to-one correspondence with the orbit set of $G(\Omega)^{\mathcal{G}_\Omega}$ in $(G(\Omega) * M_0)^{\mathcal{G}_\Omega}$.

Let us now check that this orbit set is precisely the set of $G(k)$-conjugacy classes of matrices which become $G(\Omega)$-conjugate to M_0, at least in the cases considered in the introduction. Therefore, assume until the end of this paragraph that $G = \mathbf{GL}_n$ or $\mathbf{SL}_n$. In this case, we have

(*iii*) $G(\Omega)^{\mathcal{G}_\Omega} = G(k)$.

Notice now that the action $G(\Omega)^{\mathcal{G}_\Omega} = G(k)$ on $(G(\Omega) * M_0)^{\mathcal{G}_\Omega}$ defined before Corollary II.4.5 is simply the restriction of the action of $G(k)$ on $\mathrm{M}_n(k)$ by conjugation. Indeed, if $M \in \mathrm{M}_n(k)$ has the form $M = Q * M_0$

for some $Q \in G(\Omega)$, and if $P \in G(k)$, Q is a preimage of M under the map $G(\Omega) \to G(\Omega) * M_0$, and therefore we have

$$P \cdot M = (PQ) * M_0 = P * (Q * M_0) = P * M.$$

Using (i) and (iii), it follows that the orbit set $(G(\Omega) * M_0)^{\mathcal{G}_\Omega}/G(\Omega)^{\mathcal{G}_\Omega}$ is nothing but the set of $G(k)$-conjugacy classes of matrices $M \in \mathrm{M}_n(k)$ which become $G(\Omega)$-conjugate to M_0.

Therefore, we have proved that our solution the conjugacy problem for matrices was nothing but an application of Corollary II.4.5, and we have identified three important properties which make this actually work. Notice that (i) and (iii) may seem redundant a priori, but this is only due to our specific example. In more abstract situations, both conditions may be of different nature. For example, one may replace matrices by quadratic forms of dimension n and study the Galois descent problem for isomorphism classes of quadratic forms. In this case, we see that properties (i) and (iii) do not concern the same objects.

Our next goal is now to reformulate this new approach in our functorial context, and derive a general solution for abstract Galois descent problems. In particular, we will need to find appropriate substitutes for $Z_G(M_0)(\Omega)$ and properties (i) – (iii).

III.8.2 Actions of group-valued functors

If we analyze the Galois descent problem for conjugacy classes of matrices, some key ingredients are needed in order for this question to make sense. First of all, matrices with coefficients in a field form a functor $\mathbf{M}_n : \mathfrak{C}_k \longrightarrow \mathbf{Sets}$, so that for a given Galois extension Ω/k, we may consider the sets $\mathrm{M}_n(k)$ and $\mathrm{M}_n(\Omega)$, and we have an induced map

$$\mathrm{M}_n(k) \longrightarrow \mathrm{M}_n(\Omega)$$

which allows us to extend scalars. Therefore, the algebraic objects we are going to consider will be points of a covariant functor.

Let k be any field, and let $\mathbf{F} : \mathfrak{C}_k \longrightarrow \mathbf{Sets}$ be a functor. We will write $\mathbf{F}(K)$ instead of $\mathbf{F}(K/k)$. If $K \longrightarrow L$ is a morphism of field extensions and $a \in \mathbf{F}(K)$, recall that we denote by a_L the image of a under the induced map $\mathbf{F}(K) \longrightarrow \mathbf{F}(L)$ if there is no ambiguity in the choice of the map $K \longrightarrow L$.

To set up the Galois descent problem for conjugacy classes of matrices, we also need an action of some subfunctor of $\mathbf{GL}_n$ on matrices. As we have seen in the previous paragraph, this action has some nice functorial properties. This leads to the following definition:

Definition III.8.1. Let $G : \mathfrak{C}_k \longrightarrow \mathbf{Grps}$ be a group-valued functor. We say that G **acts on F** if the following conditions hold:

(1) For every field extension K/k, the group $G(K)$ acts on the set $\mathbf{F}(K)$. This action will be denoted by $*$.

(2) For every morphism $\iota : K \longrightarrow L$ of field extensions, the following diagram is commutative:

$$\begin{array}{ccc} G(K)\times\mathbf{F}(K) & \longrightarrow & \mathbf{F}(K) \\ \Big\downarrow{\scriptstyle (G(\iota),\mathbf{F}(\iota))} & & \Big\downarrow{\scriptstyle \mathbf{F}(\iota)} \\ G(L)\times\mathbf{F}(L) & \longrightarrow & \mathbf{F}(L) \end{array}$$

that is $\mathbf{F}(\iota)(g*a) = G(\iota)(g)*\mathbf{F}(\iota)(a)$ for all $a \in \mathbf{F}(K), g \in G(K)$.

In other words, for every field extension K/k, we have a group action of $G(K)$ on $\mathbf{F}(K)$ which is functorial in K.

Notice that the last condition rewrites

$$(g * a)_L = g_L * a_L \text{ for all } a \in \mathbf{F}(K), g \in G(K)$$

for a given field extension L/K if we use the short notation recalled at the beginning of the paragraph.

Examples III.8.2.

(1) Let $G \subset \mathbf{GL}_n$ be an algebraic group-scheme and let $\mathbf{F}$ be the functor defined by $\mathbf{F}(K) = K^n$ for every field extension K/k. Then G acts by left multiplication on $\mathbf{F}$.

(2) If $G \subset \mathbf{GL}_n$ is an algebraic group-scheme and $\mathbf{F} = \mathbf{M}_n$, then G acts on $\mathbf{F}$ by conjugation.

Remark III.8.3. Let K/k be a field extension, and let Ω/K be Galois extension. Recall from Lemma III.7.15 that, given a covariant functor $\mathbf{F} : \mathfrak{C}_k \longrightarrow \mathbf{Sets}$, we have a natural action of $\mathcal{G}_\Omega$ on $\mathbf{F}(\Omega)$ defined by

$$\begin{aligned} \mathcal{G}_\Omega \times \mathbf{F}(\Omega) &\longrightarrow \mathbf{F}(\Omega) \\ (\sigma, a) &\longmapsto \sigma\cdot a = \mathbf{F}(\sigma)(a). \end{aligned}$$

If G is a group-valued functor acting on $\mathbf{F}$, we have by definition

$$\mathbf{F}(\sigma)(g * a) = G(\sigma)(g) * \mathbf{F}(\sigma)(a) \text{ for all } a \in \mathbf{F}(\Omega), \sigma \in \mathcal{G}_\Omega,$$

which rewrites as

$$\sigma\cdot(g * a) = (\sigma\cdot g) * (\sigma\cdot a) \text{ for all } a \in \mathbf{F}(\Omega), \sigma \in \mathcal{G}_\Omega.$$

We would like to continue by giving a reformulation of the general Galois descent problem. For this, we need to introduce the concept of a twisted form.

III.8.3 Twisted forms

Let G be a group-valued functor acting on a functor $\mathbf{F} : \mathfrak{C}_k \longrightarrow$ **Sets**. This action of G allows us to define an equivalence relation on the set $\mathbf{F}(K)$ for every field extension K/k by identifying two elements which are in the same $G(K)$-orbit. For example, in the case of matrices, two matrices of $\mathrm{M}_n(K)$ will be equivalent if and only if they are $G(K)$-conjugate. More precisely, we have the following definition:

Definition III.8.4. Let G be a group-valued functor defined over k acting on $\mathbf{F}$. For every field extension K/k we define an equivalence relation $\sim_K$ on $\mathbf{F}(K)$ as follows: two elements $b, b' \in \mathbf{F}(K)$ are **equivalent over** K if there exists $g \in G(K)$ such that $b = g * b'$. We will denote by $[b]$ the corresponding equivalence class.

We may now formulate a general descent problem.

Galois descent problem: let $\mathbf{F} : \mathfrak{C}_k \longrightarrow$ **Sets** and let $G : \mathfrak{C}_k \longrightarrow$ **Grps** be a group-scheme acting on $\mathbf{F}$. Finally, let Ω/k be a Galois extension and let $a, a' \in \mathbf{F}(k)$. Assume that $a_\Omega \sim_\Omega a'_\Omega$. Do we have $a \sim_k a'$?

Notice that the answer to this question only depends on the $G(k)$-equivalence class of a and a'. We now give a special name to elements of $\mathbf{F}$ which become equivalent to a fixed element $a \in \mathbf{F}(k)$ over Ω.

Definition III.8.5. Let $a \in \mathbf{F}(k)$, let K/k be a field extension and let Ω/K be a Galois extension. An element $a' \in \mathbf{F}(K)$ is called **a twisted** K**-form of** a if $a'_\Omega \sim_\Omega a_\Omega$.

Clearly, if $a' \in \mathbf{F}(K)$ is a twisted K-form of a and $a' \sim_K a''$, then a'' is also a twisted K-form of a, so the action of $G(K)$ restricts to the set of twisted K-forms of a.

We denote by $\mathbf{F}_a(\Omega/K)$ the set of K-equivalence classes of twisted K-forms of a, that is

$$\mathbf{F}_a(\Omega/K) = \{[a'] \mid a' \in \mathbf{F}(K), a'_\Omega \sim_\Omega a_\Omega\}.$$

Notice that $\mathbf{F}_a(\Omega/K)$ always contains the class of a_K, so it is natural to consider $\mathbf{F}_a(\Omega/K)$ as a pointed set, where the base point is $[a_K]$.

We now would like to define a functor $\mathbf{F}_a : \mathfrak{C}_k \longrightarrow \mathbf{Sets}^*$. Let $\iota : K \longrightarrow K'$ be a morphism of field extensions of k, let Ω/K and Ω'/K' be two Galois extension, and assume that we have an extension $\varphi : \Omega \longrightarrow \Omega'$ of ι. If $a' \in \mathbf{F}(K)$ is a twisted K-form of a, then $a'_{K'}$ is a twisted K'-form of a as well. Indeed, functorial properties of $\mathbf{F}$ imply that $(a'_{K'})_{\Omega'} = (a'_\Omega)_{\Omega'}$ (where the last scalar extension to Ω' is obtained via $\mathbf{F}(\varphi) : \mathbf{F}(\Omega) \longrightarrow \mathbf{F}(\Omega')$). Now if $g \in G(\Omega)$ satisfies $g * a'_\Omega = a_\Omega$, then we have

$$g_{\Omega'} * (a'_{K'})_{\Omega'} = g_{\Omega'} * (a'_\Omega)_{\Omega'} = (g * a'_\Omega)_{\Omega'},$$

by definition of the action of G on $\mathbf{F}$. We then get

$$g_{\Omega'} * (a'_{K'})_{\Omega'} = (a_\Omega)_{\Omega'} = a_{\Omega'},$$

showing that $a'_{K'}$ is a twisted K'-form of a. Notice that it does not depend on the choice of the extension φ of ι.

Therefore, the map $\mathbf{F}(K) \longrightarrow \mathbf{F}(K')$ induces a map

$$\begin{aligned} \mathbf{F}_a(\Omega/K) &\longrightarrow \mathbf{F}_a(\Omega'/K') \\ [a'] &\longmapsto [a'_{K'}]. \end{aligned}$$

In particular, we obtain a functor $\mathbf{F}_a : \mathfrak{C}_k \longrightarrow \mathbf{Sets}^*$ by setting $\mathbf{F}_a(K) = \mathbf{F}_a(K_s/K)$ for every field extension K/k, the map induced by a morphism of field extension $\iota : K \longrightarrow K'$ being the map

$$\mathbf{F}_a(\iota) : \begin{aligned} \mathbf{F}_a(K) &\longrightarrow \mathbf{F}_a(K') \\ [a'] &\longmapsto [a'_{K'}]. \end{aligned}$$

Example III.8.6. As pointed out before, if $\mathbf{F} = \mathbf{M}_n$, then $G \subset \mathbf{GL}_n$ acts on $\mathbf{F}$ by conjugation, and two matrices $M, M' \in \mathrm{M}_n(K)$ are then equivalent if and only if they are $G(K)$-conjugate. Moreover, if $M_0 \in \mathrm{M}_n(k)$, then $\mathbf{F}_{M_0}(\Omega/K)$ is the set of $G(K)$-conjugacy classes of matrices $M \in \mathrm{M}_n(K)$ which are $G(\Omega)$-conjugate to M_0.

Using the notation introduced previously, the Galois descent problem may be reinterpreted in terms of twisted forms as follows: given $a \in \mathbf{F}(k)$ and a Galois extension Ω/k, do we have $\mathbf{F}_a(\Omega/k) = \{[a]\}$?

We would like to describe the functor $\mathbf{F}_a$ in terms of Galois cohomology of a suitable group-scheme associated to a, under some reasonable conditions on $\mathbf{F}$ and G. To do this, we will continue to try to generalize the approach described in the first paragraph of this section.

III.8.4 The Galois descent condition

One of the crucial property we used to solve the conjugacy problem is the equality

$$\mathrm{M}_n(\Omega)^{\mathcal{G}_\Omega} = \mathrm{M}_n(k),$$

where we let $\mathcal{G}_\Omega$ act on $S \in \mathrm{M}_n(\Omega)$ coefficientwise. By Example III.7.26, this action is nothing but the action of $\mathcal{G}_\Omega$ induced by the functorial properties of M_n, as described in Lemma III.7.15. Now let us go back to our more general setting. For every field extension K/k and every Galois extension Ω/K, we have an action of $\mathcal{G}_\Omega$ on the set $\mathbf{F}(\Omega)$ given by

$$\sigma \cdot a = \mathbf{F}(\sigma)(a) \text{ for } \sigma \in \mathcal{G}_\Omega \text{ and } a \in \mathbf{F}(\Omega).$$

The second part of Lemma III.7.15, applied to the Galois extensions K/K and Ω/K, then yields

$$\sigma \cdot a_\Omega = a_\Omega \text{ for all } \sigma \in \mathcal{G}_\Omega, a \in \mathbf{F}(K).$$

However, contrary to the case of matrices, an element of $\mathbf{F}(\Omega)$ on which $\mathcal{G}_\Omega$ acts trivially does not necessarily comes from an element of $\mathbf{F}(K)$.

Example III.8.7. Let us consider the functor $\mathbf{F} : \mathfrak{C}_k \longrightarrow \mathbf{Sets}$ defined as follows: for a field extension K/k, set

$$\mathbf{F}(K) = \begin{cases} \{0\} & \text{if } [K:k] \leq 1 \\ \{0,1\} & \text{if } [K:k] \geq 2 \end{cases}$$

the map induced by a morphism $K \longrightarrow K'$ being the inclusion of sets. In particular, for every Galois extension Ω/k, the Galois group $\mathcal{G}_\Omega$ acts trivially on $\mathbf{F}(\Omega)$. However, if $[\Omega : k] > 1$, the element $1 \in \mathbf{F}(\Omega)$ does not come from an element of $\mathbf{F}(k)$.

These considerations lead to the following definition:

Definition III.8.8. We say that a functor $\mathbf{F} : \mathfrak{C}_k \longrightarrow \mathbf{Sets}$ satisfies the **Galois descent condition** if for every field extension K/k and every

Galois extension Ω/K the map $\mathbf{F}(K) \longrightarrow \mathbf{F}(\Omega)$ is injective and induces a bijection

$$\mathbf{F}(K) \simeq \mathbf{F}(\Omega)^{\mathcal{G}_\Omega}.$$

Example III.8.9. The functor $\mathbf{M}_n$ satisfies the Galois descent condition, as well as every representable functor by Lemma III.7.16, or as any Galois functor by definition.

III.8.5 Stabilizers

It follows from the considerations of the previous paragraph that it is reasonable to consider Galois descent problems for elements of a functor satisfying the Galois descent condition. Now that we have set a suitable framework for the general Galois descent problem, we need an appropriate substitute for the set $Z_G(M_0)(\Omega)$. As noticed before, denoting by $*$ the action of $G \subset \mathbf{GL}_n$ on matrices by conjugation, the subgroup $Z_G(M_0)(\Omega)$ may be reinterpreted as the stabilizer of M_0 with respect to the action of $G(\Omega)$ on $\mathrm{M}_n(\Omega)$, that is

$$Z_G(M_0)(\Omega) = \mathbf{Stab}_G(M_0)(\Omega) = \{C \in G(\Omega) \mid C * M_0 = M_0\}.$$

Since in our setting we have a group-scheme acting on $\mathbf{F}$, it seems sensible to introduce the following definition:

Definition III.8.10. Let $G : \mathfrak{C}_k \longrightarrow \mathbf{Grps}$ be a group-valued functor acting on $\mathbf{F}$. For $a \in \mathbf{F}(k)$, and every field extension K/k, we set

$$\mathbf{Stab}_G(a)(K) = \{g \in G(K) \mid g * a_K = a_K\} \text{ for all } K/k.$$

If $K \longrightarrow K'$ is a morphism of field extensions, the map $G(K) \longrightarrow G(K')$ restricts to a map $\mathbf{Stab}_G(a)(K) \longrightarrow \mathbf{Stab}_G(a)(K')$. Indeed, for every $g \in \mathbf{Stab}_G(a)(K)$, we have

$$g_{K'} * a_{K'} = (g * a_K)_{K'} = (a_K)_{K'} = a_{K'}.$$

We then get a subfunctor $\mathbf{Stab}_G(a) : \mathfrak{C}_k \longrightarrow \mathbf{Grps}$ of G, called the **stabilizer** of a.

Example III.8.11. If $\mathbf{F} = \mathbf{M}_n, G \subset \mathbf{GL}_n$ and $M_0 \in \mathrm{M}_n(k)$, then

$$\mathbf{Stab}_G(M_0)(K) = Z_G(M_0)(K) \text{ for all } K/k.$$

Remark III.8.12. Let K be a field, and let Ω/K be a Galois extension. By definition, the map $\mathbf{Stab}_G(a)(\sigma) : \mathbf{Stab}_G(a)(\Omega) \longrightarrow \mathbf{Stab}_G(a)(\Omega)$ is obtained by restriction of the map $G(\Omega) \longrightarrow G(\Omega)$. Hence, the natural action of $\mathcal{G}_\Omega$ on $G(\Omega)$ restricts to an action on $\mathbf{Stab}_G(a)(\Omega)$.

We have now to ensure that the action of $\mathcal{G}_\Omega$ on $\mathbf{Stab}_G(a)(\Omega)$ is continuous. Contrary to the case of matrices, the functor $\mathbf{Stab}_G(a)$ may not be representable even if G is, so we may not conclude to the continuity of the action of $\mathcal{G}_\Omega$ this way. However, if G is a Galois functor, so is $\mathbf{Stab}_G(a)$, as we proceed to show now.

Lemma III.8.13. *Let $G : \mathfrak{C}_k \longrightarrow$ **Grps** be a Galois functor acting on a functor $\mathbf{F}$ satisfying the Galois descent condition. Then for all $a \in \mathbf{F}(k)$, $\mathbf{Stab}_G(a)$ is a Galois functor. In particular, for every field extension K/k and every Galois extension Ω/K, $\mathbf{Stab}_G(a)(\Omega)$ is a $\mathcal{G}_\Omega$-group.*

Proof. Let K/k be a field extension and let Ω/K be a Galois extension. Since $\mathbf{Stab}_G(a)$ is a subfunctor of G, we have a commutative diagram

$$\begin{array}{ccc} \mathbf{Stab}_G(a)(K) & \longrightarrow & \mathbf{Stab}_G(a)(\Omega) \\ \downarrow & & \downarrow \\ G(K) & \longrightarrow & G(\Omega) \end{array}$$

where the vertical maps are inclusions. Therefore, the injectivity of $G(K) \longrightarrow G(\Omega)$ implies the injectivity of the map

$$\mathbf{Stab}_G(a)(K) \longrightarrow \mathbf{Stab}_G(a)(\Omega).$$

Let $g \in \mathbf{Stab}_G(a)(\Omega)^{\mathcal{G}_\Omega}$. Since G is a Galois functor, we have $g = g'_\Omega$ for some $g' \in G(K)$. We have to check that $g' \in \mathbf{Stab}_G(a)(K)$. But we have

$$(g' * a_K)_\Omega = g'_\Omega * (a_K)_\Omega = g * a_\Omega = a_\Omega = (a_K)_\Omega.$$

Since the map $\mathbf{F}(K) \longrightarrow \mathbf{F}(\Omega)$ is injective, we get $g' * a_K = a_K$, so $g' \in \mathbf{Stab}_G(a)(K)$. Hence we have a group isomorphism

$$\mathbf{Stab}_G(a)(K) \simeq \mathbf{Stab}_G(a)(\Omega)^{\mathcal{G}_\Omega}.$$

Finally, if $g \in \mathbf{Stab}_G(a)(\Omega)$, there exists a finite Galois subextension L/K of Ω/K and $g' \in G(L)$ such that $g = g'_\Omega$. Once again, we have to check that $g' \in \mathbf{Stab}_G(a)(L)$, which can be done as before. Thus we have

$$\mathbf{Stab}_G(a)(\Omega) = \bigcup_{L \subset \Omega} \mathbf{Stab}_G(a)(L).$$

Thus $\mathbf{Stab}_G(a)$ is a Galois functor. The last part of the lemma follows from Remark III.7.29. □

Remark III.8.14. To establish the result above, only the injectivity of the map $\mathbf{F}(K) \longrightarrow \mathbf{F}(\Omega)$ for all Galois extensions Ω/K was needed. However, the condition on fixed points will be essential to prove the Galois descent lemma, as for the case of matrices.

Using the results of the previous section, we then obtain a Galois cohomology set $H^1(\mathcal{G}_\Omega, \mathbf{Stab}_G(a)(\Omega))$ for any Galois extension Ω/K, as well as a functor

$$H^1(\_, \mathbf{Stab}_G(a)) : \mathfrak{C}_k \longrightarrow \mathbf{Sets}^*.$$

III.8.6 Galois descent lemma

We are now ready to state and prove the Galois descent lemma:

Theorem III.8.15 (Galois descent lemma). *Let $\mathbf{F} : \mathfrak{C}_k \longrightarrow \mathbf{Sets}$ be a functor satisfying the Galois descent condition, let $G : \mathfrak{C}_k \longrightarrow \mathbf{Grps}$ be a Galois functor acting on $\mathbf{F}$, and let $a \in \mathbf{F}(k)$. Then for every field extension K/k and every Galois extension Ω/K, we have a bijection of pointed sets*

$$\mathbf{F}_a(\Omega/K) \xrightarrow{\sim} \ker[H^1(\mathcal{G}_\Omega, \mathbf{Stab}_G(a)(\Omega)) \longrightarrow H^1(\mathcal{G}_\Omega, G(\Omega))]$$

which is functorial in Ω. More precisely, let $\iota : K \longrightarrow K'$ be a morphism of field extensions of k, let Ω/K and Ω'/K' be two Galois extensions, and assume that we have an extension $\varphi : \Omega \longrightarrow \Omega'$ of field extensions of ι. Then the diagram

$$\begin{array}{ccc}
\mathbf{F}_a(\Omega/K) & \xrightarrow{\sim} & \ker[H^1(\mathcal{G}_\Omega, \mathbf{Stab}_G(a)(\Omega)) \longrightarrow H^1(\mathcal{G}_\Omega, G(\Omega))] \\
\downarrow & & \downarrow R_\varphi \\
\mathbf{F}_a(\Omega'/K') & \xrightarrow{\sim} & \ker[H^1(\mathcal{G}_{\Omega'}, \mathbf{Stab}_G(a)(\Omega')) \longrightarrow H^1(\mathcal{G}_{\Omega'}, G(\Omega'))]
\end{array}$$

is commutative. In particular, we have an isomorphism of functors from $\mathfrak{C}_k$ to $\mathbf{Sets}^*$

$$\mathbf{F}_a \xrightarrow{\sim} \ker[H^1(\_, \mathbf{Stab}_G(a)) \longrightarrow H^1(\_, G)].$$

Therefore if $H^1(\_, G) = 1$, we have an isomorphism of functors

$$\mathbf{F}_a \xrightarrow{\sim} H^1(\_, \mathbf{Stab}_G(a)).$$

Remark III.8.16. Saying that we have a bijection of pointed sets means that it preserves the base points. Hence for every field extension K/k, the class of $[a_K]$ corresponds to the class of the trivial cocycle.

Proof of Theorem III.8.15. The key ingredient of the proof is Corollary II.4.5. First, by Remark III.8.12, the action of $\mathcal{G}_\Omega$ on $G(\Omega)$ restricts to an action on $\mathbf{Stab}_G(a)(\Omega)$, which is continuous by Lemma III.8.13. Moreover, we have an exact sequence

$$1 \longrightarrow \mathbf{Stab}_G(a)(\Omega) \longrightarrow G(\Omega) \longrightarrow G(\Omega) * a_\Omega \longrightarrow 1$$

which satisfies the conditions of § II.4.1. By Corollary II.4.5, the kernel of $H^1(\mathcal{G}_\Omega, \mathbf{Stab}_G(a)(\Omega)) \longrightarrow H^1(\mathcal{G}_\Omega, G(\Omega))$ is in one-to-one correspondence with the orbit set of $G(\Omega)^{\mathcal{G}_\Omega}$ in $(G(\Omega) * a_\Omega)^{\mathcal{G}_\Omega}$. Notice that $G(\Omega) * a_\Omega$ is simply the set of elements of $\mathbf{F}(\Omega)$ which are equivalent to a_Ω. Since $\mathbf{F}$ satisfies the Galois descent condition, it implies that $(G(\Omega) * a_\Omega)^{\mathcal{G}_\Omega}$ is equal to the set

$$\{a'_\Omega \mid a' \in \mathbf{F}(K), a'_\Omega \sim_\Omega a_\Omega\}.$$

In other words, $(G(\Omega) * a_\Omega)^{\mathcal{G}_\Omega}$ is the image of the set of twisted K-forms of a under the map $\mathbf{F}(K) \longrightarrow \mathbf{F}(\Omega)$. Now since G is Galois functor, $G(\Omega)^{\mathcal{G}_\Omega}$ is the image of $G(K)$ under the map $G(K) \longrightarrow G(\Omega)$.

Now we claim that if $g_\Omega \in G(\Omega)^{\mathcal{G}_\Omega}$ and $a'_\Omega \in (G(\Omega) * a_\Omega)^{\mathcal{G}_\Omega}$, then we have $g_\Omega \cdot a'_\Omega = (g * a')_\Omega$, where '$\cdot$' denotes here the action defined before Corollary II.4.5.

Indeed, since a' is a twisted form of a, we may write $a'_\Omega = g' * a_\Omega$ for some $g' \in G(\Omega)$. Then g' is a preimage of a'_Ω under the map $G(\Omega) \to G(\Omega) * a_\Omega$ and thus

$$\begin{aligned} g_\Omega \cdot a'_\Omega &= (g_\Omega g') * a_\Omega \\ &= g_\Omega * (g' * a_\Omega) \\ &= g_\Omega * a'_\Omega \\ &= (g * a')_\Omega. \end{aligned}$$

We then get the $G(\Omega)^{\mathcal{G}_\Omega}$-orbit of a'_Ω in $(G(\Omega) * a_\Omega)^{\mathcal{G}_\Omega}$ is the image of $G(k) * a$ under the map $\mathbf{F}(k) \to \mathbf{F}(\Omega)$. Hence the map $\mathbf{F}(k) \to \mathbf{F}(\Omega)$ induces a bijection between $\mathbf{F}_a(\Omega/k)$ and the orbit set of $G(\Omega)^{\mathcal{G}_\Omega}$ in $(G(\Omega) * a_\Omega)^{\mathcal{G}_\Omega}$. This proves the first part of the theorem.

Before proving the functorial properties of the bijection, we would like to make it a bit more explicit. If $[a'] \in \mathbf{F}_a(\Omega/K)$, the corresponding orbit of $G(\Omega)^{\mathcal{G}_\Omega}$ in $(G(\Omega) * a_\Omega)^{\mathcal{G}_\Omega}$ is the orbit of a'_Ω. By definition of a twisted K-form, we may write $g * a'_\Omega = a_\Omega$ for some $g \in G(\Omega)$. Thus g^{-1} is a preimage of a'_Ω under the map $G(\Omega) \longrightarrow G(\Omega) * a_\Omega$. By Corollary II.4.5, the corresponding cohomology class is $\delta^0(g^{-1} \cdot a_\Omega)$, that is the cohomology class represented by the cocycle

$$\alpha\colon \begin{array}{l} \mathcal{G}_\Omega \longrightarrow \mathbf{Stab}_G(a)(\Omega) \\ \sigma \longmapsto g\,\sigma\cdot g^{-1}. \end{array}$$

Conversely, if $[\alpha] \in \ker[H^1(\mathcal{G}_\Omega, \mathbf{Stab}_G(a)(\Omega)) \longrightarrow H^1(\mathcal{G}_\Omega, G(\Omega))]$, then there exists $g \in G(\Omega)$ such that

$$\alpha_\sigma = g\,\sigma\cdot g^{-1} \text{ for all } \sigma \in \mathcal{G}_\Omega.$$

In other words, we have $[\alpha] = \delta^0(g^{-1}\cdot a_\Omega)$, and the corresponding element in $\mathbf{F}_a(\Omega/K)$ is represented by the unique element $a' \in \mathbf{F}(K)$ satisfying $a'_\Omega = g^{-1} * a_\Omega$.

We now prove the functoriality of the bijection. Let $\iota : K \longrightarrow K'$ be a morphism of field extensions of k, let Ω/K and Ω'/K' be two Galois extensions, and assume that we have an extension $\varphi : \Omega \longrightarrow \Omega'$ of ι. Let $\overline{\varphi} : \mathcal{G}_{\Omega'} \longrightarrow \mathcal{G}_\Omega$ the continuous group morphism associated to φ by Corollary I.2.10.

Let us denote by η and η' the inclusions $\mathbf{Stab}_G(a)(\Omega) \subset G(\Omega)$ and $\mathbf{Stab}_G(a)(\Omega') \subset G(\Omega')$ respectively. We first show that the map

$$R_\varphi : H^1(\mathcal{G}_\Omega, \mathbf{Stab}_G(a)(\Omega)) \longrightarrow H^1(\mathcal{G}_{\Omega'}, \mathbf{Stab}_G(a)(\Omega'))$$

restricts to a map

$$R_\varphi : \ker(\eta_*) \longrightarrow \ker(\eta'_*).$$

Let $[\xi] \in \ker(\eta_*)$, so that there exists $g \in G(\Omega)$ such that

$$\xi_\sigma = g\sigma\cdot g^{-1} \text{ for all } \sigma \in \mathcal{G}_\Omega.$$

We are going to prove that $R_\varphi([\xi])$ is represented by the cocycle

$$\begin{array}{l} \mathcal{G}_{\Omega'} \longrightarrow \mathbf{Stab}_G(a)(\Omega') \\ \sigma' \longmapsto g_{\Omega'}\,\sigma'\cdot g_{\Omega'}^{-1}. \end{array}$$

In particular, this will show that $R_\varphi([\xi]) \in \ker(\eta'_*)$.

By definition, $R_\varphi([\xi])$ is represented by the cocycle

$$\xi'\colon \begin{array}{l} \mathcal{G}_{\Omega'} \longrightarrow \mathbf{Stab}_G(a)(\Omega') \\ \sigma' \longmapsto \mathbf{Stab}_G(a)(\varphi)(\xi_{\overline{\varphi}(\sigma')}). \end{array}$$

Now $\mathbf{Stab}_G(a)(\varphi)$ is the restriction of $G(\varphi)$ by definition. Hence for all $\sigma' \in \mathcal{G}_{\Omega'}$, we have

$$\xi'_{\sigma'} = G(\varphi)(g\overline{\varphi}(\sigma')\cdot g^{-1}) = [G(\varphi)(g)][G(\varphi)(\overline{\varphi}(\sigma')\cdot g)]^{-1},$$

since $G(\varphi)$ is a group morphism. By compatibility of $\overline{\varphi}$ and $G(\varphi)$, we get

$$\xi'_{\sigma'} = [G(\varphi)(g)][\sigma' \cdot G(\varphi)(g)]^{-1} = g_{\Omega'}\, \sigma' \cdot g_{\Omega'}^{-1} \text{ for all } \sigma' \in \mathcal{G}_{\Omega'}.$$

Now let $[a'] \in \mathbf{F}_a(\Omega/K)$. Then $[a']_{K'} \in \mathbf{F}_a(\Omega'/K')$ is just the class $[a'_{K'}]$ by definition. The cohomology class corresponding to $[a']$ is represented by the cocycle

$$\alpha\colon \begin{array}{l} \mathcal{G}_\Omega \longrightarrow \mathbf{Stab}_G(a)(\Omega) \\ \sigma \longmapsto g\sigma \cdot g^{-1}, \end{array}$$

where $g * a'_\Omega = a_\Omega$. Since we have

$$g_{\Omega'} * a'_{\Omega'} = (g * a'_\Omega)_{\Omega'} = (a_\Omega)_{\Omega'} = a_{\Omega'},$$

the cohomology class corresponding to $[a_{K'}]$ is represented by

$$\beta\colon \begin{array}{l} \mathcal{G}_{\Omega'} \longrightarrow \mathbf{Stab}_G(\Omega') \\ \sigma' \longmapsto g_{\Omega'}\, \sigma' \cdot g_{\Omega'}^{-1}. \end{array}$$

We have to check that $R_\varphi([\alpha])$ is represented by the cocycle β, which follows from the computations above. This proves the functoriality of the bijection. The second part of the theorem follows from an application of the previous point to $\Omega = K_s$, and from the definition of the restriction maps. □

Remark III.8.17. Quite often, it is useful in the computations to explicitly know how the correspondence works, so we describe it one more time:

(1) If $[a'] \in \mathbf{F}_a(\Omega/K)$ is the equivalence class of a twisted K-form $a' \in \mathbf{F}(K)$ of a, pick $g \in G(\Omega)$ such that $g * a'_\Omega = a_\Omega$. The corresponding cohomology class in the kernel of the map

$$H^1(\mathcal{G}_\Omega, \mathbf{Stab}_G(a)(\Omega)) \longrightarrow H^1(\mathcal{G}_\Omega, G(\Omega))$$

is the class of the cocycle

$$\alpha\colon \begin{array}{l} \mathcal{G}_\Omega \longrightarrow \mathbf{Stab}_G(a)(\Omega) \\ \alpha_\sigma \longmapsto \alpha_\sigma = g\,\sigma \cdot g^{-1}. \end{array}$$

(2) If $[\alpha] \in \ker[H^1(\mathcal{G}_\Omega, \mathbf{Stab}_G(a)(\Omega)) \longrightarrow H^1(\mathcal{G}_\Omega, G(\Omega))]$, pick $g \in G(\Omega)$ such that $\alpha_\sigma = g\,\sigma \cdot g^{-1}$ for all $\sigma \in \mathcal{G}_\Omega$; the corresponding class in $\mathbf{F}_a(\Omega/K)$ is the equivalence class of the unique $a' \in \mathbf{F}(K)$ satisfying $a'_\Omega = g^{-1} * a_\Omega$.

Remark III.8.18. Let L/K be a finite Galois extension, and let

$$\rho_L : H^1(\mathcal{G}_L, \mathbf{Stab}_G(a)(L)) \longrightarrow H^1(K, \mathbf{Stab}_G(a))$$

be the corresponding map of pointed sets. Applying the last part of the Galois descent lemma to $K = K'$, $\Omega = L$ and $\Omega' = K_s$ shows that, if $[\alpha] \in \ker[H^1(\mathcal{G}_L, \mathbf{Stab}_G(a)(L)) \longrightarrow H^1(\mathcal{G}_L, G(L))]$ corresponds to $[a'] \in \mathbf{F}_a(L/K)$, then $\rho_L([\alpha])$ corresponds to $[a'] \in \mathbf{F}_a(K)$.

To apply the Galois Descent Lemma in a more convenient way, we need examples of Galois functors G satisfying $H^1(\_, G) = 1$. This will be provided by Hilbert 90 and this is the topic of the next section.

III.8.7 Hilbert's Theorem 90

To prove the so-called Hilbert's Theorem 90, we will need some preliminary results on semi-linear actions.

Definition III.8.19. Let Ω/k be a Galois extension, and let U be a (right) vector space over Ω with an action $*$ of $\mathcal{G}_\Omega$ on U. We will denote by '$\cdot$' the standard linear action of $\mathcal{G}_\Omega$ on Ω. We say that $\mathcal{G}_\Omega$ acts **by semi-linear automorphisms** on U if we have for all $u, u' \in U, \lambda \in \Omega$

$$\begin{aligned} \sigma * (u + u') &= \sigma * u + \sigma * u'; \\ \sigma * (u\lambda) &= (\sigma * u)(\sigma \cdot \lambda). \end{aligned}$$

Examples III.8.20.

(1) Let V be a k-vector space, and let $U = V_\Omega$. The action of $\mathcal{G}_\Omega$ on U defined on elementary tensors by

$$\sigma * (v \otimes \lambda) = v \otimes (\sigma \cdot \lambda) \text{ for all } v \in V, \lambda \in \Omega$$

is a continuous action by semi-linear automorphisms.

(2) Let $U = \Omega^n$, and let $\mathcal{G}_\Omega$ act in an obvious way on each coordinate. We obtain that way a continuous action by semi-linear automorphisms. Morever, $U^{\mathcal{G}_\Omega} = k^n$, and we have a canonical isomorphism of Ω-vector spaces

$$U^{\mathcal{G}_\Omega} \otimes_k \Omega \xrightarrow{\sim} U,$$

which sends $u \otimes \lambda$ onto $u\lambda$. This isomorphism is also an isomorphism of $\mathcal{G}_\Omega$-modules, as the reader may check.

The following lemma generalizes the previous example.

Lemma III.8.21 (Galois descent of vector spaces). *Let U be a vector space over Ω. If $\mathcal{G}_\Omega$ acts continuously on U by semi-linear automorphisms, then $U^{\mathcal{G}_\Omega} = \{u \in U \mid \sigma * u = u \text{ for all } \sigma \in \mathcal{G}_\Omega\}$ is a k-vector space and the map*

$$f\colon \begin{array}{rcl} U^{\mathcal{G}_\Omega} \otimes_k \Omega & \longrightarrow & U \\ u \otimes \lambda & \longmapsto & u\lambda \end{array}$$

is an isomorphism of Ω-vector spaces.

Proof (borrowed from [30]): It is clear that $U^{\mathcal{G}_\Omega}$ is a k-vector space. We first prove the surjectivity of f. Let $u \in U$. Since the action of $\mathcal{G}_\Omega$ on U is continuous, the stabilizer of u under the action $*$ is open, hence has the form $\mathrm{Gal}(\Omega/K)$ for some finite extension K/k. Let L/k be the Galois closure of K/k in Ω. Then L/k is a finite Galois extension such that $\mathrm{Gal}(\Omega/L)$ acts trivially on u. Let $(\lambda_i)_{1\leq i\leq n}$ be a k-basis of L and let $\sigma_1 = \mathrm{Id}_\Omega, \sigma_2, \ldots, \sigma_n$ be a set of representatives of the left cosets of $\mathrm{Gal}(\Omega/L)$ in $\mathcal{G}_\Omega$, so that the orbit of u in U consists of $\sigma_1 * u = u, \sigma_2 * u, \ldots, \sigma_n * u$ (we have exactly $n = [L:k]$ cosets, since $\mathcal{G}_\Omega/\mathrm{Gal}(\Omega/L) \simeq \mathcal{G}_L$).

Let

$$u_i = \sum_j \sigma_j * (u\lambda_i).$$

We are going to show that $u_i \in U^{\mathcal{G}_\Omega}$. For any $\sigma \in \mathcal{G}_\Omega$, we have $\sigma\sigma_j = \sigma_\ell\sigma'$ for some $\ell \in \{1, \ldots, n\}$ and some $\sigma' \in \mathrm{Gal}(\Omega/L)$ by choice of the σ_j's. Hence we have

$$(\sigma\sigma_j) * (u\lambda_i) = (\sigma_\ell\sigma') * (u\lambda_i) = \sigma_\ell * (\sigma' * (u\lambda_i)).$$

Since $\sigma' \in \mathrm{Gal}(\Omega/L)$ acts trivially on $u \in U$ by choice of L and $\lambda_i \in L$, we have

$$\sigma' * (u\lambda_i) = (\sigma' * u)(\sigma' \cdot \lambda_i) = u\lambda_i.$$

Thus $(\sigma\sigma_j) * (u\lambda_i) = \sigma_\ell * (u\lambda_i)$. The action of σ on $u_i = \sum\limits_j \sigma_j * (u\lambda_i)$ then just permutes the terms of the sum, so $u_i \in U^{\mathcal{G}_\Omega}$.

Since $(\sigma_1)_{|L}, \ldots, (\sigma_n)_{|L}$ are precisely the n k-automorphisms of L/k, they are linearly independent over L in $\mathrm{End}_k(L)$ (this is Dedekind's Lemma; see [42] for instance). Hence the matrix $M = (\sigma_j \cdot \lambda_i)_{i,j}$ lies in $\mathrm{GL}_n(L)$. By definition of u_j, we have $u_j = \sum\limits_k (\sigma_k * u)(\sigma_k \cdot \lambda_j)$. Now if

$M^{-1} = (m'_{ij})$, from the equation $M^{-1}M = I_n$, we get

$$\sum_j m'_{1j}(\sigma_k \cdot \lambda_j) = \delta_{1k} \text{ for all } k = 1, \dots n,$$

by comparing first rows. Hence we have

$$\sum_j u_j m'_{1j} = \sum_j \sum_k (\sigma_k * u)(\sigma_k \cdot \lambda_j) m'_{1j} = \sum_k (\sigma_k * u)\delta_{1k} = \sigma_1 * u = u,$$

the last equality coming from the fact that $\sigma_1 = \mathrm{Id}_\Omega$. Therefore, we have

$$u = \sum_j u_j m'_{1j} = f\left(\sum_j u_j \otimes m'_{1j}\right),$$

which proves the surjectivity of f.

To prove its injectivity, it is enough to prove the following:

Claim: Any vectors $u_1, \dots, u_r \in U^{\mathcal{G}_\Omega}$ which k-linearly independent remain Ω-linearly independent in U.

Indeed, assume that the claim is proved, and let $x \in \ker(f)$. One may write

$$x = u_1 \otimes \mu_1 + \dots + u_r \otimes \mu_r,$$

for some $\mu_1, \dots, \mu_r \in \Omega$ and some $u_1, \dots, u_r \in U^{\mathcal{G}_\Omega}$ which are linearly independent. By assumption, $f(x) = 0 = u_1\mu_1 + \dots + u_r\mu_r$. Now the claim implies that $\mu_1 = \dots = \mu_r = 0$, and thus $x = 0$, proving the injectivity of f.

It remains to prove the claim. We are going to do it by a way of contradiction. Assume that we have k-linearly independent vectors $u_1, \dots, u_r \in U^{\mathcal{G}_\Omega}$ for which there exist $\mu_1, \dots, \mu_r \in \Omega$ which are not all zero, such that

$$u_1\mu_1 + \dots + u_r\mu_r = 0.$$

We may assume that r is minimal, $r > 1$ and $\mu_1 = 1$. By assumption, the μ_i's are not all in k, so we may also assume that $\mu_2 \notin k$. Choose $\sigma \in \mathcal{G}_\Omega$ such that $\sigma \cdot \mu_2 \neq \mu_2$. We have

$$\sigma * \left(\sum_i u_i \mu_i\right) = \sum_i (\sigma * u_i)(\sigma \cdot \mu_i) = \sum_i u_i(\sigma \cdot \mu_i) = 0$$

and therefore we get $\sum_{i \geq 2} u_i(\sigma \cdot \mu_i - \mu_i) = 0$, a non-trivial relation with fewer terms. This is a contradiction, and this concludes the proof. $\square$

Remark III.8.22. If we endow $U^{\mathcal{G}_\Omega} \otimes_k \Omega$ with the natural semi-linear action as in Example III.8.20 (1), we claim that the isomorphism f above is an isomorphism of $\mathcal{G}_\Omega$-modules, that is f is equivariant with respect to the two semi-linear actions.

To check this, it is enough to do it on elementary tensors. Now for all $u \in U^{\mathcal{G}_\Omega}, \lambda \in \Omega$ and $\sigma \in \mathcal{G}_\Omega$, we have

$$f(\sigma * (u \otimes \lambda)) = u(\sigma \cdot \lambda) = (\sigma * u)(\sigma \cdot \lambda) = \sigma * (u\lambda) = \sigma * f(u \otimes \lambda),$$

hence the claim.

We may then rephrase the lemma above by saying that any Ω-vector space U endowed with a semi-linear action of $\mathcal{G}_\Omega$ is 'defined over k', hence the name of 'Galois descent of vector spaces'.

Before continuing, we need to recall a few definitions.

Definition III.8.23. A **simple** k-algebra is an associative finite dimensional unital k-algebra which has no proper two-sided ideals. A **semi-simple** k-algebra is a k-algebra which is isomorphic to the direct product of finitely many simple algebras. A **separable k-algebra** is a k-algebra A such that A_K is semi-simple for every field extension K/k. For example, a finite separable extension L/k is a separable k-algebra in that sense.

We are now ready to state Hilbert's theorem 90.

Proposition III.8.24 (Hilbert 90). *Let A be a semi-simple k-algebra. For every Galois extension Ω/k, we have $H^1(\mathcal{G}_\Omega, \mathbf{GL}_1(A)(\Omega)) = 1$.*

In particular, the following properties hold:

(1) *For any* **separable** *k-algebra A, we have $H^1(\_, \mathbf{GL}_1(A)) = 1$.*

(2) *For any finite dimensional k-vector space V, we have*

$$H^1(\_, \mathbf{GL}(V)) = 1.$$

In particular, $H^1(\_, \mathbb{G}_m) = 1$.

Proof (much inspired from [30]): Since A is semi-simple, we may write $A \simeq A_1 \times \cdots \times A_r$, where $A_1, \ldots, A_r$ are simple k-algebras. Then we have an isomorphism of $\mathcal{G}_\Omega$-groups

$$A_\Omega^\times \simeq (A_1)_\Omega^\times \times \cdots \times (A_r)_\Omega^\times,$$

and therefore

$$H^1(\mathcal{G}_\Omega, A_\Omega^\times) \simeq H^1(\mathcal{G}_\Omega, (A_1)_\Omega^\times) \times \cdots \times H^1(\mathcal{G}_\Omega, (A_r)_\Omega^\times).$$

Thus we may assume without any loss of generality that A is a simple algebra.

Let $\alpha \in Z^1(\mathcal{G}_\Omega, \mathbf{GL}_1(A)(\Omega))$. The Galois group $\mathcal{G}_\Omega$ of Ω/k acts continuously on A_Ω as follows:

$$\sigma \cdot (v \otimes \lambda) = v \otimes (\sigma \cdot \lambda) \text{ for all } a \in A, \lambda \in \Omega.$$

We now twist the action in a continuous action by semi-linear automorphisms as follows:

$$\sigma * a = \alpha_\sigma(\sigma \cdot a) \text{ for all } a \in A_\Omega, \sigma \in \mathcal{G}_\Omega.$$

Set $U = \{u \in A_\Omega \mid \sigma * u = u \text{ for all } \sigma \in \mathcal{G}_\Omega\}$. In our particular case, scalar multiplication by an element $\lambda \in \Omega$ in the vector space A_Ω is just right multiplication by $1 \otimes \lambda$. We then get an isomorphism

$$f\colon \begin{array}{l} U_\Omega \xrightarrow{\sim} A_\Omega \\ u \otimes \lambda \longmapsto u(1 \otimes \lambda) \end{array}$$

from the previous lemma. Notice that for all $a \in A_\Omega, a_0 \in A$ and $\sigma \in \mathcal{G}_\Omega$, we have

$$\sigma * (a(a_0 \otimes 1)) = (\sigma * a)(a_0 \otimes 1),$$

as we may see by checking it on elementary tensors. In particular, if $a \in U$ then $a(a_0 \otimes 1) \in U$. Thus the external law

$$\begin{array}{l} U \times A \longrightarrow U \\ (u, a_0) \longmapsto u \bullet a_0 = u(a_0 \otimes 1) \end{array}$$

endows U with a structure of a right A-module. This endows U_Ω with a structure of a right A_Ω-module, and f turns out to be an isomorphism of A_Ω-modules. Indeed, for all $u \in U, a \in A$ and $\lambda, \lambda' \in \Omega$, we have

$$\begin{aligned} f((u \otimes \lambda) \bullet (a \otimes \lambda')) &= f(u \bullet a \otimes \lambda\lambda') \\ &= (u \bullet a)(1 \otimes \lambda\lambda') \\ &= u(a \otimes 1)(1 \otimes \lambda\lambda') \\ &= u(1 \otimes \lambda)(a \otimes \lambda') \\ &= f(u \otimes \lambda) \bullet (a \otimes \lambda), \end{aligned}$$

which is enough to prove A_Ω-linearity using a distributivity argument. Let I be a minimal right ideal of A. Since U and A are non-trivial finitely generated A-modules (since they are finite dimensional over k), we have $U \simeq I^r$ and $A \simeq I^s$ as A-modules (see for example [53] for more details). Since $U_\Omega \simeq A_\Omega$, we have $\dim_k(U) = \dim_k(A)$ and thus $r = s$, meaning that U is isomorphic to A as a right A-module; in particular, $U = u_0 \bullet A$ for some $u_0 \in U$.

The map

$$\varphi\colon \begin{array}{l} A_\Omega \longrightarrow A_\Omega \\ a \otimes \lambda \longmapsto f((u_0 \bullet a) \otimes \lambda) \end{array}$$

is then an automorphism of A_Ω-modules. Since φ is A_Ω-linear, φ is simply the right multiplication by $\varphi(1) = f(u_0 \otimes 1) = u_0$. The bijectivity of φ then implies that $u_0 \in A_\Omega^\times$. Now since $u_0 \in U$, we have by definition of the twisted action that

$$u_0 = \alpha_\sigma \sigma \cdot u_0 \text{ for all } \sigma \in \mathcal{G}_\Omega.$$

Hence $\alpha_\sigma = u_0 \sigma \cdot u_0^{-1}$ for all $\sigma \in \mathcal{G}_\Omega$. This shows that α is cohomologous to the trivial cocycle. This proves the first part of the proposition. The fact that $H^1(\_, \mathbf{GL}_1(A))$ is the trivial functor whenever A is separable follows from the previous point and from the fact that A_K is a semi-simple K-algebra for any field extension K/k. Applying this to $A = \operatorname{End}_k(V)$ yields $H^1(\_, \mathbf{GL}(V)) = 1$. □

Remark III.8.25. Assume that Ω/k is a finite cyclic extension of degree n, and let γ be a generator of $\mathcal{G}_\Omega$. If $\alpha \in Z^1(\mathcal{G}_\Omega, \Omega^\times)$ is a 1-cocycle, we have

$$\alpha_{\gamma^n} = \alpha_{\gamma^{n-1}} \gamma^{n-1} \cdot \alpha_\gamma = \ldots = N_{\Omega/k}(\alpha_\gamma).$$

Since $\gamma^n = 1$, we get $N_{\Omega/k}(\alpha_\gamma) = 1$. Conversely, any element $x \in \Omega^\times$ of norm 1 determines completely a cocycle with values in $\Omega^\times$ by the formula

$$\alpha_{\gamma^m} = \prod_{i=0}^{m-1} \gamma^i \cdot x \text{ for } m = 0, \ldots, n-1.$$

Now let $x \in \Omega^\times$ satisfying $N_{\Omega/k}(x) = 1$, and let α be the corresponding cocycle. By Hilbert 90, we know that α is cohomologous to the trivial cocycle, so there exists $z \in \Omega^\times$ such that $\alpha_\sigma = \dfrac{\sigma(z)}{z}$ for all $\sigma \in \mathcal{G}_\Omega$. Applying this equality to $\sigma = \gamma$, we get $x = \dfrac{\gamma(z)}{z}$, which is the classical version of Hilbert 90.

Corollary III.8.26. *Let V be a finite dimensional k-vector space. For every field extension K/k and every Galois extension Ω/K, we have* $H^1(\mathcal{G}_\Omega, \mathbf{SL}(V)(\Omega)) = 1$.

In particular, $H^1(\_, \mathbf{SL}(V)) = 1$.

Proof. We have an exact sequence of $\mathcal{G}_\Omega$-groups

$$1 \longrightarrow \mathrm{SL}(V_\Omega) \longrightarrow \mathrm{GL}(V_\Omega) \longrightarrow \Omega^\times \longrightarrow 1\,,$$

where the last map is the determinant. By Theorem III.7.39, we have an exact sequence in cohomology

$$\mathrm{GL}(V) \longrightarrow k^\times \xrightarrow{\delta^0} H^1(\mathcal{G}_\Omega, \mathbf{SL}(V)(\Omega)) \longrightarrow H^1(\mathcal{G}_\Omega, \mathbf{GL}(V)(\Omega))\,,$$

where the first map is the determinant. By Hilbert 90, we get an exact sequence

$$\mathrm{GL}(V) \longrightarrow k^\times \xrightarrow{\delta^0} H^1(\mathcal{G}_\Omega, \mathbf{SL}(V)(\Omega)) \longrightarrow 1\,.$$

Since the determinant map is surjective, and since the sequence above is exact at $k^\times$, it follows that the 0th connecting map is trivial, hence we get $H^1(\mathcal{G}_\Omega, \mathbf{SL}(V)(\Omega)) = 1$. □

Let $n \geq 2$ be an integer. Recall that the algebraic group-scheme μ_n is defined by $\mu_n(R) = \{r \in R \mid r^n = 1\}$.

Proposition III.8.27. *Assume that* $\mathrm{char}(k)$ *does not divide* n. *Then we have a group isomorphism*

$$H^1(k, \mu_n) \simeq k^\times / k^{\times n}.$$

Proof. Since $\mathrm{char}(k)$ does not divide n, the polynomial $X^n - a \in k_s[X]$ is separable for all $a \in k_s^\times$. Therefore, the map

$$\begin{aligned} k_s^\times &\longrightarrow k_s^\times \\ x &\longmapsto x^n \end{aligned}$$

is surjective and we have an exact sequence of $\mathcal{G}_{k_s}$-groups

$$1 \longrightarrow \mu_n(k_s) \longrightarrow k_s^\times \xrightarrow{\cdot^n} k_s^\times \longrightarrow 1\,,$$

which induces an exact sequence in cohomology by Theorem III.7.39:

$$k^\times \xrightarrow{\cdot^n} k^\times \xrightarrow{\delta^0} H^1(k, \mu_n) \longrightarrow H^1(k, \mathbb{G}_m)\,.$$

Using Hilbert 90 and the exactness of this sequence, we get the desired isomorphism. □

Remark III.8.28. The proof shows that the isomorphism is given by

$$\begin{aligned} k^n/k^{\times n} &\xrightarrow{\sim} H^1(k,\mu_n) \\ \overline{a} &\longmapsto \delta^0(a). \end{aligned},$$

where δ^0 is the 0th connecting map associated to the exact sequence above. Therefore, the bijection can be made explicit as follows:

if $\bar{a} \in k^\times/k^{\times n}$, let $x \in k_s$ such that $x^n = a$ and set $\alpha_\sigma = \dfrac{\sigma(x)}{x}$. Then the map

$$\alpha : \mathcal{G}_{k_s} \longrightarrow \mu_n(k_s)$$

is a cocycle, whose cohomology class does not depend on the choice of a.

Conversely, if α is a cocycle with values in $\mu_n(k_s) \subset k_s^\times$, then there exists $x \in k_s^\times$ such that $\alpha_\sigma = \dfrac{\sigma(x)}{x}$ for all $\sigma \in \mathcal{G}_{k_s}$ by Hilbert 90. Now since $\alpha_\sigma^n = 1$, we get that $\sigma \cdot x^n = x^n$ for all $\sigma \in \mathcal{G}_{k_s}$, so $a = x^n \in k^\times$. The class of a modulo $k^{\times n}$ does not depend on the choice of x.

§III.9 First applications of Galois descent

III.9.1 Galois descent of algebras

Let $\underline{A}$ be a finite dimensional k-vector space. For any field extension K/k, let $\mathbf{F}(K)$ be the set of unital associative K-algebras with underlying K-vector space $\underline{A}_K$. If $\iota : K \longrightarrow L$ is a morphism of field extensions of k, we define a map $\mathbf{F}(\iota)$ by

$$\mathbf{F}(\iota) \colon \begin{aligned} \mathbf{F}(K) &\longrightarrow \mathbf{F}(L) \\ A &\longmapsto A_L. \end{aligned}$$

We then obtain a functor $\mathbf{F} : \mathfrak{C}_k \longrightarrow$ **Sets**. Now let $f \in \mathrm{GL}(\underline{A}_K)$, and let A be a unital associative K-algebra. We will write $x \cdot_A y$ for the product of two elements $x, y \in A$. The map

$$\begin{aligned} \underline{A}_K \times \underline{A}_K &\longrightarrow \underline{A}_K \\ (x,y) &\longmapsto f(f^{-1}(x) \cdot_A f^{-1}(y)) \end{aligned}$$

endows $\underline{A}_K$ with a structure of a unital associative K-algebra, that we will denote by $f{\cdot}A$. Straightforward computations show that this induces an action of $\mathbf{GL}(\underline{A})$ on $\mathbf{F}$. Notice that by definition, we have

$$f(x) \cdot_{f\cdot A} f(y) = f(x \cdot_A y) \text{ for all } x, y \in A,$$

so that f is an isomorphism of K-algebras from A onto $f \cdot A$. It easily follows that two unital associative K-algebras A and B are equivalent if and only if there are isomorphic as K-algebras.

Now fix a k-algebra $A \in \mathbf{F}(k)$. Then $\mathbf{Stab}_{\mathbf{GL}(\underline{A})}(A)$ is nothing but $\mathbf{Aut}_{alg}(A)$. It is not difficult to check that $\mathbf{F}$ satisfies the conditions of the Galois descent lemma. Hence we get

Proposition III.9.1. *Let K/k be a field extension, and let Ω/K be a Galois field extension.*

For any k-algebra A, the pointed set $H^1(\mathcal{G}_\Omega, \mathbf{Aut}_{alg}(A)(\Omega))$ classifies the isomorphism classes of K-algebras which become isomorphic to A over Ω. Moreover, the class of the trivial cocycle corresponds to the isomorphism class of A_K.

Remark III.9.2. Let A be a k-algebra and let Ω/k be a Galois extension. If B is a k-algebra such that there exists an isomorphism $f : B_\Omega \xrightarrow{\sim} A_\Omega$ of Ω-algebras, the corresponding cohomology class is represented by the cocycle

$$\alpha\colon \begin{array}{l} \mathcal{G}_\Omega \longrightarrow \mathrm{Aut}_\Omega(A_\Omega) \\ \sigma \longmapsto f \circ \sigma \cdot f^{-1}. \end{array}$$

Indeed, since f is a Ω-algebra isomorphism, we have

$$x \cdot_{f \cdot B_\Omega} y = f(f^{-1}(x) \cdot_{A_\Omega} f^{-1}(y)) = x \cdot_{A_\Omega} y \text{ for all } x, y \in A_\Omega,$$

and therefore $f \cdot B_\Omega = A_\Omega$. Remark III.8.17 then yields the result.

Conversely, the k-algebra corresponding to $[\alpha] \in H^1(\mathcal{G}_\Omega, \mathbf{Aut}_{\Omega-alg}(A_\Omega))$ is the isomorphism class of

$$B = \{a \in A_\Omega \mid \alpha_\sigma(\sigma \cdot a) = a \text{ for all } \sigma \in \mathcal{G}_\Omega\},$$

where the k-algebra structure is given by restriction of the algebra structure on A_Ω.

To see this, notice first that the isomorphism of Ω-vectors spaces

$$f : B_\Omega \xrightarrow{\sim} A_\Omega$$

given by Lemma III.8.21 is in fact an isomorphism of Ω-algebras, so that $f \cdot B_\Omega = A_\Omega$. In view of Remark III.8.17, it is therefore enough to show that we have

$$\alpha_\sigma = f \circ \sigma \cdot f^{-1} \text{ for all } \sigma \in \mathcal{G}_\Omega,$$

that is

$$\alpha_\sigma \circ \sigma \cdot f = f \text{ for all } \sigma \in \mathcal{G}_\Omega.$$

Since the elements of $B \otimes_k 1$ span B_Ω as an Ω-vector space, it is enough to check this equality on the elements of the form $x \otimes 1, x \in B$. Now for all $x \in B$ and $\sigma \in \mathcal{G}_\Omega$, we have

$$\begin{aligned} \alpha_\sigma((\sigma \cdot f)(x \otimes 1)) &= \alpha_\sigma(\sigma \cdot (f(\sigma^{-1} \cdot (x \otimes 1)))) \\ &= \alpha_\sigma(\sigma \cdot (f(x \otimes 1))) \\ &= \alpha_\sigma(\sigma \cdot x) \\ &= x \\ &= f(x \otimes 1), \end{aligned}$$

which is the result we were looking for.

We will use this remark in the sequel without any further reference. We now give an application of Galois descent to the case of central simple algebras.

Definition III.9.3. Let K be a field. A **central simple k-algebra** is a simple k-algebra with center k. One can show that a k-algebra A is central simple if it becomes isomorphic to a matrix algebra over k_s (resp. over a finite Galois extension Ω/k). See [19] for more details for example.

In view of the definition, the dimension of a central simple algebra A over its center k is always the square of an integer and central simple algebras define a functor. We will denote by $\mathbf{CSA}_n : \mathfrak{C}_k \longrightarrow \mathbf{Sets}^*$ the functor of isomorphism classes of central simple algebras of dimension n^2 (where the base point is the matrix algebra). The integer n is called the **degree** of A, and is denoted by $\deg_k(A)$.

Denoting by $\mathbf{PGL}_n$ the group-scheme $\mathbf{Aut}_{alg}(\mathrm{M}_n(k))$, we get by Proposition III.9.1:

Proposition III.9.4. *For any field extension K/k, and every Galois extension Ω/K, the pointed set $H^1(\mathcal{G}_\Omega, \mathbf{PGL}_n(\Omega))$ classifies the isomorphism classes of central simple K-algebras of degree n which become isomorphic to $\mathrm{M}_n(k)$ over Ω.*

In particular, we have an isomorphism of functors from $\mathfrak{C}_k$ to $\mathbf{Sets}^*$

$$\mathbf{CSA}_n \simeq H^1(\_, \mathbf{PGL}_n).$$

Matrix algebras are mapped onto the class of the trivial cocycle, and the restriction map $\mathrm{Res}_{L/K}$ corresponds to the tensor product $\otimes_K L$.

Remark III.9.5. The notation $\mathbf{PGL}_n$ is deliberate, and is consistent with the definition of the group $\mathrm{PGL}_n(K)$ where K is a field. Indeed, by definition $\mathbf{PGL}_n(K) = \mathrm{Aut}_{K-alg}(\mathrm{M}_n(K))$. It is well-known that every K-algebra automorphism of $\mathrm{M}_n(K)$ has the form

$$\mathrm{Int}(M)\colon \begin{array}{rcl} \mathrm{M}_n(K) & \longrightarrow & \mathrm{M}_n(K) \\ M' & \longmapsto & MM'M^{-1} \end{array}$$

for some $M \in \mathrm{GL}_n(K)$, and therefore we have

$$\mathbf{PGL}_n(K) = \{\mathrm{Int}(M) \mid M \in \mathrm{GL}_n(K)\}.$$

Now $\mathrm{Int}(M)$ is the identity map if and only if M commutes with any matrix, which means that $M = \lambda I_n$ for some $\lambda \in k^\times$. Therefore, we get a canonical isomorphism

$$\mathbf{PGL}_n(K) \simeq \mathrm{GL}_n(K)/K^\times = \mathrm{PGL}_n(K).$$

Notice also that $\mathbf{PGL}_n(K_s) \simeq \mathrm{SL}(K_s)/\mu_n(K_s)$ if n is prime to $\mathrm{char}(K)$.

Let us now consider the case of G-algebras.

Definition III.9.6. Let k be a field and let G be an abstract group. A G-**algebra** over k is a k-algebra on which G acts faithfully by k-algebra automorphisms. Two G-algebras over k are **isomorphic** if there exists an isomorphism of k-algebras which commutes with the actions of G. It will be denoted by $\simeq_G$.

Let $\underline{A}$ be a finite dimensional k-vector space and let G be an abstract finite group. For any field extension K/k, let $\mathbf{F}(K)$ be the set of G-algebras over K with underlying vector space $\underline{A}_K$. If $\iota : K \longrightarrow L$ is a morphism of field extensions of k, we define a map $\mathbf{F}(\iota)$ by

$$\mathbf{F}(\iota)\colon \begin{array}{rcl} \mathbf{F}(K) & \longrightarrow & \mathbf{F}(L) \\ A & \longmapsto & A_L, \end{array}$$

where the structure of G-algebra on A_L is given on elementary tensors by

$$g\cdot(a \otimes \lambda) = (g\cdot a) \otimes \lambda \text{ for all } g \in G, a \in A, \lambda \in L.$$

Now let $f \in \mathrm{GL}(\underline{A}_K)$, and let A be a G-algebra over K. Consider the K-algebra $f\cdot A$ as defined before. The map

$$\begin{array}{rcl} G \times f\cdot A & \longrightarrow & f\cdot A \\ (g, x) & \longmapsto & f(g\cdot f^{-1}(x)) \end{array}$$

endows $f \cdot A$ with a structure of a G-algebra over K (details are left to the reader as an exercise).

We then get an action of $\mathbf{GL}(\underline{A})$ on $\mathbf{F}$. Moreover, two G-algebras over K are equivalent if and only if they are isomorphic as G-algebras. Once again, all the conditions of the Galois descent lemma are fulfilled, and we get:

Proposition III.9.7. *Let G be a finite abstract group and let A be a G-algebra. For any field extension K/k and every Galois extension Ω/K, the pointed set $H^1(\mathcal{G}_\Omega, \mathbf{Aut}_{G-alg}(A)(\Omega))$ classifies the isomorphism classes of G-algebras over K which become G-isomorphic to A over Ω. Moreover, the class of the trivial cocycle corresponds to the isomorphism class of A_K.*

Remark III.9.8. Galois descent of algebras still works perfectly if we ask for the multiplication law to satisfy additional properties, such as commutativity or associativity. Il also works if the algebras are not unital nor associative.

III.9.2 The conjugacy problem

We now study the following conjugacy problem: let $G \subset \mathbf{GL}_n$ be a Galois functor and let $M, M_0 \in \mathrm{M}_n(k)$ satisfying $QMQ^{-1} = M_0$ for some $Q \in G(k_s)$. Does there exist $P \in G(k)$ such that $PMP^{-1} = M_0$?

Let us denote by $Z_G(M_0)$ the centralizer of M_0 in G, that is the group-scheme defined by

$$Z_G(M_0)(K) = \{M \in G(K) \mid MM_0 = M_0 M\}$$

for all field extension K/k. By Galois descent, the $G(k)$-conjugacy classes over k of matrices M which are $G(k_s)$-conjugate to M_0 are in one-to-one correspondence with $\ker[H^1(k, Z_G(M_0)) \longrightarrow H^1(k, G)]$. Moreover, the $G(k)$-conjugacy class of M_0 corresponds to the trivial cocycle.

Therefore, the conjugacy problem has a positive answer for all matrices M if and only if the map $H^1(k, Z_G(M_0)) \longrightarrow H^1(k, G)$ has trivial kernel. In particular, if $H^1(k, G) = 1$, the total obstruction to this problem is measured by $H^1(k, Z_G(M_0))$.

We have seen in the introduction that the conjugacy problem has a negative answer for $G = \mathbf{SL}_n$. We would like to recover this fact by using Galois cohomology.

Definition III.9.9. Let E be a finite dimensional k-algebra, and let R be a commutative k-algebra. For all $x \in E \otimes_k R$, we set

$$N_{E_R/R}(x) = \det(\ell_x),$$

where $\ell_x : E_R \longrightarrow E_R$ is the R-linear map induced by left multiplication by x. Notice that the definition above makes sense since E_R is a free R-module of finite rank.

The map $N_{E_R/R} : E_R \longrightarrow R$ is called the **norm map** of E_R.

Example III.9.10. Let $E = k^n, n \geq 1$. If $x = (x_1, \ldots, x_n)$, then we have

$$N_{E/k}(x) = x_1 \cdots x_n,$$

since the representative matrix of ℓ_x in the canonical basis of E is simply the diagonal matrix whose diagonal entries are $x_1, \ldots, x_n$.

Definition III.9.11. If L is a semi-simple commutative k-algebra, we denote by $\mathbb{G}_{m,L}^{(1)}$ the algebraic group-scheme over k defined by

$$\mathbb{G}_{m,L}^{(1)}(R) = \{x \in L_R^\times \mid N_{L_R/R}(x) = 1\},$$

for every k-algebra R.

We now compute $H^1(k, \mathbb{G}_{m,L}^{(1)})$ in a special case.

Lemma III.9.12. *Assume that we have an isomorphism of k_s-algebras*

$$L_{k_s} \simeq k_s^n \text{ for some } n \geq 1.$$

Then we have

$$H^1(k, \mathbb{G}_{m,L}^{(1)}) \simeq k^\times / N_{L/k}(L^\times).$$

Proof. The idea of course is to fit $\mathbb{G}_{m,L}^{(1)}(k_s)$ into an exact sequence of $\mathcal{G}_{k_s}$-modules. We first prove that the norm map

$$N_{L_{k_s}/k_s} : L_{k_s}^\times \longrightarrow k_s^\times$$

is surjective. Let $\varphi : L_{k_s} \xrightarrow{\sim} k_s^n$ be an isomorphism of k_s-algebras. We claim that we have $N_{L_{k_s}/k_s}(x) = N_{k_s^n/k_s}(\varphi(x))$ for all $x \in L_{k_s}$.

Indeed, if $\mathbf{e} = (e_1, \ldots, e_n)$ is a k_s-basis of L_{k_s}, then we have easily

$$\mathrm{Mat}(\ell_{\varphi(x)}, \varphi(\mathbf{e})) = \mathrm{Mat}(\ell_x, \mathbf{e}),$$

where $\varphi(\mathbf{e}) = (\varphi(e_1), \dots, \varphi(e_n))$. The desired equality then follows immediately. Now for $\lambda \in k_s^\times$, set $x_\lambda = \varphi^{-1}((\lambda, 1, \dots, 1))$. The equality above and Example III.9.10 then yield

$$N_{L\otimes_k k_s/k_s}(x_\lambda) = N_{k_s^n/k_s}((\lambda, 1, \dots, 1)) = \lambda.$$

Thus $N_{L_{k_s}/k_s}$ is surjective and we have an exact sequence of $\mathcal{G}_{k_s}$-modules

$$1 \longrightarrow \mathbb{G}_{m,L}^{(1)}(k_s) \longrightarrow L_{k_s}^\times \longrightarrow k_s^\times \longrightarrow 1\,,$$

where the last map is the norm map $N_{L_{k_s}/k_s}$. Notice now the assumption on L implies easily that L has no nilpotent elements, so L is a semi-simple k-algebra by [8, §7, Proposition 5]. Since $L_{k_s}^\times = \mathbf{GL}_1(L)(k_s)$, applying Galois cohomology to this sequence and using Hilbert 90 yield the exact sequence

$$(L \otimes_k 1)^\times \longrightarrow k^\times \longrightarrow H^1(k, \mathbb{G}_{m,L}^{(1)}) \longrightarrow 1,$$

the first map being $N_{L_{k_s}/k_s}$. Now it is obvious from the properties of the determinant that we have

$$N_{L_{k_s}/k_s}(x \otimes 1) = N_{L/k}(x) \text{ for all } x \in L.$$

The exactness of the sequence above then gives the desired result. □

Remark III.9.13. The proof above shows that the 0th-connecting map $\delta^0 : k^\times \to H^1(k, \mathbb{G}_{m,L}^{(1)})$ associated to the exact sequence of $\mathcal{G}_{k_s}$-modules

$$1 \longrightarrow \mathbb{G}_{m,L}^{(1)}(k_s) \longrightarrow L_{k_s}^\times \longrightarrow k_s^\times \longrightarrow 1$$

is surjective with kernel $N_{L/k}(k^\times)$. Thus, the correspondence works as follows:

if $\bar{a} \in k^\times/N_{L/k}(L^\times)$, let $z \in L_{k_s}^\times$ such that $a = N_{L_{k_s}/k_s}(z)$. Then the cohomology class corresponding to $\bar{a}$ is represented by the cocycle

$$\alpha\colon \begin{array}{rcl} \mathcal{G}_{k_s} & \longrightarrow & \mathbb{G}_{m,L}^{(1)}(k_s) \\ \sigma & \longmapsto & \dfrac{\sigma(z)}{z}. \end{array}$$

Conversely, if α is a cocycle with values in $\mathbb{G}_{m,L}^{(1)}(k_s)$, then there exists $z \in L_{k_s}^\times$ such that $\alpha_\sigma = \dfrac{\sigma(z)}{z}$ for all $\sigma \in \mathcal{G}_{k_s}$. Now set $a = N_{L_{k_s}/k_s}(z)$. We have $a \in k^\times$, and the class of a modulo $N_{L/k}(L^\times)$ is the one corresponding to $[\alpha]$.

Remark III.9.14. Algebras satisfying the condition of the previous lemma are called **étale**, and will be studied in detail in a forthcoming chapter.

Now let us go back to the conjugacy problem of matrices. If $G = \mathbf{GL}_n$, it is well-known that the conjugacy problem has an affirmative answer in this case, and this can be proved without any use of cohomology. The cohomological reinterpretation of the problem implies that the functor $H^1(\_, Z_G(M_0))$ should be trivial, which is actually the case. We will not prove this in full generality, but just in a particular example.

Assume that $M_0 = C_\chi \in \mathrm{M}_n(k)$ is a companion matrix of some monic polynomial $\chi \in k[X]$ of degree $n \geq 1$. In this case, it is known that every matrix commuting with M_0 is a polynomial in M_0, so $Z_G(M_0)(k_s) = k_s[M_0] \cap G(k_s)$. Moreover, the minimal polynomial and the characteristic polynomial are both equal to χ. Set $L = k[X]/(\chi)$, so that we have an isomorphism of k-algebras

$$\begin{aligned} L &\xrightarrow{\sim} k[M_0] \\ \overline{P} &\longmapsto P(M_0), \end{aligned}$$

which induces in turn a Galois equivariant isomorphism of k_s-algebras

$$f\colon \begin{aligned} L_{k_s} &\xrightarrow{\sim} k_s[M_0] \\ \overline{X} \otimes \lambda &\longmapsto \lambda M_0. \end{aligned}$$

In particular, f induces an isomorphism of $\mathcal{G}_{k_s}$-modules

$$L_{k_s}^\times \simeq k_s[M_0]^\times.$$

Notice now that if $C \in \mathrm{GL}_n(k_s)$ commutes with M_0, then C^{-1} also commutes with M_0. Therefore, we have the equalities

$$Z_{\mathbf{GL}_n}(M_0)(k_s) = k_s[M_0] \cap \mathrm{GL}_n(k_s) = k_s[M_0]^\times,$$

so f induces an isomorphism of $\mathcal{G}_{k_s}$-modules

$$\mathbf{GL}_1(L)(k_s) = L_{k_s}^\times \simeq Z_{\mathbf{GL}_n}(M_0)(k_s).$$

By Hilbert 90, we then get $H^1(\mathcal{G}_\Omega, Z_{\mathbf{GL}_n}(M_0)(\Omega)) = 1$, as expected.

Now let us identify $Z_{\mathbf{SL}_n}(M_0)(\Omega)$.

Claim: We have $\det(f(x)) = N_{L_{k_s}/k_s}(x)$ for all $x \in L_{k_s}$.

To see this, set $\alpha = \overline{X} \in L$. Then $\mathbf{e} = (1 \otimes 1, \alpha \otimes 1, \ldots, \alpha^{n-1} \otimes 1)$ is a k_s-basis of L_{k_s}. Let $x = \sum_{i=0}^{n-1} \alpha^i \otimes \lambda_i \in L_{k_s}$, and let $P = \sum_{i=0}^{n-1} \lambda_i X^i$. Clearly, we have $\ell_x = P(\ell_{\alpha\otimes 1})$. Now the matrix of $\ell_{\alpha\otimes 1}$ in the basis $\mathbf{e}$ is easily seen to be $C_\chi = M_0$, and so the matrix of ℓ_x in the basis $\mathbf{e}$ is $P(M_0) = f(x)$. Therefore $\det(\ell_x) = \det(f(x))$, and we are done.

We then get an isomorphism of $\mathcal{G}_{k_s}$-modules

$$Z_{\mathbf{SL}_n}(M_0)(k_s) \simeq \mathbb{G}_{m,L}^{(1)}(k_s).$$

In particular, if χ is separable, we have

$$L_{k_s} \simeq k_s[X]/(\chi) \simeq k_s^n,$$

and therefore by Lemma III.9.12 we have

$$H^1(k, Z_{\mathbf{SL}_n}(M_0)) \simeq k^\times / N_{L/k}(L^\times),$$

which is not trivial in general.

For example, assume that $\mathrm{char}(k) \neq 2$. Let $M_0 = \begin{pmatrix} 0 & d \\ 1 & 0 \end{pmatrix}$ and let $M = \begin{pmatrix} 0 & -d \\ -1 & 0 \end{pmatrix}, d \in k^\times$.

Then M_0 is the companion matrix of $\chi = X^2 - d$ and thus $L = k(\sqrt{d})$. Moreover, we have $QMQ^{-1} = M_0$, with $Q = \begin{pmatrix} i & 0 \\ 0 & -i \end{pmatrix}$ (where i is a square root of -1 in k_s), so M and M_0 are conjugate by an element of $\mathrm{SL}_2(k_s)$. However, they are not conjugate by an element of $\mathrm{SL}_2(k)$ in general.

To see this, let us compute the class in $k^\times / N_{L/k}(L^\times)$ corresponding to the conjugacy class of M. Notice first that $Q\sigma{\cdot}Q^{-1}$ is the identity matrix I_2 if $\sigma(i) = i$ and is -I_2 otherwise. In other words, we have

$$\alpha_\sigma^Q = (iI_2)^{-1}\sigma{\cdot}(iI_2) \text{ for all } \sigma \in \mathcal{G}_{k_s}.$$

Via the isomorphism $H^1(\mathcal{G}_{k_s}, \mathbb{G}_{m,L}^{(1)}(k_s)) \simeq H^1(\mathcal{G}_{k_s}, Z_{\mathbf{SL}_n}(M_0)(k_s))$ induced by f_*, the cohomology class $[\alpha^{(Q)}]$ correspond to the cohomology class of the cocycle

$$\beta^{(Q)}: \begin{array}{rcl} \mathcal{G}_{k_s} & \longrightarrow & \mathbb{G}_{m,L}^{(1)}(k_s) \\ \sigma & \longmapsto & (1\otimes i)^{-1}\sigma{\cdot}1\otimes i. \end{array}$$

Now $N_{L_{k_s}/k_s}(1\otimes i) = (1\otimes i)^2 = -1$, and thus the conjugacy class of M

corresponds to the class of -1 in $k^\times/N_{L/k}(L^\times)$ by Remark III.9.13. In particular, M and M_0 are conjugate over k if and only if $-1 \in N_{L/k}(L^\times)$. Therefore, to produce counterexamples, one may take for k any subfield of $\mathbb{R}$ and $d < 0$, as we did in the introduction.

III.9.3 Cup-products with values in μ_2

Throughout this section, k will be a field of characteristic different from 2. Since $\mu_2(k_s)$ is a $\mathcal{G}_{k_s}$-module, we have cohomology groups $H^m(k, \mu_2)$ for all $m \geq 1$. These groups are abelian groups and we will write the group law additively. However, the group of coefficients $\mu_2(k_s)$ will be written multiplicatively as usual.

Therefore, if $[\alpha], [\beta] \in H^m(k, \mu_2)$, we have by definition

$$[\alpha] + [\beta] = [\alpha\beta].$$

In particular, $2[\alpha] = 0$ for all $[\alpha] \in H^m(k, \mu_2)$.

If $a \in k^\times$, we will denote by (a) the cohomology class in $H^1(k, \mu_2)$ representing the square-class $\overline{a} \in k^\times/k^{\times 2}$. Let $x_a \in k_s^\times$ satisfying $x_a^2 = a$. Recall from Remark III.8.28 that (a) is represented by the cocycle

$$\alpha_a \colon \begin{array}{ccc} \mathcal{G}_{k_s} & \longrightarrow & \mu_2(k_s) \\ \sigma & \longmapsto & \dfrac{\sigma(x_a)}{x_a}. \end{array}$$

If now $b \in k^\times$ and $x_b \in k_s^\times$ satisfies $x_b^2 = b$, then the element $x_a x_b \in k_s^\times$ satisfies $(x_a x_b)^2 = ab$. Therefore, (ab) is represented by the cocycle $\alpha_a \alpha_b$, and we have the equality

$$(ab) = (a) + (b) \text{ for all } a, b \in k^\times.$$

Since $\mu_2(k_s)$ is an abelian group, hence a $\mathbb{Z}$-module, we can form the tensor product $\mu_2(k_s) \otimes_{\mathbb{Z}} \mu_2(k_s)$. The map

$$\varphi \colon \begin{array}{ccc} \mu_2(k_s) \times \mu_2(k_s) & \longrightarrow & \mu_2(k_s) \\ ((-1)^m, (-1)^n) & \longmapsto & (-1)^{nm} \end{array}$$

is $\mathbb{Z}$-bilinear, so we may consider the cup-product $\cup_\varphi$, that we will simply denote by $\cup$. We refer to Chapter 1, Section II.6 for the definition of the cup-product.

Let us describe the cup product $(a) \cup (b) \in H^2(k, \mu_2)$, for $a, b \in k_s^\times$. For any $\sigma \in \mathcal{G}_{k_s}$, let $\varepsilon_a(\sigma), \varepsilon_b(\sigma) \in \{0,1\}$ defined by

$$\frac{\sigma(x_a)}{x_a} = (-1)^{\varepsilon_a(\sigma)}, \frac{\sigma(x_b)}{x_b} = (-1)^{\varepsilon_b(\sigma)}.$$

It follows from the definition that the cup-product is represented by the cocycle

$$\alpha_a \cup \alpha_b \colon \begin{array}{c} \mathcal{G}_{k_s} \times \mathcal{G}_{k_s} \longrightarrow \mu_2(k_s) \\ \sigma, \tau \longmapsto (-1)^{\varepsilon_a(\sigma)\varepsilon_b(\tau)}. \end{array}$$

We are now going to establish some useful properties of the cup-product:

Proposition III.9.15. *For all $a, b \in k^\times$, the following properties hold:*

(1) $(a) \cup (b) = (b) \cup (a)$.

(2) $(a) \cup (b) = 0 \iff$ *b is a norm of $k(\sqrt{a})/k$ (where $k(\sqrt{a}) = k$ if a is a square).*

(3) $(a) \cup (1-a) = 0$.

(4) $(a) \cup (-a) = 0$.

(5) $(a) \cup (a) = (a) \cup (-1)$.

Remark III.9.16. The reader with an expert eye has certainly noticed the similarity between these properties and those of the quaternion algebra (a, b). This is not a coincidence. One can show that we have an isomorphism $\mathrm{Br}_2(k) \simeq H^2(k, \mu_2)$, where $\mathrm{Br}_2(k)$ denotes the subgroup of the Brauer group of k consisting of elements of exponent at most 2 and that, under this isomorphism, the class of the quaternion algebra (a, b) corresponds to $(a) \cup (b)$. This isomorphism will be fully proved in Chapter VIII. Of course, theses relations may be established directly. However, as an illustration of Galois descent, we would like to supply another proof based on the properties of quaternion algebras.

Let us first recall the definition of (a, b).

Definition III.9.17. If $a, b \in k^\times$, we define (a, b) to be the k-algebra generated by two elements i, j and subject to the relations

$$i^2 = a, j^2 = b \text{ and } ij = -ji.$$

One can check that it is a central simple k-algebra of degree 2 over k, called a **quaternion algebra**.

Remark III.9.18. One can show that every central simple algebra of degree 2 is isomorphic to a quaternion algebra. We will not need this fact here.

Proposition III.9.19. *The connecting map*

$$\delta^1 : H^1(k, \mathbf{PGL}_2) \longrightarrow H^2(k, \mu_2)$$

associated to the sequence

$$1 \longrightarrow \mu_2(k_s) \longrightarrow \mathbf{SL}_2(k_s) \longrightarrow \mathbf{PGL}_2(k_s) \longrightarrow 1$$

has trivial kernel and maps the isomorphism class of the quaternion algebra (a,b) to $(a) \cup (b)$.

Proof. The fact that δ^1 has trivial kernel comes from the fact that $H^1(k, \mathbf{SL}_2) = 1$. Now, to compute the image of $Q = (a,b)$, we need first to find a cocycle $\alpha \in Z^1(k, \mathbf{PGL}_2)$ corresponding to Q, and therefore we need an explicit isomorphism between Q_{k_s} and $\mathrm{M}_2(k_s)$. One can check that the map $f : Q_{k_s} \longrightarrow \mathrm{M}_2(k_s)$ defined by

$$f(x+yi+zj+tij) = \begin{pmatrix} x+y\sqrt{a} & b(z+t\sqrt{a}) \\ z-t\sqrt{a} & x-y\sqrt{a} \end{pmatrix}$$

is an isomorphism of k_s-algebras; this is a possible way to prove that (a,b) is actually a central simple algebra.

If E_{11}, E_{21}, E_{12} and E_{22} denote the elementary matrices of $\mathrm{M}_2(k)$, the inverse map $f^{-1} : \mathrm{M}_2(k_s) \longrightarrow Q_{k_s}$ is given by

$$\begin{aligned} f^{-1}(E_{11}) &= \frac{1}{2}\left(1+\frac{1}{\sqrt{a}}i\right), & f^{-1}(E_{22}) &= \frac{1}{2}\left(1-\frac{1}{\sqrt{a}}i\right) \\ f^{-1}(E_{12}) &= \frac{1}{2b}\left(j+\frac{1}{\sqrt{a}}ij\right), & f^{-1}(E_{21}) &= \frac{1}{2}\left(j-\frac{1}{\sqrt{a}}ij\right). \end{aligned}$$

For $\sigma \in \mathcal{G}_{k_s}$, let $x_a, x_{-b}, \in k_s$ satisfying $x_a^2 = a, x_{-b}^2 = -b$, and for any $\sigma \in \mathcal{G}_{k_s}$, let $\varepsilon_a(\sigma), \varepsilon_{-b}(\sigma) \in \{0,1\}$ defined by

$$\frac{\sigma(x_a)}{x_a} = (-1)^{\varepsilon_a(\sigma)}, \frac{\sigma(x_{-b})}{x_{-b}} = (-1)^{\varepsilon_{-b}(\sigma)}.$$

We now compute a matrix $M_\sigma \in \mathbf{SL}_2(k_s)$ such that

$$\alpha_\sigma = f\sigma{\cdot}f^{-1} = \mathrm{Int}(M_\sigma) \text{ for all } \sigma \in \mathcal{G}_{k_s}.$$

Assume first that $\varepsilon_a(\sigma) = 0$, that is $\sigma(x_a) = x_a$. Then $\sigma \cdot f^{-1}(E_{ij}) = f^{-1}(E_{ij})$ for $1 \le i,j \le 2$, so $M_\sigma = I_2$ in this case. Now if $\varepsilon_a(\sigma) = 1$, that is $\sigma(x_a) = -x_a$, then we have

$$\sigma\cdot f^{-1}(E_{11}) = \frac{1}{2}\left(1-\frac{1}{\sqrt{a}}i\right), \quad \sigma\cdot f^{-1}(E_{22}) = \frac{1}{2}\left(1+\frac{1}{\sqrt{a}}i\right)$$
$$\sigma\cdot f^{-1}(E_{12}) = \frac{1}{2b}\left(j-\frac{1}{\sqrt{a}}ij\right), \quad \sigma\cdot f^{-1}(E_{21}) = \frac{1}{2}\left(j+\frac{1}{\sqrt{a}}ij\right).$$

Therefore we have

$$\alpha_\sigma(E_{11}) = E_{22}, \alpha_\sigma(E_{22}) = E_{11}, \alpha_\sigma(E_{12}) = \frac{1}{b}E_{21}, \alpha_\sigma(E_{21}) = bE_{12}.$$

One can check that we can take $M_\sigma = \begin{pmatrix} 0 & x_{-b} \\ -x_{-b}^{-1} & 0 \end{pmatrix} \in \mathbf{SL}_2(k_s)$ in this case.

Thus, setting $M = \begin{pmatrix} 0 & x_{-b} \\ -x_{-b}^{-1} & 0 \end{pmatrix}$, we have $M_\sigma = M^{\varepsilon_a(\sigma)}$ for all $\sigma \in \mathcal{G}_{k_s}$. Then the cocycle

$$\alpha\colon \begin{array}{rcl} \mathcal{G}_{k_s} & \longrightarrow & \mathbf{PGL}_2(k_s) \\ \sigma & \longmapsto & \alpha_\sigma = \mathrm{Int}(M^{\varepsilon_a(\sigma)}) \end{array}$$

represents the isomorphism class of Q. For all $\sigma \in \mathcal{G}_{k_s}$, a preimage of α_σ in $\mathbf{SL}_2(k_s)$ is then $M^{\varepsilon_a(\sigma)}$. By definition of the first connecting map, $\delta^1(Q)$ is then represented by the cocycle $\gamma : \mathcal{G}_{k_s} \times \mathcal{G}_{k_s} \longrightarrow \mu_2(k_s)$ defined by

$$M_\sigma \sigma\cdot M_\tau M_{\sigma\tau}^{-1} = \gamma_{\sigma,\tau}\mathrm{Id}_2 \text{ for all } \sigma, \tau \in \mathcal{G}_{k_s}.$$

Notice that $\sigma\cdot M = (-1)^{\varepsilon_{-b}(\sigma)}M$ for all $\sigma \in \mathcal{G}_{k_s}$. Therefore, we get

$$\gamma_{\sigma,\tau}I_2 = (-1)^{\varepsilon_{-b}(\sigma)\varepsilon_a(\tau)}M^{\varepsilon_a(\sigma)+\varepsilon_a(\tau)-\varepsilon_a(\sigma\tau)}.$$

Since the map

$$\begin{array}{rcl} \mathcal{G}_{k_s} & \longrightarrow & \mu_2(k_s) \\ \sigma & \longmapsto & (-1)^{\varepsilon_a(\sigma)} \end{array}$$

is a cocycle and $\mathcal{G}_{k_s}$ acts trivially on $\mu_2(k_s)$, we get that $\varepsilon_a(\sigma)+\varepsilon_a(\tau)-\varepsilon_a(\sigma\tau)$ is even for all $\sigma, \tau \in \mathcal{G}_{k_s}$. Taking into account that $M^2 = -I_2$, we get

$$\gamma_{\sigma,\tau} = (-1)^{\frac{\varepsilon_a(\sigma)+\varepsilon_a(\tau)-\varepsilon_a(\sigma\tau)}{2}+\varepsilon_{-b}(\sigma)\varepsilon_a(\tau)}.$$

Now if σ and τ both fix x_a or both map x_a onto $-x_a$, then $\sigma\tau$ fixes x_a. If only one of them fixes x_a, then $\sigma\tau$ maps x_a onto $-x_a$. Therefore in all cases, we have $\dfrac{\varepsilon_a(\sigma)+\varepsilon_a(\tau)-\varepsilon_a(\sigma\tau)}{2} = \varepsilon_a(\sigma)\varepsilon_a(\tau)$. Finally, we get

$$\gamma_{\sigma,\tau} = (-1)^{\varepsilon_a(\sigma)\varepsilon_a(\tau)+\varepsilon_{-b}(\sigma)\varepsilon_a(\tau)},$$

that is

$$\delta^1(Q) = (a) \cup (a) + (-b) \cup (a).$$

It follows from the identity $(u)+(v) = (uv) \in H^1(k, \mu_2)$ for all $u, v \in k^\times$ and the bilinearity of the cup product that we have

$$\delta^1(Q) = (a) \cup (a) + (-1) \cup (a) + (b) \cup (a) = (-a) \cup (a) + (b) \cup (a).$$

Since $(a, 1) \simeq \mathrm{M}_2(k)$, which represents the trivial cohomology class, applying this formula to $b = 1$ leads to $0 = (-a) \cup (a)$, so $\delta^1(Q) = (b) \cup (a)$. Now $(a, b) \simeq (b, a)$, so we get $\delta^1(Q) = \delta^1((b, a)) = (a) \cup (b)$. □

Proof of Proposition III.9.15. Since $(a, b) \simeq (b, a)$, we get (1) by applying δ^1. Now δ^1 has trivial kernel, so $(a) \cup (b) = 0 \iff (a, b) \simeq \mathrm{M}_2(k)$, which is equivalent to say that b is a norm of $k(\sqrt{a})$ by the well-known properties of quaternion algebras (see [53], for example), hence (2). A direct application of (2) gives (3) and (4). Now by bilinearity of the cup-product and the fact that $(-a) = (-1) + (a)$, (5) follows from (4). □

Remark III.9.20. Applying Example II.6.7 and the definition of the restriction map, we see that for every field extension L/K, we have

$$\mathrm{Res}_{L/K}([\alpha] \cup [\beta]) = \mathrm{Res}_{L/K}([\alpha]) \cup \mathrm{Res}_{L/K}([\beta])$$

for all $[\alpha] \in H^p(K, \mu_2), [\beta] \in H^q(K, \mu_2)$.

Notes

The proof of the fact that $H^n(k, G)$ is independent from the choice of an algebraic closure and the construction of the restriction map follow the arguments outlined in [58, Chapter II,§1.1]. Our exposition of Galois descent is more general than the versions that may be found in [58] or [30]. However, we are convinced that it follows from the general descent theory of Grothendieck. The reader may find other interesting applications of Galois descent in [30], where the first cohomology sets associated to classical algebraic groups are described in terms of various algebraic structures.

Exercises

1. Show that $H^1(\_, \mathbf{Sp}_{2n}) = 1$.
   Hint: What is the automorphism group of a non-degenerate alternating bilinear form?

2. Let k be a field, and let $n, d \geq 1$ be two integers. For every field extension K/k, let us denote by $\mathcal{H}_{n,d}(K)$ the set of homogeneous polynomials of degree d in n variables with coefficients in K. We let act $K^\times \times \mathrm{GL}_n(K)$ on $\mathcal{H}_{n,d}(K)$ by

$$((\lambda, M) \cdot P)(X_1, \ldots, X_n) = \lambda P(M^{-1}\begin{pmatrix} X_1 \\ \vdots \\ X_n \end{pmatrix}).$$

If $P, Q \in \mathcal{H}_{n,d}(K)$, we say that P and Q are **similar** if they are in the same orbit under the action of $K^\times \times \mathrm{GL}_n(K)$. We set

$$\mathbf{Sim}(P)(K) = \{(\lambda, M) \in K^\times \times \mathrm{GL}_n(K) \mid (\lambda, M) \cdot P = P\}.$$

If $P_0 \in \mathcal{H}_{n,d}(k)$, we denote by $\mathbf{G}_{P_0}$ the functor of similarity classes of twisted forms of P_0.

(a) Show that $\mathcal{H}_{n,d}$ is a functor on which $\mathbb{G}_m \times \mathbf{GL}_n$ acts.

(b) Let $P_0 \in \mathcal{H}_{n,d}(k)$. Show that we have an isomorphism of functors

$$H^1(\_, \mathbf{Sim}(P_0)) \simeq \mathbf{G}_{P_0}.$$

We now assume until the end that $n = d = 3$ and $P_0 = X_1^3$.

(c) For every field extension K/k, check that

$$\mathbf{Sim}(P_0)(K) = \left\{ (a^3, \begin{pmatrix} a & 0 & 0 \\ b & c & d \\ e & f & g \end{pmatrix}) \in K^\times \times \mathrm{GL}_3(K) \right\}.$$

(d) Compute $H^1(k, \mathbf{Sim}(P_0))$.
Hint: Show that the obvious morphism

$$\mathbf{Sim}(P_0)(k_s) \longrightarrow k_s^\times \times \mathrm{GL}_2(k_s)$$

is surjective, with kernel isomorphic to $k_s \times k_s$.

(e) What are the twisted forms of P_0, up to similarity?

3. We keep the notation of the previous exercise. We now let $\mathbf{GL}_n$ act on $\mathcal{H}_{n,d}$ by

$$(M \cdot P)(X_1, \ldots, X_n) = P(M^{-1}\begin{pmatrix} X_1 \\ \vdots \\ X_n \end{pmatrix}).$$

If $P, Q \in \mathcal{H}_{n,d}(K)$, we say that P and Q are **isomorphic** if they are in the same orbit under the action of $\mathrm{GL}_n(K)$. We set

$$\mathbf{Aut}(P)(K) = \{M \in \mathrm{GL}_n(K) \mid M \cdot P = P\}.$$

If $P_0 \in \mathcal{H}_{n,d}(k)$, we denote by $\mathbf{F}_{P_0}$ the functor of isomorphism classes of twisted forms of P_0.

(a) Let $P_0 \in \mathcal{H}_{n,d}(k)$. Show that we have an isomorphism of functors

$$H^1(\_, \mathbf{Aut}(P_0)) \simeq \mathbf{F}_{P_0}.$$

We now assume until the end that $n = d = 3$ and $P_0 = X_1^3$.

(b) For all field extension K/k, check that

$$\mathbf{Aut}(P_0)(K) = \left\{ \begin{pmatrix} a & 0 & 0 \\ b & c & d \\ e & f & g \end{pmatrix} \in \mathrm{GL}_3(K) \mid a \in \mu_3(K) \right\}.$$

(c) Check that we have a **split** exact sequence of $\mathcal{G}_{k_s}$-groups

$$1 \to k_s \times k_s \to \mathbf{Aut}(P_0)(k_s) \to \mu_3(k_s) \times \mathrm{GL}_2(k_s) \to 1.$$

Deduce that the map

$$H^1(k, \mathbf{Aut}(P_0)) \longrightarrow H^1(k, \mu_3 \times \mathbf{GL}_2)$$

is surjective.

The goal of the next questions is to prove that this map is also injective. Let $\beta : \mathcal{G}_{k_s} \to \mathbf{Aut}(P_0)(k_s)$, let $\gamma = \pi_*(\beta)$ and let $\alpha : \mathcal{G}_{k_s} \to k_s \times k_s$ be the image of β by the map induced by the conjugation map $\mathbf{Aut}(P_0)(k_s) \longrightarrow \mathrm{Aut}(k_s \times k_s)$.

For all $\sigma \in \mathcal{G}_{k_s}$, write $\beta = \begin{pmatrix} a_\sigma & \mathbf{0} \\ X_\sigma & M_\sigma \end{pmatrix}$, where $a_\sigma \in \mu_3(k_s), X \in k_s \times k_s$ and $M_\sigma \in \mathrm{GL}_2(k_s)$.

(d) Check that for all $X \in (k_s \times k_s)_\alpha$ and all $\sigma \in \mathcal{G}_{k_s}$, we have

$$\sigma * X = a_\sigma^{-1} M_\sigma \sigma \cdot X.$$

(e) Justify the existence of $P \in \mathrm{GL}_2(k_s)$ such that $a_\sigma^{-1} M_\sigma = P\sigma \cdot P^{-1}$ for all $\sigma \in \mathcal{G}_{k_s}$, and check that the map

$$\varphi : \begin{array}{rcl} k_s \times k_s & \longrightarrow & (k_s \times k_s)_\alpha \\ X & \longmapsto & PX \end{array}$$

is an isomorphism of $\mathcal{G}_{k_s}$-modules.

(f) Conclude using Chapter II Exercise 5 (*c*).

(g) Show that we have a bijection of pointed sets

$$H^1(k, \mathbf{Aut}(P_0)) \simeq k^\times/k^{\times 3}.$$

If $\overline{a} \in k^\times/k^{\times 3}$, what is the corresponding twisted form of P_0 (up to isomorphim) ?

4. Let Ω/k be a Galois extension, and let $B(\Omega)$ be the subgroup of upper triangular matrices of $\mathrm{GL}_2(\Omega)$.

(a) Check that $\mathbb{P}^1(\Omega)$ is a $\mathcal{G}_\Omega$-set for the obvious action of $\mathcal{G}_\Omega$, and that $\mathbb{P}^1(\Omega)^{\mathcal{G}_\Omega} = \mathbb{P}^1(k)$.

(b) Show that the pointed $\mathcal{G}_\Omega$-set of left cosets $\mathrm{GL}_2(\Omega)/B(\Omega)$ is isomorphic to $\mathbb{P}^1(\Omega)$.

(c) Show that $H^1(\mathcal{G}_\Omega, B(\Omega)) = 1$.

(d) Check that we have an exact sequence of $\mathcal{G}_\Omega$-groups

$$1 \longrightarrow \Omega \longrightarrow B(\Omega) \longrightarrow \Omega^\times \times \Omega^\times \longrightarrow 1.$$

(e) Recover the fact that the group $H^1(\mathcal{G}_\Omega, \Omega)$ is trivial.

IV

Galois cohomology of quadratic forms

In this chapter, we define some algebraic group-schemes associated to quadratic forms. We also define some classical invariants of quadratic forms and give a cohomological interpretation of these invariants using Galois cohomology.

§IV.10 Algebraic group-schemes associated to quadratic forms

IV.10.1 Quadratic forms over rings

Throughout this section, R is a commutative ring with unit satisfying $2 \in R^\times$.

We start with some definitions and results on quadratic forms over rings. We will not need the full theory, so we only define quadratic forms over free R-modules of finite rank. In this setting, the basic results which are well-known when R is a field remain true. We will refer to [53, Chapter 1, § 6] for proofs.

A **quadratic form** over R is a pair (M, φ), where M is a free R-module of finite rank and a map $\varphi : M \to R$ satisfying the following conditions:

(1) $\varphi(\lambda x) = \lambda^2 \varphi(x)$ for all $\lambda \in R, x \in M$.

(2) The map

$$b_\varphi : \begin{aligned} M \times M &\longrightarrow R \\ (x, y) &\longmapsto \frac{1}{2}(\varphi(x+y) - \varphi(x) - \varphi(y)) \end{aligned}$$

is R-bilinear. Notice that b_φ completely determines φ, since $b_\varphi(x, x) = \varphi(x)$.

Two quadratic spaces (M, φ) and (M', φ') are **isomorphic** if there exists an isomorphism $f : M \to M'$ of R-modules such that $\varphi'(f(x)) = \varphi(x)$ for all $x \in M$. An automorphism $f : M \to M$ of R-modules such that

$$\varphi(f(x)) = \varphi(x) \text{ for all } x \in M$$

is called an **isometry**.

We say that (M, φ) is **regular** (or **non-singular**) if the adjoint map $b_\varphi^* : x \in M \mapsto b_\varphi(x, \cdot) \in M^*$ is an isomorphism of R-modules.

Finally, if $\rho : R \to S$ is a ring morphism, we may consider the S-module $M_S = M \otimes_R S$. We then define $\varphi_S : M_S \to S$ to be the quadratic form associated to the bilinear form

$$b_{\varphi_S}(x \otimes \lambda, y \otimes \mu) = \lambda\mu\rho(b_\varphi(x, y)), \text{ for all } \lambda, \mu \in S, x, y \in M.$$

It is straightforward to check that φ is regular if and only if φ_S is regular for some extension $S \supset R$.

If $\mathbf{e} = (e_1, \ldots, e_n)$ is an R-basis of M, the **representative matrix** of φ in this basis is by definition the matrix $\mathrm{Mat}(\varphi, \mathbf{e}) = (b_\varphi(e_i, e_j))$; if $x = x_1e_1 + \ldots + x_ne_n$, then $\varphi(x) = X^t\mathrm{Mat}(\varphi, \mathbf{e})X$, where X is the column vector $\begin{bmatrix} x_1 \\ \vdots \\ x_n \end{bmatrix}$. If $f \in \mathrm{GL}(V)$, then the representative matrix of φ with respect to the basis $f(\mathbf{e})$ is $M^t\mathrm{Mat}(\varphi, \mathbf{e})M$, where $M = \mathrm{Mat}(f, \mathbf{e})$.

It is easy to check that φ is regular if and only if $\det(\mathrm{Mat}(\varphi, \mathbf{e})) \in R^\times$. If φ is regular, we define the **determinant** of φ to be the square-class

$$\det(\varphi) = \det(\mathrm{Mat}(\varphi, \mathbf{e})) \in R^\times/R^{\times 2}.$$

It does not depend on the choice of $\mathbf{e}$, and only depends on the isomorphism class of φ. Moreover, for every ring morphism $\rho : R \to S$, we have $\det(\varphi_S) = \rho(\det(\varphi)) \in S^\times/S^{\times 2}$.

Two elements $x, y \in M$ are **orthogonal** with respect to φ if $b_\varphi(x, y) = 0$. An **orthogonal basis** is a basis of M consisting of pairwise orthogonal elements. If R is a field, every quadratic form has an orthogonal basis.

If (M, φ) is a regular quadratic space, the rank of the R-module M is called the **dimension of** φ, and is denoted by $\dim(\varphi)$.

Notation: We denote by $\langle a_1, \ldots, a_n \rangle$ the quadratic form

$$\begin{aligned} R^n &\longrightarrow R \\ (x_1, \ldots, x_n) &\longmapsto a_1x_1^2 + \ldots + a_nx_n^2. \end{aligned}$$

If $a \in R$ and $n \geq 0$ is an integer, we will denote by $n \times \langle a \rangle$ then n-dimensional quadratic form $\langle a, \ldots, a \rangle$.

If (M, φ) and (M', φ') are two quadratic spaces, we define a quadratic form $\varphi \perp \varphi'$ on $M \times M'$ by

$$\varphi \perp \varphi'((x, x')) = \varphi(x) + \varphi(x') \text{ for all } x \in M, x' \in M'.$$

The quadratic space $(M \times M', \varphi \perp \varphi')$ is called the **orthogonal sum** of (M, φ) and (M', φ'). It is easy to check that the orthogonal sum is an associative operation on the set of quadratic spaces. If $n \geq 0$ is an integer, we denote by $n \times \varphi$ the orthogonal sum of n copies of φ.

There is also a unique quadratic form $\varphi \otimes \varphi'$ on $M \otimes_R M'$ satisfying

$$b_{\varphi \otimes \varphi'}(x \otimes x', y \otimes y') = b_\varphi(x, y) b_{\varphi'}(x', y') \text{ for all } x, y \in M, x', y' \in M'.$$

The quadratic space $(M \otimes_R M', \varphi \otimes \varphi')$ is called the **tensor product** of (M, φ) and (M', φ'). If $\lambda \in R$, the tensor product of $(R, \langle \lambda \rangle)$ and (M, φ) is canonically isomorphic to the quadratic space $(M, \lambda\varphi)$, where $\lambda\varphi$ is defined by

$$(\lambda\varphi)(x) = \lambda\varphi(x) \text{ for all } x \in M.$$

If (M, φ) and (M', φ) are regular and $\lambda \in R^\times$, so are the quadratic spaces $(M \times M', \varphi \perp \varphi'), (M \otimes_R M', \varphi \otimes \varphi')$ and $(M, \lambda\varphi)$.

IV.10.2 Orthogonal groups

From now on, k will denote a field of characteristic different from 2, and (V, q) will denote a regular quadratic form of dimension n over k. We define the algebraic group-scheme $\mathbf{O}(q)$ as follows:

$$\mathbf{O}(q)(R) = \{f \in \mathbf{GL}(V)(R) \mid q_R \circ f = q_R\}.$$

Definition IV.10.1. The algebraic group-scheme $\mathbf{O}(q)$ is called the **orthogonal group** of q.

From the definition of the orthogonal group $\mathbf{O}(q)$, it is easy to see that for every k-algebra R and every $f \in \mathbf{O}(q)(R)$, we have $\det(f) \in \mu_2(R)$ (be careful, it does **not** imply that $\det(f) = \pm 1$). We therefore get a morphism of algebraic group-schemes

$$\det : \mathbf{O}(q) \to \mu_2.$$

Therefore it is quite natural to consider the kernel of this morphism, that is the group-scheme whose set of R-points is $\ker(\det_R)$:

Definition IV.10.2. The **special orthogonal group** of q is the algebraic group-scheme $\mathbf{O}^+(q)$ defined by

$$\mathbf{O}^+(q)(R) = \{f \in \mathbf{GL}(V)(R) \mid q_R \circ f = q_R, \det(f) = 1\}.$$

We now define a particular type of isometry, which will play an important role in the sequel.

Definition IV.10.3. Let $x \in V$ such that $q(x) \neq 0$. Let H be the hyperplane $H = \{x\}^\perp = \{y \in V \mid b_q(x, y) = 0\}$.

Let $\tau_x : M \to M$ be the unique k-linear map satisfying

$$\tau_x(x) = -x \text{ and } \tau(y) = y \text{ for all } y \in H.$$

The map τ_x is called a **reflection** relative to the hyperplane H. One can check that we have

$$\tau_x(v) = v - 2\frac{b_q(x, v)}{b_q(x, x)}x \text{ for all } v \in V.$$

Moreover, from the definition, it is clear that $\tau_x \in \mathbf{O}(q)(k)$ and that $\det(\tau_x) = -1$.

Remark IV.10.4. A vector $x \in V, x \neq 0$ is called an **isotropic** vector if $q(x) = 0$, and is called **anisotropic** otherwise. A quadratic form is called **isotropic** if q has at least one isotropic vector, and **anisotropic** otherwise. If q is isotropic, one can show that we have

$$q \simeq q' \perp \langle 1, -1 \rangle,$$

for some suitable quadratic form q'. We refer the reader to [53] for more details.

The following proposition is classical:

Proposition IV.10.5. *Let (V, q) be a regular quadratic form over k. Then every isometry f is a product of reflections.*

Proof. We will prove the theorem by induction on $n = \dim V$. If $n = 1$, then there exists $a \in k^\times$ such that $q(x) = ax^2$. Therefore, the only isometries of q are $\pm\mathrm{Id}$, and the result is clear. Now assume that the result is true for $n \geq 1$ and assume that $\dim V = n + 1$. Let f be an isometry of q, and choose $x \in V$ such that $q(x) \neq 0$, which is possible since q is regular, hence not identically zero.

Let us assume first that $f(x) = x$, and let $W = \{x\}^{\perp}$. Since $q(x) \neq 0$, V is the orthogonal sum of kx and W (it comes from the general theory of quadratic forms over fields, or may be proved directly). Then it is easy to check that $f(W) \subset W$ and that $f' = f_{|W}$ is an isometry of $q_{|W}$. Since $\dim(W) = n$, we may apply the induction hypothesis. Hence we may write $f' = \tau'_{w_1} \circ \cdots \circ \tau'_{w_r}$, where $\tau'_{w_1}, \ldots, \tau'_{w_r}$ are reflections of the quadratic space $(W, q_{|W})$. Now we have for all $\lambda \in k$ and all $w \in W$

$$\tau_{w_i}(\lambda x + w) = \lambda \tau_{w_i}(x) + \tau_{w_i}(w) = \lambda x + \tau'_{w_i}(w),$$

the last equality coming from the fact that x is orthogonal to every $w \in W$. Hence we get

$$\begin{aligned}(\tau_{w_1} \circ \cdots \circ \tau_{w_r})(\lambda x + w) &= \lambda x + (\tau'_{w_1} \circ \cdots \circ \tau'_{w_r})(w) \\ &= \lambda x + f'(w) \\ &= f(\lambda x + w),\end{aligned}$$

since f' is the restriction of f to W and $f(x) = x$. Hence we get $f = \tau_{w_1} \circ \cdots \circ \tau_{w_r}$ and we are done.

Let us go back to the general case and set $y = f(x)$. Since f is an isometry, we have $q(y) = q(f(x)) = q(x) \neq 0$. Using the bilinearity of b_q, it is easy to check that $q(x+y) + q(x-y) = 2(q(x) + q(y))$. Hence $q(x+y) + q(x-y) = 4q(x) \neq 0$, so either $q(x+y)$ or $q(x-y)$ is different from 0. Notice also that $x+y$ and $x-y$ are orthogonal, since we have

$$b_q(x+y, x-y) = b_q(x,x) - b_q(y,y) = q(x) - q(y) = 0.$$

Assume first that $q(x+y) \neq 0$. Then we have

$$\tau_{x+y}(y) = \frac{1}{2}(\tau_{x+y}(x+y) - \tau_{x+y}(x-y)) = \frac{1}{2}(-(x+y) - (x-y)) = -x$$

and thus

$$(\tau_x \circ \tau_{x+y} \circ f)(x) = \tau_x(-x) = x.$$

By the previous case, $\tau_x \circ \tau_{x+y} \circ f$ is a product of reflections, and therefore so is f. Now if $q(x-y) \neq 0$, we have

$$\tau_{x-y}(y) = \frac{1}{2}(\tau_{x-y}(x+y) - \tau_{x-y}(x-y)) = \frac{1}{2}((x+y) + (x-y)) = x.$$

Hence we have

$$(\tau_{x-y} \circ f)(x) = \tau_{x-y}(y) = x$$

and we conclude as before. □

We end this section with an easy lemma, whose proof is left to the reader:

Lemma IV.10.6. *We have an exact sequence of $\mathcal{G}_{k_s}$-groups*

$$1 \longrightarrow \mathbf{O}^+(q)(k_s) \longrightarrow \mathbf{O}(q)(k_s) \xrightarrow{\det} \mu_2(k_s) \longrightarrow 1 .$$

IV.10.3 Clifford groups and spinors

Definition IV.10.7. Let R be a ring, and let M be a free R-module of finite rank. Set $T_0(M) = R, T_n(M) = M^{\otimes n}$ if $n \geq 1$. The **tensor algebra** $T(M)$ is the R-algebra

$$T(M) = \bigoplus_{n \geq 0} T_n(M),$$

where the product is defined on elementary tensors by

$$(x_1 \otimes \cdots \otimes x_n) \cdot (y_1 \otimes \cdots \otimes y_m) = x_1 \otimes \cdots \otimes x_n \otimes y_1 \otimes \cdots \otimes y_m.$$

Definition IV.10.8. Let (M, φ) be a quadratic form over a ring R $(2 \in R^\times)$ and let $I(\varphi)$ be the two-sided ideal of $T(M)$ generated by the set $\{x \otimes x - \varphi(x) \mid x \in M\}$. The **Clifford algebra** of (M, φ), denoted by $C(M, \varphi)$ is defined by

$$C(M, \varphi) = T(M)/I(\varphi).$$

The image of a vector $x \in M$ under the canonical projection

$$T(M) \to C(M, \varphi)$$

is denoted by $\overline{x}$. Clearly, $C(M, \varphi)$ is generated by $\overline{M} = \{\overline{x} \mid x \in M\}$ as an R-algebra. There is a canonical involution ^{t} on $T(M)$ defined on elementary tensors by

$$(x_1 \otimes x_2 \otimes \cdots \otimes x_n)^t = x_n \otimes \cdots \otimes x_2 \otimes x_1.$$

This involution stabilizes $I(\varphi)$, and therefore induces an involution on $C(M, \varphi)$, still denoted by ^{t}. The automorphism

$$\begin{aligned} M &\longrightarrow M \\ x &\longmapsto -x \end{aligned}$$

induces an automorphism on $T(M)$ which stabilizes $I(\varphi)$. Therefore, it also induces an automorphism

$$\gamma : C(M, \varphi) \to C(M, \varphi).$$

Notice that γ and ^{t} commute with scalar extensions. We then get a $\mathbb{Z}/2\mathbb{Z}$-grading of $C(M,\varphi)$ by setting

$$\begin{aligned} C_0(M,\varphi) &= \{s \in C(M,\varphi) \mid \gamma(s) = s\}; \\ C_1(M,\varphi) &= \{s \in C(M,\varphi) \mid \gamma(s) = -s\}. \end{aligned}$$

Let (V,q) be a quadratic form over a field k. The following proposition collects the standard properties of the Clifford algebra. We refer to [53] once again for more details.

Proposition IV.10.9. *Let (V,q) be a regular quadratic space.*

(1) *For any k-algebra R, we have a canonical isomorphism*

$$C(V,q)_R \simeq C(V_R, q_R)$$

which respects the grading.

(2) *If $x,y \in V$, then $b_q(x,y) = \dfrac{1}{2}(\overline{x}\cdot\overline{y} + \overline{y}\cdot\overline{x})$.*
In particular, if x and y are orthogonal (with respect to q), then $\overline{x}\cdot\overline{y} = -\overline{y}\cdot\overline{x}$ in $C(V,q)$.

(3) *If $q(x) \neq 0$, then $\overline{x}$ is invertible in $C(V,q)$.*

(4) *If $e_1,\ldots,e_n$ is an orthogonal basis for q, the elements $\overline{e}_1^{m_1}\cdots\overline{e}_n^{m_n}$, where $m_i = 0,1$ for all i, form a k-basis of $C(V,q)$. In particular, the map $V \to C(V,q)$ is injective.*

In view of (4), we will omit the bar notation from now on. We now define the Clifford group of q. First, we need a lemma:

Lemma IV.10.10. *Let (V,q) be a quadratic form over a field k, and let $f : V \to V$ be an isometry of q. Then there exists an invertible element $s_f \in C(V,q)$ such that $f(x) = \gamma(s_f)xs_f^{-1}$ for all $x \in V$.*

Proof. We start with the case of reflections. If $f = \tau_x$, for some $x \in V$ satisfying $q(x) \neq 0$, then for all $y \in V$, we have

$$\tau_x(y) = y - 2\frac{b_q(x,y)}{b_q(x,x)}x = y - (xy+yx)x^{-2}x = -xyx^{-1} = \gamma(x)yx^{-1}.$$

Now if f is any isometry of q, then $f = \tau_{x_1}\circ\cdots\circ\tau_{x_r}$ by Proposition IV.10.5. Setting $s_f = x_1\cdots x_r$, the previous case leads immediately to $f(y) = \gamma(s_f)ys_f^{-1}$ for all $y \in V$. □

We now define a group scheme associated to (V,q).

Definition IV.10.11. Let R be a k-algebra. The **Clifford group** of (V, q) is the affine algebraic group scheme defined by

$$\mathbf{\Gamma}(V,q)(R) = \{s \in C(V,q)_R^\times \mid \gamma(s)V_R s^{-1} = V_R\}.$$

For any element $s \in \mathbf{\Gamma}(V,q)(R)$, we can define an automorphism

$$\alpha_s : \begin{array}{rcl} V_R & \longrightarrow & V_R \\ x & \longmapsto & \gamma(s)xs^{-1}, \end{array}$$

and then we get a morphism $\alpha_R : \mathbf{\Gamma}(R) \to \mathbf{GL}(V)(R)$. We therefore get a group-scheme morphism

$$\alpha : \mathbf{\Gamma}(V,q) \to \mathbf{GL}(V).$$

Notice that we have a canonical isomorphism

$$\mathbf{\Gamma}(V,q)(R) \simeq \Gamma(V_R, q_R).$$

We now study more carefully the morphism α. We start with the following:

Lemma IV.10.12. *For every k-algebra R, we have* $\ker(\alpha_R) = R^\times$.

Proof. The inclusion $R^\times \subset \ker(\alpha_R)$ is clear. Now assume that $s \in C(V,q)_R^\times$ satisfies $\gamma(s)xs^{-1} = x$ for all $x \in V_R$, and write $s = s_0 + s_1, s_i \in C_i(V,q)_R$. We get easily

$$\begin{array}{rcl} s_0 x & = & x s_0 \\ s_1 x & = & -x s_1. \end{array}$$

Let us introduce some notation. Let $e_1, \ldots, e_n$ be an orthogonal basis for q. Hence $e_1 \otimes 1_R, \ldots, e_n \otimes 1_R$ is an orthogonal basis for q_R. We will still write $e_1, \ldots, e_n$ for short. Notice that e_i is invertible, since $e_i^2 = q(e_i) \neq 0$ (since q is non-singular).

For any subset I of $\{1, \ldots, n\}$, set $e_I = e_{i_1} \cdots e_{i_k}$, where $i_1 < \cdots < i_k$ are the elements of I if $I \neq \emptyset$, and $e_\emptyset = 1$. Then it is clear that the family of elements e_I where I has an even (resp. odd) number of elements is a k-basis of $C_0(V,q)_R$ (resp. $C_1(V,q)_R$). Hence we may write

$$s_0 = \sum_{I, |I| \text{ even}} \lambda_I e_I.$$

Let $i \in \{1, \ldots, n\}$. If $i \notin I$, then $e_I e_i = e_i e_I$, since I has even cardinality. If $i \in I$, then e_i commutes with itself, and anticommutes with the other

$e_j, j \in I - \{i\}$; since $I - \{i\}$ has odd cardinality, we get $e_I e_i = -e_i e_I$. From the equality $s_0 e_i = e_i s_0$, we get (after dividing by e_i)

$$\sum_{I \ni i, |I| \text{ even}} \lambda_I e_I = 0,$$

and then $\lambda_I = 0$ for all I containing i. Since any non-empty subset I contains at least one element, the only remaining term is the constant term, so $s_0 \in R$.

Let us show that $s_1 = 0$. Write $s_1 = \sum_{I, |I| \text{ odd}} \lambda_I e_I$, and let $i \in \{1, \ldots, n\}$. If $i \notin I$, then $e_I e_i = -e_i e_I$ since I has odd cardinality. If $i \in I$, then e_i commutes with itself, and anticommutes with the other $e_j, j \in I - \{i\}$; since $I - \{i\}$ has even cardinality, we get $e_I e_i = e_i e_I$. From the equality $s_1 e_i = -e_i s_1$, we get (after dividing by e_i)

$$\sum_{I \ni i, |I| \text{ odd}} \lambda_I e_I = 0,$$

and then $\lambda_I = 0$ for all I containing i. Since any non-empty subset I contains at least one element, we get $s_1 = 0$. Finally, $s \in R$, and since s is invertible, we get $s \in R^\times$. □

Definition IV.10.13. Let (M, φ) be a quadratic form over a ring R. If $s \in C(M, \varphi)$, we set $N_R(s) = s^t s \in C(M, \varphi)$. The map

$$N_R : C(M, \varphi) \to C(M, \varphi)$$

is called the **norm** of $C(M, \varphi)$. If $x \in M$, we have $N_R(x) = x^2 = \varphi(x)$.

Notice that N_R commutes with scalar extensions (since the involution ^{t} does).

Lemma IV.10.14. *Let (V, q) be a quadratic form over a field k, and let R be a k-algebra. If $s \in \mathbf{\Gamma}(V, q)(R)$, then $N_R(s) = s^t s \in R^\times$. The norm defines a group-scheme morphism $N : \mathbf{\Gamma}(V, q) \to \mathbb{G}_m$, and $N_R(\gamma(s)) = N_R(s)$.*

Proof. If $s \in \mathbf{\Gamma}(V, q)(R)$, then $\gamma(s) V_R = V_R s$. Applying the involution ^{t} to this equality, we get $V_R \gamma(s)^t = s^t V_R$. It is easy to see from the definitions that γ and ^{t} commute, so we get $V_R \gamma(s^t) = s^t V_R$. Applying the automorphism γ to this equality, we get $V_R s^t = \gamma(s^t) V_R$. Since s^t is invertible whenever s is invertible, we obtain that $s^t \in \mathbf{\Gamma}(V, q)(R)$. Hence $N_R(s) = s^t s \in \mathbf{\Gamma}(V, q)(R)$.

To prove that $N_R(s) \in R^\times$, it is enough to show that $\alpha_R(s^t s) = \mathrm{Id}$ by the previous lemma. Notice first that if $v \in V_R$, we have

$$\gamma(v^t) = \gamma(v) = -v.$$

Now let $s \in \mathbf{\Gamma}(V, q)(R)$. Then for all $x \in V_R$, we have $\gamma(s)xs^{-1} \in V_R$ and therefore

$$\gamma(s)xs^{-1} = -\gamma((\gamma(s)xs^{-1})^t) = -\gamma(s^{-t}x\gamma(s^t)) = \gamma(s^{-t})xs^t.$$

We then get

$$\alpha_{s^t s}(x) = \gamma(s^t)\gamma(s)xs^{-1}s^{-t} = \gamma(s^t)\gamma(s^{-t})xs^t s^{-t} = x.$$

Thus $\alpha(s^t s) = \mathrm{Id}$ as claimed, proving that $N_R(s) \in R^\times$.

It remains to show the last part. If $s, s' \in \mathbf{\Gamma}(V, q)(R)$, we have

$$N_R(ss') = (ss')^t(ss') = s'^t s^t ss'.$$

Since $N_R(s) = s^t s \in R^\times$ by the previous point, we get

$$N_R(ss') = s'^t s^t ss' = s^t ss'^t s' = N_R(s)N_R(s').$$

Therefore N is a group-scheme morphism. Now for all $s \in \mathbf{\Gamma}(V, q)(R)$, we have

$$N_R(\gamma(s)) = \gamma(s)^t\gamma(s) = \gamma(s^t s) = \gamma(N_R(s)) = N_R(s),$$

the last equality coming from the fact that γ is the identity on R. □

Lemma IV.10.15. *For all $s \in \mathbf{\Gamma}(V, q)(R)$, α_s is an isometry of q_R.*

Proof. We have $q_R(\alpha_s(x)) = \alpha_s(x)^t\alpha_s(x)$, since ^{t} is the identity on V_R by definition. Hence

$$q_R(\alpha_s(x)) = s^{-t}x^t\gamma(s)^t(\gamma(s)xs^{-1}) = s^{-t}xN_R(s)xs^{-1}.$$

Thus we get

$$q_R(\alpha_s(x)) = N_R(s)s^{-t}x^2s^{-1} = q_R(x)N_R(s)N_R(s^{-1}) = q_R(x),$$

since $N_R(s^{-1}) = N_R(s)^{-1}$. □

Hence we get a group-scheme morphism $\alpha : \mathbf{\Gamma}(V, q) \to \mathbf{O}(q)$ with kernel $\mathbb{G}_m$. We then have the following result.

Theorem IV.10.16. *For any field extension K/k, we have an exact sequence*

$$1 \longrightarrow K^\times \longrightarrow \mathbf{\Gamma}(V,q)(K) \xrightarrow{\alpha_K} \mathbf{O}(q)(K) \longrightarrow 1\,.$$

Moreover, for any $x \in V$ satisfying $q(x) \neq 0$, the preimages of τ_x in $\mathbf{\Gamma}(V,q)(K)$ are the elements $\lambda x, \lambda \in K^\times$.

Proof. Let us prove that α_K is surjective. By Lemma IV.10.10 and its proof, we have $\alpha_K(x) = \tau_x$. Since $\mathbf{O}(q)(K)$ is generated by reflections, the surjectivity follows. The equality $\ker(\alpha_K) = K^\times$ is a particular case of Lemma IV.10.12. The last part of the theorem is then clear. □

If K/k is a field extension, it follows from the definition of the norm that we have

$$N_K(\lambda) = \lambda^2 \text{ for all } \lambda \in K^\times.$$

Therefore the previous theorem implies that the norm induces a morphism $SN_K : \mathbf{O}(q)(K) \to K^\times/K^{\times 2}$ satisfying

$$SN_K(\tau_x) = q(x) \in K^\times/K^{\times 2} \text{ for all } x \in V \text{ such that } q(x) \neq 0.$$

Definition IV.10.17. The morphism SN_K is called the **spinor norm**.

We now define another group-scheme associated to q.

Definition IV.10.18. We denote by $\mathbf{Pin}(q)$ the algebraic group-scheme defined by

$$\mathbf{Pin}(q)(R) = \{x \in \mathbf{\Gamma}(V,q)(R) \mid N_R(x) = 1\}.$$

Corollary IV.10.19. *We have an exact sequence of $\mathcal{G}_{k_s}$-groups*

$$1 \longrightarrow \mu_2(k_s) \longrightarrow \mathbf{Pin}(q)(k_s) \xrightarrow{\alpha_{k_s}} \mathbf{O}(q)(k_s) \longrightarrow 1\,,$$

where the map $\mu_2(k_s) \longrightarrow \mathbf{Pin}(q)(k_s)$ is the inclusion. If $x \in V_{k_s}$ satisfies $q_{k_s}(x) \neq 0$, the preimages of τ_x in $\mathbf{Pin}(q)(k_s)$ are $\pm \dfrac{1}{\sqrt{q(x)}} x$.

Proof. If $s \in \mathbf{Pin}(q)(k_s)$ satisfies $\alpha_{k_s}(s) = 1$, then $s \in k_s^\times$ by Lemma IV.10.12. But we have in this case $1 = N_{K_s}(s) = s^2$, so $s = \pm 1$. Now let $x \in V$ and let $f \in \mathbf{O}(q)(k_s)$ be an isometry of q_{k_s}. By the computations made in the proof of Lemma IV.10.10, we have $\tau_x(y) = \gamma(x)yx^{-1}$ for all $y \in V_{k_s}$. We have $N_{k_s}(x) = x^2 = q_{k_s}(x)$, so $N_{k_s}\left(\dfrac{1}{\sqrt{q(x)}} x\right) = 1$ and

$\frac{1}{\sqrt{q(x)}}x \in \mathbf{Pin}(q)(k_s)$. Moreover, we have

$$\tau_x(y) = \gamma\left(\frac{1}{\sqrt{q(x)}}x\right) y \left(\frac{1}{\sqrt{q(x)}}x\right)^{-1} \text{ for all } y \in V_{k_s}.$$

Hence $\tau_x = \alpha_{k_s}\left(\frac{1}{\sqrt{q(x)}}x\right)$. It follows that $\frac{1}{\sqrt{q(x)}}x$ is a preimage of τ_x in $\mathbf{Pin}(q)k_s$, and since the kernel of α_{k_s} is $\{\pm 1\}$, there is only one other preimage, which is $-\frac{1}{\sqrt{q(x)}}x$. Surjectivity follows from the fact that $\mathbf{O}(q)(k_s)$ is generated by reflections (cf. Proposition IV.10.5). The fact that the maps preserve the $\mathcal{G}_{k_s}$-actions is left to the reader. □

Notation: If $q = \langle 1, \dots, 1\rangle$, we denote $\mathbf{O}(q)$ and $\mathbf{Pin}(q)$ simply by $\mathbf{O}_n$ and $\mathbf{Pin}_n$ respectively.

Definition IV.10.20. The inverse image of $\mathbf{O}^+(q)$ in $\mathbf{\Gamma}(V, q)$ by the morphism α is called the **special Clifford group of** q, and is denoted by $\mathbf{\Gamma}^+(V, q)$.

The **spinor group of** q is the algebraic group-scheme

$$\mathbf{Spin}(q) = \mathbf{Pin}(q) \cap \mathbf{\Gamma}^+(V, q).$$

If $q = \langle 1, \dots, 1\rangle$, we simply denote it by $\mathbf{Spin}_n$.

The next lemma directly follows from the previous corollary and the definition of the spinor group.

Lemma IV.10.21. *We have an exact sequence of $\mathcal{G}_{k_s}$-groups*

$$1 \longrightarrow \mu_2(k_s) \longrightarrow \mathbf{Spin}(q)(k_s) \xrightarrow{\alpha_{k_s}} \mathbf{O}^+(q)(k_s) \longrightarrow 1\,.$$

§IV.11 Galois cohomology of quadratic forms

IV.11.1 Galois cohomology of orthogonal groups

Let V be a k-vector space of dimension n. For every field extension K/k, denote by $\mathbf{F}(K)$ the set of quadratic forms on V_K. If $\iota : K \to K'$ is a morphism of field extensions, let

$$\mathbf{F}(\iota) : \begin{array}{rcl} \mathbf{F}(K) & \longrightarrow & \mathbf{F}(K') \\ q' & \longmapsto & q'_{K'} \end{array}$$

be the corresponding scalar extension map. We then obtain a functor

$$\mathbf{F} : \mathfrak{C}_k \longrightarrow \mathbf{Sets},$$

on which the functor $\mathbf{GL}(V)$ acts as follows: for every field extension K/k, every $f \in \mathbf{GL}(V)(K)$ and every $q' \in \mathbf{F}(K)$, set

$$f \cdot q' = q' \circ f^{-1}.$$

Now let (V, q) be a regular quadratic form on V, and let $\mathbf{Quad}_n(L)$ be the pointed set of isomorphism classes of regular quadratic forms on V_L, the base point being the isomorphism class of q_L. We then obtain a functor $\mathbf{Quad}_n : \mathfrak{C}_k \to \mathbf{Sets}^*$. By definition of the action of $\mathbf{GL}(V)$ on $\mathbf{F}$, it is clear that the stabilizer of q is $\mathbf{O}(q)$. Since all regular quadratic forms are isomorphic over a separably closed field, the functor of twisted forms of q is $\mathbf{Quad}_n$, and an immediate application of the Galois descent lemma gives:

Proposition IV.11.1. *We have an isomorphism of functors from* $\mathfrak{C}_k$ *to* $\mathbf{Sets}^*$

$$\mathbf{Quad}_n \simeq H^1(\_, \mathbf{O}(q)).$$

We continue with a cohomological interpretation of the determinant.

Proposition IV.11.2. *The map* $\det_* : H^1(k, \mathbf{O}(q)) \to H^1(k, \mu_2)$ *sends the isomorphism class of a quadratic form* q' *onto* $\det(q)\det(q')^{-1}$.

Proof. Let $f \in \mathrm{GL}(V_{k_s})$ such that $f \cdot q'_{k_s} = q_{k_s}$. A cocycle corresponding to q' is given by

$$\alpha : \begin{array}{l} \mathcal{G}_{k_s} \longrightarrow \mathbf{O}(q)(k_s) \\ \quad \sigma \longmapsto f \circ \sigma \cdot f^{-1}. \end{array}$$

Let B, B' denote the representative matrices of q, q' in a fixed basis of V. By choice of f, we have $q'_{k_s} = q_{k_s} \circ f$, that is

$$q'_{k_s}(x) = q_{k_s}(f(x)) \text{ for all } x \in V_{k_s}.$$

From this equality, we get

$$\det(B') = \det(B)\det(f)^2.$$

Hence $(\det(f)^{-1})^2 = \det(B)\det(B')^{-1}$. Now a cocycle representing $\det_*([\alpha])$ is

$$\begin{array}{l} \mathcal{G}_{k_s} \longrightarrow \mu_2(k_s) \\ \quad \sigma \longmapsto \det(\alpha_\sigma) = \dfrac{\sigma(\det(f)^{-1})}{\det(f)^{-1}}. \end{array}$$

This cocycle corresponds to the square-class of $\det(B)\det(B')^{-1}$ by Remark III.8.28, which is also the square-class of $\det(q)\det(q')^{-1}$. □

Corollary IV.11.3. *The pointed set $H^1(k, \mathbf{O}^+(q))$ is in 1-1 correspondence with the set of isomorphism classes of quadratic forms (V, q') satisfying* $\det(q') = \det(q)$.

Proof. The image of $H^1(k, \mathbf{O}^+(q)) \longrightarrow H^1(k, \mathbf{O}(q))$ is equal to the kernel of $\det_*$, so it is equal to the set of isomorphism classes of quadratic forms (V, q') satisfying $\det(q') = \det(q)$. Now we want to prove that the map $H^1(k, \mathbf{O}^+(q)) \longrightarrow H^1(k, \mathbf{O}(q))$ is injective. Let $[\alpha] \in H^1(k, \mathbf{O}^+(q))$ and let $[\beta] \in H^1(k, \mathbf{O}(q))$ be its image via the map induced by the inclusion $\mathbf{O}^+(q)(k_s) \subset \mathbf{O}(q)(k_s)$. By Lemma II.5.5, it is enough to show that the kernel of $H^1(k, \mathbf{O}^+(q)_\alpha) \longrightarrow H^1(k, \mathbf{O}(q)_\beta)$ is trivial for every $[\alpha]$. Now the sequence

$$1 \longrightarrow \mathbf{O}^+(q)(k_s)_\alpha \longrightarrow \mathbf{O}(q)(k_s)_\beta \longrightarrow \mu_2(k_s) \longrightarrow 1$$

is easily seen to be exact, so we have an exact sequence

$$\mu_2(k) \longrightarrow H^1(k, \mathbf{O}^+(q)(k_s)_\alpha) \longrightarrow H^1(k, \mathbf{O}(q)(k_s)_\beta).$$

Let us compute the connecting map $\delta^0 : \mu_2(k) \longrightarrow H^1(k, \mathbf{O}^+(q)(k_s)_\alpha)$. A preimage of $\varepsilon \in \mu_2(k_s)$ in $\mathbf{O}(q)(k_s)$ is $\varepsilon\mathrm{Id}$. Now we have

$$\sigma * \varepsilon\mathrm{Id} = \beta_\sigma(\sigma\cdot\varepsilon\mathrm{Id})\beta_\sigma^{-1} = \varepsilon\mathrm{Id},$$

since $\varepsilon \in k$. Hence $\delta^0(\varepsilon) = 1$, so δ^0 is trivial. Therefore the map $H^1(k, \mathbf{O}^+(q)(k_s)_\alpha) \longrightarrow H^1(k, \mathbf{O}(q)(k_s)_\beta)$ has trivial kernel, which concludes the proof. □

IV.11.2 Galois cohomology of spinors

IV.11.2.1 The Hasse invariant of a quadratic form

In this section, we define a cohomology class of degree 2 attached to a quadratic form q over k. Assume that $q \simeq \langle a_1, \ldots, a_n\rangle$. We set

$$w_2(q) = \sum_{1\leq i<j\leq n} (a_i) \cup (a_j) \in H^2(k, \mu_2).$$

One can show that $w_2(q)$ does not depend on the choice of a diagonalization of q, but only depends on the isomorphism class of q. This follows from [53, Chapter 2, Remark 12.5] and the fact that the map

$$\begin{aligned} k^\times \times k^\times &\longrightarrow H^2(k, \mu_2) \\ (a,b) &\longmapsto (a) \cup (b) \end{aligned}$$

is a Steinberg symbol (see [53, Chapter 2, §12] for a definition).

Definition IV.11.4. The cohomology class $w_2(q)$ is called the **Hasse invariant of** q.

The following lemma may be proved by direct computation.

Lemma IV.11.5. *Let q, q' be two quadratic forms. Then we have*

$$w_2(q \perp q') = w_2(q) + (\det(q)) \cup (\det(q')) + w_2(q')$$

and

$$w_2(\lambda q) = w_2(q) + (n-1)(\lambda) \cup (\det(q)) + \frac{n(n-1)}{2}(-1) \cup (\lambda),$$

where $n = \dim(q)$.

The notation $w_2(q)$, which may appear strange at first, is justified by the fact that one may define a family of cohomology classes $(w_m(q))_{m \geq 0}$, called Stiefel-Whitney classes. We refer to [18] for more information.

IV.11.2.2 Galois cohomology of spinor groups

We now give a cohomological interpretation of the Hasse invariant.

Theorem IV.11.6. *The first connecting map associated to the exact sequence*

$$1 \longrightarrow \mu_2(k_s) \longrightarrow \mathbf{Pin}(q)(k_s) \xrightarrow{\alpha_{k_s}} \mathbf{O}(q)(k_s) \longrightarrow 1$$

is given by

$$\delta^1(q') = w_2(q) + w_2(q') + (\det(q)) \cup (-\det(q')).$$

Proof. Let (V, q') be a quadratic form on V. Let $\mathbf{e} = (e_1, \ldots, e_n), \mathbf{e}' = (e'_1, \ldots, e'_n)$ be two orthogonal bases of V with respect to q and q' respectively, and set

$$a_i = q(e_i), b_i = q'(e'_i), c_i = b_i a_i^{-1}.$$

In order to simplify notation, we will still denote by $e_1, \ldots, e_n$ and

$e'_1, \ldots, e'_n$ the corresponding orthogonal bases of V_{k_s} with respect to q_{k_s} and q'_{k_s} respectively.
We will denote by $\tau_i \in \mathbf{O}(q)(k_s)$ the reflection τ_{e_i}. Finally, write

$$\frac{\sigma(\sqrt{c_i})}{c_i} = (-1)^{s_i(\sigma)} \text{ and } \frac{\sigma(\sqrt{a_i})}{a_i} = (-1)^{\varepsilon_i(\sigma)} \text{ for all } \sigma \in \mathcal{G}_{k_s}.$$

We first show that a cocycle $\xi : \mathcal{G}_{k_s} \longrightarrow \mathbf{O}(q)(k_s)$ representing q' is given by

$$\xi_\sigma = \tau_1^{s_1(\sigma)} \circ \cdots \circ \tau_n^{s_n(\sigma)} \text{ for all } \sigma \in \mathcal{G}_{k_s}.$$

The map

$$f : \begin{array}{rcl} V_{k_s} & \longrightarrow & V_{k_s} \\ e'_i & \longmapsto & \sqrt{c_i} e_i \end{array}$$

is easily seen to satisfy

$$f \cdot q'_{k_s} = q_{k_s}.$$

Then a cocycle representing q' is

$$\xi : \begin{array}{rcl} \mathcal{G}_{k_s} & \longrightarrow & \mathbf{O}(q)(k_s) \\ \sigma & \longmapsto & \xi_\sigma = f \circ \sigma \cdot f^{-1}. \end{array}$$

For $i = 1, \ldots, n$, we have

$$\begin{aligned} \xi_\sigma(e_i) &= f(\sigma \cdot (f^{-1}(\sigma^{-1} \cdot e_i))) \\ &= f(\sigma \cdot f^{-1}(e_i)) \\ &= f\left(\sigma \cdot \frac{1}{\sqrt{c_i}} e'_i\right) \\ &= f\left((-1)^{-s_i(\sigma)} \frac{1}{\sqrt{c_i}} e'_i\right) \\ &= (-1)^{s_i(\sigma)} e_i. \end{aligned}$$

The fact that $\tau_i(e_i) = -e_i$ and that the e_j's are mutually orthogonal lead easily to the desired conclusion. We now compute a cocycle representing $\delta^1(q')$. By Corollary IV.10.19, a preimage of ξ_σ in $\mathbf{Pin}(q)(k_s)$ is given by

$$\tilde{\xi}_\sigma = \frac{1}{\sqrt{a_1}^{s_1(\sigma)} \cdots \sqrt{a_n}^{s_n(\sigma)}} e_1^{s_1(\sigma)} \cdots e_n^{s_n(\sigma)},$$

and we have

$$\sigma \cdot \tilde{\xi}_\tau = \frac{(-1)^{\sum\limits_i \varepsilon_i(\sigma) s_i(\tau)}}{\sqrt{a_1}^{s_1(\tau)} \cdots \sqrt{a_n}^{s_n(\tau)}} e_1^{s_1(\tau)} \cdots e_n^{s_n(\tau)}.$$

Taking into account that e_i and e_j anticommute if $i \neq j$, we get that

$$\tilde{\xi}_\sigma \sigma \cdot \tilde{\xi}_\tau = \frac{(-1)^{\sum_i \varepsilon_i(\sigma)s_i(\tau) + \sum_{i<j} s_i(\sigma)s_j(\tau)}}{\sqrt{a_1}^{s_1(\sigma)+s_1(\tau)} \cdots \sqrt{a_n}^{s_n(\sigma)+s_n(\tau)}} e_1^{s_1(\sigma)+s_1(\tau)} \cdots e_n^{s_n(\sigma)+s_n(\tau)}.$$

Now recall that the map $\mathcal{G}_{k_s} \longrightarrow \mu_2(k_s), \sigma \mapsto (-1)^{s_i(\sigma)}$ is a cocycle with values in $\mu_2(k_s)$. Since $\mu_2(k_s)$ is abelian and $\mathcal{G}_{k_s}$ acts trivially on it, it follows that this map is a group morphism. Therefore, we have

$$s_i(\sigma) + s_i(\tau) - s_i(\sigma\tau) \in 2\mathbb{Z}$$

for all $\sigma, \tau \in \mathcal{G}_{k_s}$. Set $m^{(i)}_{\sigma\tau} = \dfrac{1}{2}(s_i(\sigma) + s_i(\tau) - s_i(\sigma\tau))$.
Then we have

$$e_i^{s_i(\sigma)+s_i(\tau)} = e_i^{2m^{(i)}_{\sigma,\tau}+s_i(\sigma\tau)} = a_i^{m^{(i)}_{\sigma,\tau}} e_i^{s_i(\sigma\tau)}.$$

Since $s_i(\sigma) + s_i(\tau) = 2m^{(i)}_{\sigma,\tau} + s_i(\sigma\tau)$, we have

$$\sqrt{a_i}^{s_i(\sigma)+s_i(\tau)} = a_i^{m^{(i)}_{\sigma,\tau}} \sqrt{a_i}^{s_i(\sigma\tau)},$$

and thus

$$\tilde{\xi}_\sigma \sigma \cdot \tilde{\xi}_\tau = \frac{(-1)^{\sum_i \varepsilon_i(\sigma)s_i(\tau) + \sum_{i<j} s_i(\sigma)s_j(\tau)}}{\sqrt{a_1}^{s_1(\sigma\tau)} \cdots \sqrt{a_n}^{s_n(\sigma\tau)}} e_1^{s_1(\sigma\tau)} \cdots e_n^{s_n(\sigma\tau)},$$

that is

$$\tilde{\xi}_\sigma \sigma \cdot \tilde{\xi}_\tau \tilde{\xi}_{\sigma\tau}^{-1} = (-1)^{\sum_i \varepsilon_i(\sigma)s_i(\tau) + \sum_{i<j} s_i(\sigma)s_j(\tau)}.$$

Finally, we get that $\delta^1(q')$ is represented by the cocycle

$$\mathcal{G}_{k_s} \times \mathcal{G}_{k_s} \longrightarrow \mu_2(k_s), (\sigma, \tau) \mapsto (-1)^{\sum_i \varepsilon_i(\sigma)s_i(\tau) + \sum_{i<j} s_i(\sigma)s_j(\tau)}.$$

Hence

$$\delta^1(q') = \sum_i (a_i) \cup (b_i a_i^{-1}) + \sum_{i<j} (b_i a_i^{-1}) \cup (b_j a_j^{-1}).$$

Taking into account that we work up to squares, we obtain

$$\delta^1(q') = \sum_i (a_i) \cup (a_i b_i) + \sum_{i<j} (a_i b_i) \cup (a_j b_j).$$

Using the bilinearity and the commutativity of the cup-product, we get

$$\delta^1(q') = \sum_i (a_i) \cup (a_i) + \sum_{i<j} (a_i) \cup (a_j) + \sum_{i<j} (b_i) \cup (b_j) + \sum_{i,j} (a_i) \cup (b_j).$$

Now we have

$$\sum_i (a_i) \cup (a_i) = \sum_i (a_i) \cup (-1) = (a_1 \cdots a_n) \cup (-1) = (\det(q)) \cup (-1).$$

Morever we have

$$\sum_{i,j} (a_i) \cup (b_j) = \left(\sum_i (a_i)\right) \cup \left(\sum_j (b_j)\right) = (\det(q)) \cup (\det(q')).$$

Hence, using bilinearity again, we finally get

$$\delta^1(q') = w_2(q) + w_2(q') + (\det(q)) \cup (-\det(q')).$$

This completes the proof of the theorem. □

This result was originally proved by Springer in [63]. Notice however that Springer's definition of the Hasse invariant is slightly different, and that the minus sign is missing in his formula. Applying the previous theorem to $q = \langle 1, \ldots, 1 \rangle$, we get:

Corollary IV.11.7. *The connecting map $\delta^1 : H^1(k, \mathbf{O}_n) \longrightarrow H^2(k, \mu_2)$ associated to the exact sequence*

$$1 \longrightarrow \mu_2(k_s) \longrightarrow \mathbf{Pin}_n(k_s) \xrightarrow{\alpha_{k_s}} \mathbf{O}_n(k_s) \longrightarrow 1$$

is given by

$$\delta^1(q) = w_2(q).$$

Remark IV.11.8. The previous corollary provides another way to show that the Hasse invariant is well-defined, since we know that $\delta^1(q)$ only depends on the isomorphism class of q.

Remark IV.11.9. For every morphism $K \longrightarrow L$ of field extensions of k, the diagram

$$\begin{array}{ccccccccc} 1 & \longrightarrow & \mu_2(K_s) & \longrightarrow & \mathbf{Pin}_n(K_s) & \longrightarrow & \mathbf{O}_n(K_s) & \longrightarrow & 1 \\ & & \downarrow & & \downarrow & & \downarrow & & \\ 1 & \longrightarrow & \mu_2(L_s) & \longrightarrow & \mathbf{Pin}_n(L_s) & \longrightarrow & \mathbf{O}_n(L_s) & \longrightarrow & 1 \end{array}$$

is commutative. Hence Theorem III.7.39 (5) and Corollary IV.11.7 imply that we have

$$w_2(q_L) = \mathrm{Res}_{L/K}(w_2(q)),$$

for every quadratic form q over K.

Corollary IV.11.10. *The first connecting map associated to the exact sequence*

$$1 \longrightarrow \mu_2(k_s) \longrightarrow \mathbf{Spin}(q)(k_s) \xrightarrow{\alpha_{k_s}} \mathbf{O}^+(q)(k_s) \longrightarrow 1$$

is given by

$$\delta^1(q') = w_2(q) + w_2(q').$$

Proof. We have a commutative diagram

$$\begin{array}{ccccccccc} 1 & \longrightarrow & \mu_2(k_s) & \longrightarrow & \mathbf{Spin}(q)(k_s) & \xrightarrow{\alpha_{k_s}} & \mathbf{O}^+(q)(k_s) & \longrightarrow & 1 \\ & & \| & & \downarrow & & \downarrow & & \\ 1 & \longrightarrow & \mu_2(k_s) & \longrightarrow & \mathbf{Pin}(q)(k_s) & \xrightarrow{\alpha_{k_s}} & \mathbf{O}(q)(k_s) & \longrightarrow & 1 \end{array}$$

which induces a commutative square in cohomology

$$\begin{array}{ccc} H^1(k, \mathbf{O}^+(q)) & \longrightarrow & H^2(k, \mu_2) \\ \downarrow & & \| \\ H^1(k, \mathbf{O}(q)) & \longrightarrow & H^2(k, \mu_2) \end{array}$$

by Theorem III.7.39 (4). Now it suffices to apply the previous theorem and the fact that the pointed set $H^1(k, \mathbf{O}^+(q))$ classifies quadratic forms with same dimension and determinant as q (cf. Corollary IV.11.3), taking into account that $(a) \cup (-a) = 0$ for all $a \in k^\times$. □

§IV.12 Cohomological invariants of quadratic forms

IV.12.1 Classification of quadratic forms over $\mathbb{Q}$

We would like now to give a complete classification result of quadratic forms over $\mathbb{Q}$ in terms of invariants. Before doing this, we need to define the signature of a quadratic form.

Lemma IV.12.1. *Let (V, q) be a (regular) quadratic space over $\mathbb{R}$. Then there exist unique integers $r, s \geq 0$ such that*

$$q \simeq r \times \langle 1 \rangle \perp s \times \langle -1 \rangle.$$

Proof. We know that $q \simeq \langle a_1, \cdots, a_n \rangle$, for some $a_i \in \mathbb{R}^\times$. Now each a_i is congruent to ± 1 modulo $\mathbb{R}^{\times 2}$. This proves the existence part. Now assume that we have

$$q \simeq r \times \langle 1 \rangle \perp s \times \langle -1 \rangle \simeq r' \times \langle 1 \rangle \perp s' \times \langle -1 \rangle.$$

Let $e_1, \ldots, e_r, e_{r+1}, \ldots, e_n$ be the orthogonal basis corresponding to the first diagonalization, and let $e'_1, \ldots, e'_{r'}, e'_{r'+1}, \ldots, e'_n$ be the orthogonal basis corresponding to the second one. Let V_+ be the subspace of V generated by $e_1, \ldots, e_r$ and let V'_- be the subspace of V generated by $e'_{r'+1}, \ldots, e'_n$. By definition, we have $\dim V_+ = r$ and $\dim V'_- = n - r'$. It is easy to check that we have

$$q(x) > 0 \text{ for all } x \in V_+, x \neq 0$$

and that

$$q(x) < 0 \text{ for all } x \in V'_-, x \neq 0.$$

In particular, $V_+ \cap V'_- = \{0\}$. Now assume that $r \neq r'$, that is for example $r > r'$. In this case, we would have

$$\dim V_+ + \dim V'_- = r + n - r' > n,$$

and therefore $V_+ \cap V'_- \neq \{0\}$, a contradiction. Hence $r = r'$ and therefore $s = s'$ since $r + s = r' + s' = n$. □

Definition IV.12.2. Let q be a quadratic form over $\mathbb{R}$, and write

$$q \simeq r \times \langle 1 \rangle \perp s \times \langle -1 \rangle.$$

The **signature** of q is the integer $r - s \in \mathbb{Z}$, and is denoted by $\mathrm{sign}(q)$. If q is a quadratic form over $\mathbb{Q}$, the signature of q is by definition the signature of $q_{\mathbb{R}}$, and is still denoted by $\mathrm{sign}(q)$. We say that q is **indefinite** if we have

$$|\mathrm{sign}(q)| < \dim(q).$$

Remark IV.12.3. Notice that if $q \simeq q'$, then $q_{\mathbb{R}} \simeq q'_{\mathbb{R}}$, and by uniqueness of the integers r and s defined in the previous theorem, we get $\mathrm{sign}(q) = \mathrm{sign}(q')$. In particular, the signature only depends on the isomorphism class of a given quadratic form.

Remark IV.12.4. Given two quadratic forms q, q' over $\mathbb{Q}$ or $\mathbb{R}$, it is straightforward to check that we have

$$\begin{aligned} \mathrm{sign}(-q) &= -\mathrm{sign}(q), \\ \mathrm{sign}(q \perp q') &= \mathrm{sign}(q) + \mathrm{sign}(q'). \end{aligned}$$

We are now ready to state our classification theorem.

Theorem IV.12.5. *Two quadratic forms over $\mathbb{Q}$ are isomorphic if and only if they have same dimension, determinant, Hasse invariant and signature.*

We refer the reader to [53, Chapter 6, Corollary 6.6] for a proof. Notice that in [53] the Hasse invariant is defined in terms of Brauer classes of quaternion algebras. However, in view of the group isomorphism $\mathrm{Br}_2(k) \simeq H^2(k, \mu_2)$ for every field k, which maps the Brauer class of (a, b) onto $(a) \cup (b)$, we see that the theorem above is an equivalent formulation of the result proved in [53].

IV.12.2 Higher cohomological invariants

We now would like to briefly mention the existence of other invariants of quadratic forms with values in $H^d(\_, \mu_2)$.

Definition IV.12.6. A **cohomological invariant of degree** d of $\mathbf{F} : \mathfrak{C}_k \longrightarrow$ **Sets** with values in μ_2 is a natural transformation

$$\iota : \mathbf{F} \longrightarrow H^d(\_, \mu_2)$$

of functors from $\mathfrak{C}_k$ to **Sets**.

A cohomological invariant ι of degree d is **constant** if there exists $[\alpha] \in H^d(k, \mu_2)$ such that for all K/k and all $[\xi] \in H^1(K, G)$ we have

$$\iota_K([\xi]) = \mathrm{Res}_{K/k}([\alpha]).$$

If $\mathbf{F} : \mathfrak{C}_k \longrightarrow \mathbf{Sets}^*$, a cohomological invariant $\iota : \mathbf{F} \longrightarrow H^d(\_, \mu_2)$ is **normalized** if, for every field extension K/k, ι_K maps the base point of $\mathbf{F}(K)$ onto 0.

Example IV.12.7. The Hasse invariant is a cohomological invariant of degree 2 with values in μ_2, in view of Remark IV.11.9.

If K/k is a field extension, we define an equivalence relation on the set of regular quadratic spaces as follows: if q, q' are two regular quadratic forms on K, we say that q and q' are Witt equivalent if there exist two integers $r, r' \geq 0$ such that

$$q \perp r \times \langle 1, -1 \rangle \simeq q' \perp r' \times \langle 1, -1 \rangle.$$

We denote it by $q \sim q'$. We denote by $[q]$ the corresponding equivalence class. The set of Witt equivalence classes is denoted by $W(K)$. One can show that the operations

$$[q] + [q'] = [q \perp q'], [q] \cdot [q'] = [q \otimes q'] \text{ and } -[q] = [-q]$$

are well-defined, and induce a structure of a commutative ring on $W(K)$, with neutral elements [0] and $[\langle 1 \rangle]$ (see [53], for example).

It is then easy to check that the map

$$e_0 : \begin{array}{rcl} W(K) & \longrightarrow & \mathbb{Z}/2\mathbb{Z} \\ [q] & \longmapsto & \overline{\dim(q)} \end{array}$$

is a well-defined ring morphism. Its kernel $I(K)$ is called the **fundamental ideal** of $W(K)$. In other words, we have

$$I(K) = \{[q] \in W(K) \mid \dim(q) \text{ is even }\}.$$

If $n \geq 1$ is an integer, and $a_1, \ldots, a_n \in K^\times$, we set

$$\langle\langle a_1, \ldots, a_n \rangle\rangle = \langle 1, -a_1 \rangle \otimes \cdots \otimes \langle 1, -a_n \rangle.$$

Such a quadratic form is called a **n-fold Pfister form.**

Since $[\langle a, b \rangle] = [\langle\langle -a \rangle\rangle] - [\langle\langle b \rangle\rangle]$, $I(K)$ is additively generated by the classes of 1-fold Pfister forms, and its nth-power $I^n(K)$ is additively generated by the classes of n-fold Pfister forms for all $n \geq 1$. By convention, we will set $I^0(K) = W(K)$.

If $K \to L$ is a morphism of field extensions of k, the maps $W(K) \longrightarrow W(L), [q] \mapsto [q_L]$ and $I^n(K) \longrightarrow I^n(L), [q] \mapsto [q_L]$ are respectively well-defined ring and group morphisms, and we get functors

$$W(\_) : \mathfrak{C}_k \longrightarrow \mathbf{Rings} \text{ and } I^n(\_) : \mathfrak{C}_k \longrightarrow \mathbf{Abgrps}.$$

Let K/k be a field extension. For $n \geq 1$, and for $a_1, \ldots, a_n \in K^\times$, we set

$$e_n([\langle\langle a_1, \ldots, a_n \rangle\rangle]) = (a_1) \cup \cdots \cup (a_n) \in H^n(K, \mu_2).$$

Theorem IV.12.8. *For all $n \geq 1$, the map e_n extends in a unique way to a group morphism*

$$e_n : I^n(K) \longrightarrow H^n(K, \mu_2)$$

and induces a group isomorphism

$$I^n(K)/I^{n+1}(K) \simeq H^n(K, \mu_2).$$

This is an easy exercise for $n = 1$ and is well-known for $n = 2$ (see [53, Chapter 2, § 12] for more details). The case $n = 3$ has been proved by Arason in [1], and by Jacob and Rost in [27] for $n = 4$. For arbitrary n, this follows from the work of Voevodsky et al. (see [45], [66] and [43]). The proof of this theorem is extremely difficult and earned Voevodsky a Fields medal in 2002. In the next chapters, we will only need the existence of e_n for $n = 0, 1, 2$ and 3.

Remark IV.12.9. This theorem says in particular that we have

$$e_n(\lambda q) = e_n(q) \text{ for all } \lambda \in K^\times, q \in I^n(K).$$

Remark IV.12.10. The group morphisms e_n are easily seen to be compatible with scalar extension maps. In other words, we get comohological invariants

$$e_n : I^n(\_) \longrightarrow H^n(\_, \mu_2).$$

The isomorphism described in Theorem IV.12.8 shows that a class $[q] \in W(k)$ lies in $I^n(k)$ if and only if $e_j([q]) = 0$ for $j = 0, \dots, n-1$. This was proved for $n = 3$ by Merkurjev in [39].

Since $\bigcap\limits_{n \geq 1} I^n(k) = (0)$ (see [53] p.156), we get the following general classification theorem for quadratic forms:

Theorem IV.12.11. *Let k be a field of characteristic different from 2, and let q, q' be two regular quadratic forms over k. Then $q \simeq q'$ if and only if $e_n([q \perp -q']) = 0$ for all $n \geq 0$.*

We end this section by some elementary results on the invariants e_n.

Lemma IV.12.12. *Let k be a field of characteristic different from 2, and let $n, m \geq 0$ be two integers. If $[q] \in I^n(k)$ and $[q'] \in I^m(k)$, then $[q] \cdot [q'] \in I^{n+m}(k)$ and we have*

$$e_{n+m}([q] \cdot [q']) = e_n([q]) \cup e_m([q']).$$

Proof. Since e_n, e_m and e_{n+m} are group morphisms, a distributivity argument shows that it is enough to prove it when q and q' are Pfister forms. The result being clear in this case, we get the desired result. □

Lemma IV.12.13. *Let q be a regular quadratic form of dimension $2m$. Then we have*

$$e_1([q]) = ((-1)^m \det(q)) \in H^1(k, \mu_2).$$

Proof. Write $q \simeq \langle a_1, b_1, \dots, a_m, b_m \rangle$. For all $a, b \in k^\times$, we have

$$[\langle a, b \rangle] = [\langle\langle -a \rangle\rangle] - [\langle\langle b \rangle\rangle].$$

The result follows easily. □

Remark IV.12.14. The invariant e_n may be difficult to compute for $n \geq 2$, since an explicit decomposition of $[q]$ into sum of classes of the form $\pm[\langle\langle a_1, \ldots, a_n\rangle\rangle]$ is needed. However for $n = 2$, this is not necessary. Indeed, one may define another invariant attached to a quadratic form q as follows: if $n = \dim(q)$, set

$$c(q) = w_2(q) + \begin{cases} 0 & \text{if } n \equiv 1, 2 \mod 8 \\ (-1) \cup (-\det(q)) & \text{if } n \equiv 3, 4 \mod 8 \\ (-1) \cup (-1) & \text{if } n \equiv 5, 6 \mod 8 \\ (-1) \cup (\det(q)) & \text{if } n \equiv 7, 8 \mod 8. \end{cases}$$

The invariant $c(q)$ is called the **Clifford invariant** (or Witt invariant) of q. One can show that $c(q)$ only depends on the Witt class of q, and that its restriction to $I^2(k)$ induces a group morphism

$$c : I^2(k) \longrightarrow H^2(k, \mu_2).$$

We refer to [53, Chapter 2, p.81] for more details. Easy computations show that we have

$$c([\langle\langle a, b\rangle\rangle]) = (a) \cup (b).$$

Hence the restriction of c to I^2 is nothing but the invariant e_2.

Let L/k be a finite field extension and let $s : L \longrightarrow k$ be a non-zero k-linear map. If $z \in L$, recall that $N_{L/k}(z)$ is the determinant of the endomorphism of k-vector space

$$\ell_z : \begin{array}{l} L \longrightarrow L \\ x \longmapsto zx. \end{array}$$

If (V, q) is a regular quadratic space, the map

$$s_*(q) : \begin{array}{l} V \longrightarrow k \\ x \longmapsto s(q(x)) \end{array}$$

is a regular quadratic form over k (where V is viewed as a k-vector space). Moreover, this induces a well-defined group morphism

$$s_* : \begin{array}{l} W(L) \longrightarrow W(k) \\ [q] \longmapsto [s_*(q)]. \end{array}$$

See [53] for more details for example. We can now state the next result.

Lemma IV.12.15. *Let L/k be a finite extension, and let $z \in L^\times$. Then we have*

$$e_1(s_*(\langle 1, -z\rangle)) = (N_{L/k}(z)) \in H^1(k, \mu_2).$$

Proof. Let $v_1, \ldots, v_n$ be a k-basis of L, and let $B = (s(v_iv_j))$ the corresponding representative matrix of $s_*(\langle 1\rangle)$. Let $M = (m_{ij})$ be the matrix of ℓ_z in the basis $v_1, \ldots, v_n$. In other words, we have

$$zv_j = \sum_{i=1}^{n} m_{ij}v_i.$$

Notice that by definition, we have $\det(M) = N_{L/k}(z)$. We have

$$s(zv_iv_j) = \sum_{r=1}^{n} m_{ri}s(v_rv_j),$$

so the representative matrix of $s_*(\langle z\rangle)$ is M^tB. Hence, we get

$$\det(s_*(\langle z\rangle)) = \det(M)\det(B) = N_{L/k}(z)\det(s_*(\langle 1\rangle)).$$

If $[L : k] = m$, we get

$$\begin{aligned}\det(s_*(\langle 1, -z\rangle)) &= \det(s_*(\langle 1\rangle) \perp -s_*(\langle z\rangle)) \\ &= (-1)^m \det(s_*(\langle 1\rangle))\det(s_*(\langle z\rangle)) \\ &= (-1)^m N_{L/k}(z).\end{aligned}$$

Using Lemma IV.12.13, we get the desired result. □

Corollary IV.12.16. *Let $[q] \in I^n(k)$, and let L/k be a finite field extension such that $[q_L] = 0 \in W(L)$. Then for all $z \in L^\times$, we have*

$$(N_{L/k}(z)) \cup e_n(q) = 0 \in H^{n+1}(k, \mu_2).$$

Proof. Let $s : L \longrightarrow k$ be any non-zero linear map. Since $[q_L] = 0$, we have $[\langle 1, -z\rangle \otimes q_L] = [\langle 1, -z\rangle]\cdot[q_L] = 0$, and therefore $[s_*(\langle 1, -z\rangle \otimes q_L)] = 0$. It is easy to check using definitions that we have

$$s_*(\langle 1, -z\rangle \otimes q_L) \simeq s_*(\langle 1, -z\rangle) \otimes q.$$

Therefore, we get $e_{n+1}([s_*(\langle 1, -z\rangle)] \cdot [q]) = 0$. Using Lemma IV.12.12 and Lemma IV.12.15, we get the desired result. □

Exercises

1. Let q be a quadratic form over k. Show that the connecting map $\delta^0 : \mathbf{O}^+(q)(k) \longrightarrow H^1(k, \mu_2)$ associated to the exact sequence

$$1 \longrightarrow \mu_2(k_s) \longrightarrow \mathbf{Spin}(q)(k_s) \longrightarrow \mathbf{O}^+(q)(k_s) \longrightarrow 1$$

is given by the spinor norm SN.

2. Let q be a quadratic form over k.
   (a) Show that $H^1(k, \mathbf{Spin}(q)) = 1$ if and only if the spinor norm is surjective and every quadratic form q' satisfying
   $$\dim(q') = \dim(q), \det(q') = \det(q) \text{ and } w_2(q') = w_2(q)$$
   is isomorphic to q.
   (b) Show that the spinor norm is surjective in the following cases:
   (i) q is universal, i.e. q represents every element of $k^\times$
   (ii) q represents 0.
   (c) Assume that the spinor norm is surjective. Show that the map $H^1(k, \mathbf{Spin}(q)) \longrightarrow H^1(k, \mathbf{O}^+(q))$ has trivial kernel, but is not injective in general.

3. Prove Lemma IV.11.5.

4. Let $r \in \mathbb{Q}^{+*}$. By comparing their invariants, show that we have an isomorphism
$$\langle r, r, r, r \rangle \simeq \langle 1, 1, 1, 1 \rangle,$$
and deduce that every positive rational number is a sum of 4 squares in $\mathbb{Q}$.

5. Let (V, q) be a quadratic space. We keep the definitions of Chapter II, Exercise 7.
   (a) For $a \in k^\times$, show that
   $$X_a = \{x \in \mathbf{\Gamma}^+(V, q)(k_s) \mid N_{k_s}(x) = a\}$$
   is a principal homogeneous space over $\mathbf{Spin}(q)(k_s)$, whose isomorphism class only depends on the square-class of a.
   (b) Show that $X_a \simeq X_b$ if and only if $b \equiv a \mod k^{\times 2}$.
   (c) Show that the map $H^1(k, \mu_2) \longrightarrow H^1(k, \mathbf{Spin}(q))$ sends (a) onto the isomorphism class of X_a.

6. Describe $H^1(k, \mathbf{Spin}(q))$ in terms of twisted forms of tensors.

7. Describe the map $H^1(k, \mu_2^n) \longrightarrow H^1(k, \mathbf{O}_n)$ induced by the natural injection $\mu_2(k_s)^n \hookrightarrow \mathbf{O}_n(k_s)$.

V

Étale and Galois algebras

§V.13 Étale algebras

The goal of this section is to give an interpretation of separable field extensions in terms of Galois cohomology. First of all, notice that if E/k is a separable field extension, and if L/k is a field extension, E_L is not necessarily a field.

For example, if $k = \mathbb{R}, E = L = \mathbb{C}$, we have $E_L \simeq \mathbb{C} \times \mathbb{C}$. Hence, the collection of separable extensions is not a functor, so we have no chance to classify them using Galois cohomology. To do so, we need a broader class of algebras. The following result is really classical, and is proven in [8].

Theorem V.13.1. *For a finite dimensional commutative k-algebra E, the following conditions are equivalent:*

(1) $E \simeq E_1 \times \cdots \times E_r$, *where* $E_1, \ldots, E_r$ *are finite separable extensions of* k.

(2) $E_{k_s} \simeq k_s \times \cdots \times k_s$.

(3) $|X(E)| = \dim_k E$, *where* $X(E) = \mathrm{Hom}_{k-alg}(E, k_s)$.

If k is infinite, the conditions above are also equivalent to:

(4) $E \simeq k[X]/(f)$ *for some separable polynomial* $f \in k[X]$.

Definition V.13.2. A finite dimensional commutative k-algebra satisfying the equivalent conditions above is called **étale**.

Examples V.13.3.

(1) The k-algebra k^n is an étale algebra, called the **split** étale algebra.

(2) A finite separable extension E/k is an étale algebra.

It follows directly from the definition that if E is étale over K and $K \longrightarrow L$ is a morphism of field extensions, then E_L is an étale algebra over L. We denote by $\mathbf{\acute{E}t}_n : \mathfrak{C}_k \longrightarrow \mathbf{Sets}^*$ the functor of isomorphism classes of étale algebras of dimension n, the base point being the isomorphism class of the split étale algebra. Notice the twisted K-forms of K^n are by definition étale K-algebras of dimension n, so by Galois descent, étale algebras of dimension n are classified by $H^1(\_, \mathbf{Aut}_{alg}(k^n))$.

Lemma V.13.4. *For any field K, we have an isomorphism of abstract groups*

$$\mathrm{Aut}_{K-alg}(K^n) \simeq S_n.$$

Proof. If $f : K^n \longrightarrow K^n$ is an automorphism, then it maps an idempotent to an idempotent. Denote by e_i the element of K^n defined by $e_i = (\delta_{ij})_{1\leq j\leq n}$, where δ_{ij} is the Kronecker symbol. For $I \subset \{1, \ldots, n\}$, let $e_I = \sum_{i\in I} e_i$ (so $e_\emptyset = 0$). It is clear that the idempotents of K^n are the e_I's. From the relation $e_i e_j = \delta_{ij} e_i$, it follows that for two subsets I, J of $\{1, \ldots, n\}$, we have

$$e_I e_J = e_{I\cap J}.$$

Now let $f(e_i) = e_{I_i}$. Each I_i is non-empty since f is injective and $f(0) = 0$. Moreover, the previous relation implies that the subsets $I_i \subset \{1, \ldots, n\}$ are pairwise disjoint. Hence $I_1, \ldots, I_n$ is a partition of $\{1, \ldots, n\}$ and each I_i contains exactly one element, so f permutes the e_i's. Conversely, any permutation of the e_i's defines a K-algebra automorphism of K^n. It is then easy to check that we get a group isomorphism $S_n \simeq \mathrm{Aut}_{K-alg}(K^n)$. □

Now we have $\mathbf{Aut}_{alg}(k^n)(K_s) = \mathrm{Aut}_{K_s-alg}(K_s^n) \simeq S_n$; the reader will easily check that $\mathcal{G}_{K_s}$ acts trivially on $\mathrm{Aut}_{K_s-alg}(K_s^n)$, so using Proposition III.9.1 we deduce:

Proposition V.13.5. *We have an isomorphism of functors from $\mathfrak{C}_k$ to* **Sets***

$$\mathbf{\acute{E}t}_n \simeq H^1(\_, S_n),$$

where S_n is considered as a trivial $\mathcal{G}_{K_s}$-group for every field extension K/k.

We now take a closer look at this correspondence. By Remark III.9.2, a cocycle α whose class in $H^1(k, S_n)$ corresponds to a given étale k-algebra

E may be constructed as follows: take any isomorphism of k_s-algebras $f : E_{k_s} \xrightarrow{\sim} k_s^n$, and for every $\sigma \in \mathcal{G}_{k_s}$, set

$$\alpha_\sigma = f \circ \sigma \cdot f^{-1}.$$

However, we are going to see that in the case of étale k-algebras, we have a natural choice for f. For any $\chi \in X(E)$, we denote by $\ell_\chi : E_{k_s} \longrightarrow k_s$ the morphism of k_s-algebras defined by

$$\ell_\chi(u \otimes \lambda) = \chi(u)\lambda \text{ for all } \lambda \in k_s \text{ and all } u \in E.$$

Notice that the absolute Galois group $\mathcal{G}_{k_s}$ acts naturally on $X(E)$ by composition on the left.

Lemma V.13.6. *For all $\chi \in X(E)$ and all $\sigma \in \mathcal{G}_{k_s}$, we have*

$$\ell_{\sigma \cdot \chi}(\sigma \cdot x) = \sigma(\ell_\chi(x)) \textit{ for all } x \in E_{k_s}.$$

Proof. It is enough to prove the equality for $x = u \otimes \lambda, u \in E, \lambda \in k_s$. The equality follows in this case by direct computation. □

Lemma V.13.7. *Let E be an étale k-algebra and $f : E_{k_s} \longrightarrow k_s^{X(E)}$ the canonical map defined by*

$$f(x) = (\ell_\chi(x))_{\chi \in X(E)} \textit{ for all } x \in E_{k_s}.$$

Then f is an isomorphism of k_s-algebras.

Proof. It is clear that f is an morphism of k_s-algebras. In order to show that f is a bijection, it is enough to prove that f is injective. Now $\ker(f) = \bigcap_{\chi \in X(E)} \ker(\ell_\chi)$. It is well-known that this intersection is reduced to 0 if and only if the linear forms $\ell_\chi, \chi \in X(E)$ span the dual space $E^*_{k_s}$ of k_s-linear forms on E_{k_s}. Since we have

$$\dim_{k_s}(E^*_{k_s}) = \dim_{k_s}(E_{k_s}) = \dim_k(E) = |X(E)|,$$

this is equivalent to saying that the linear forms $\ell_\chi, \chi \in X(E)$ are linearly independent over k_s. Assume the contrary and consider a dependence relation of minimal length $r \geq 1$:

$$a_1\ell_{\chi_1} + \ldots + a_r\ell_{\chi_r} = 0,$$

where $\chi_1, \ldots, \chi_r$ are distinct elements of $X(E)$ and $a_1, \ldots, a_r$ are non-zero elements of k_s. Then we have necessarily $r \geq 2$. Let $u_0 \in E$ such that $\chi_1(u_0) \neq \chi_r(u_0)$ (such a u_0 exists since $\chi_1 \neq \chi_r$). Since the

ℓ_{χ_i}'s are algebra morphisms, applying $(u_0 \otimes 1)x$ for all $x \in E_{k_s}$ to this relation gives

$$a_1\chi_1(u_0)\ell_{\chi_1} + \ldots + a_r\chi_r(u_0)\ell_{\chi_r} = 0.$$

Multiplying by $\chi_r(u_0)$ each side of the initial dependence relation then yields

$$a_1\chi_r(u_0)\ell_{\chi_1} + \ldots + a_r\chi_r(u_0)\ell_{\chi_r} = 0.$$

Subtracting the two relations gives

$$a_1(\chi_1(u_0) - \chi_r(u_0))\ell_{\chi_2} + \ldots + a_{r-1}(\chi_{r-1}(u_0) - \chi_r(u_0))\ell_{\chi_{r-1}} = 0.$$

We then produce another dependence relation of length $< r$, which contradicts the minimality of r. □

Proposition V.13.8. *Let E be an étale k-algebra, $\dim_k(E) = n$, and let $\varphi : \mathcal{G}_{k_s} \longrightarrow S_n$ be a continuous group morphism representing E. Let $N \subset k_s$ be the compositum of the subfields $\chi(E)$ of k_s, $\chi \in X(E)$. Then N/k is a finite Galois extension and we have*

$$\ker(\varphi) = \mathrm{Gal}(k_s/N).$$

In particular, $\mathrm{Im}(\varphi) \simeq \mathcal{G}_N$.

Proof. Let us prove the first part. Assume that $E = E_1 \times \cdots \times E_r$, where E_i is a finite separable field extension of k. Clearly the map

$$\theta : X(E_1) \times \ldots \times X(E_r) \longrightarrow X(E)$$

defined by

$$\theta(\chi_1, \ldots, \chi_r)(x_1, \ldots, x_r) = (\chi_1(x_1), \ldots, \chi_r(x_r)) \text{ for all } x_i \in E_i$$

is bijective, with inverse map

$$\begin{aligned} X(E) &\longrightarrow X(E_1) \times \ldots \times X(E_r) \\ \chi &\longmapsto (\iota_1 \circ \chi, \ldots, \iota_r \circ \chi), \end{aligned}$$

where $\iota_i : E_i \longrightarrow E$ is the canonical injection. Let N_i be the compositum of all the subfields $\chi'(E_i)$, where χ' describes $X(E_i)$. Since E_i/k is a separable field extension, N_i is just the Galois closure of E_i in k_s. It then follows from the description of $X(E)$ that the compositum N of the subfields $\chi(E), \chi \in X(E)$ is the compositum of $N_1, \ldots, N_r$. Since the compositum of Galois extensions of k is a Galois extension of k, we are done.

Now set $X = X(E)$. Let $f : E_{k_s} \longrightarrow k_s^X$ be the canonical isomorphism of k_s-algebras defined in Lemma V.13.7, and denote by $\alpha : \mathcal{G}_{k_s} \longrightarrow \mathrm{Aut}_{k_s-alg}(k_s^X)$ the continuous morphism defined by

$$\alpha_\sigma = f \circ (\sigma \cdot f^{-1}), \text{ for all } \sigma \in \mathcal{G}_{k_s}.$$

Composing α on the left by an isomorphism $\mathrm{Aut}_{k_s-alg}(k_s^X) \simeq S_n$, we obtain a continuous morphism $\mathcal{G}_{k_s} \longrightarrow S_n$ which represents E. Since two morphisms representing E are conjugate, they have same kernel, so we have $\ker(\varphi) = \ker(\alpha)$. Hence it is enough to prove that $\ker(\alpha) = \mathrm{Gal}(k_s/N)$.

Let $\sigma \in \mathcal{G}_{k_s}$. Then $\sigma \in \ker(\alpha)$ if and only if $\sigma \cdot f = f$. Since these two maps are k_s-linear, this is equivalent to

$$(\sigma \cdot f)(u \otimes 1) = f(u \otimes 1) \text{ for all } u \in E.$$

Now we have

$$(\sigma \cdot f)(u \otimes 1) = \sigma \cdot (f(\sigma^{-1} \cdot (u \otimes 1))) = \sigma \cdot f(u \otimes 1).$$

By definition $f(u \otimes 1) = (\chi(u))_{\chi \in X}$, so we obtain that $\sigma \in \ker(\alpha)$ if and only if $\sigma \cdot \chi(u) = \chi(u)$ for all $u \in E$. Since N is the compositum of the subfields $\chi(E), \chi \in X$, this is equivalent to say that $\sigma \in \mathrm{Gal}(k_s/N)$. The last part of the proposition comes from the first isomorphism theorem and the isomorphism

$$\mathcal{G}_{k_s}/\mathrm{Gal}(k_s/N) \simeq \mathcal{G}_N.$$

This completes the proof. □

§V.14 Galois algebras

In this section, we would like to describe a cohomological interpretation of finite Galois extensions with a given Galois group G. As previously, the difficulty is that the category of Galois extensions is not closed under scalar extension, so we need first to generalize the notion of Galois extension in an appropriate way. The right notion appears to be that of a Galois G-algebra.

V.14.1 Definition and first properties

Recall that a G-algebra over k is a k-algebra on which G acts faithfully by k-algebra automorphisms. Notice that if L is a G-algebra over k, then

G acts naturally on $X(L) = \mathrm{Hom}_{k-alg}(L, k_s)$ on the right as follows: for all $\chi \in X(L)$, all $g \in G$ and all $u \in L$, we set

$$(\chi \cdot g)(u) = \chi(g \cdot u).$$

We start with a little lemma.

Lemma V.14.1. *Let G be an abstract finite group and let L/k be a G-algebra. Assume that L/k is a finite separable field extension and that $|G| = \dim_k(L)$. Then L/k is a Galois extension if and only if the natural right action of G on $X(L)$ is transitive. In this case, we have $\mathcal{G}_L \simeq G$.*

Proof. Notice first that, since L/k is separable of degree n, $X(L)$ has n elements, where $n = \dim_k L$. Let $\iota \in X(L)$ be the inclusion $L \subset k_s$, and consider the map

$$\rho\colon \begin{array}{l} \mathcal{G}_L \hookrightarrow X(L) \\ \sigma \longmapsto \iota \circ \sigma. \end{array}$$

Notice that by construction, the image of ρ is the set of morphisms $\chi \in X(L)$ satisfying $\chi(L) \subset L$.

Let us consider the group morphism

$$\varphi\colon \begin{array}{l} G \longrightarrow \mathcal{G}_L \\ g \longmapsto [L \longrightarrow L, x \longmapsto g \cdot x] \end{array}$$

induced by the action of G on L, and consider the map

$$\psi = \rho \circ \varphi : G \longrightarrow X(L).$$

By definition, we have

$$\psi(g)(x) = \iota(g \cdot x) \text{ for all } g \in G, x \in L.$$

In other words, we have $\psi(g) = \iota \cdot g$.

Assume first that G acts transitively on $X(L)$, that is ψ is surjective. Since $|G| = |X(L)|$ by assumption, ψ is then bijective and thus ρ is bijective as well. Therefore, $\chi(L) \subset L$ for all $\chi \in X(L)$ and L/k is a Galois extension. Now the bijectivity of ψ also implies the bijectivity of φ. Hence φ induces a group isomorphism

$$G \simeq \mathcal{G}_L.$$

Thus, we have proved that L/k is a Galois extension with Galois group isomorphic to G.

Conversely, assume that L/k is Galois. In this case, $|\mathcal{G}_L| = n = |G|$

and φ is an isomorphism. Now let $\chi, \chi' \in X(L)$. By assumption, there exist $\sigma, \sigma' \in \mathcal{G}_L$ such that $\chi = \iota \circ \sigma$ and $\chi' = \iota \circ \sigma'$. We then have $\chi' = \chi \circ (\sigma^{-1} \circ \sigma')$. Hence, $\mathcal{G}_L \simeq G$ acts transitively on $X(L)$. □

Definition V.14.2. Let G be an abstract group of order n, and let L be a G-algebra of dimension n over k. We say that L is a **Galois G-algebra** if L is étale and the right action of G on $X(L)$ is transitive.

Remark V.14.3. If L is a Galois G-algebra, then the action of G on $X(L)$ is simply transitive since $|G| = \dim_k L = |X(L)|$.

Remark V.14.4. In view of Lemma V.14.1, every Galois extension L/k is a Galois $\mathcal{G}_L$-algebra, and there is only one group G (up to isomorphism) for which L has a Galois G-algebra structure.

We continue by giving another standard example of a Galois G-algebra.

Example V.14.5. Let k be a field, and let G be a finite group of order n. Let us index the coordinates of the elements of k^n with the elements of G, and let $(e_g)_{g \in G}$ the corresponding n idempotents. If $g \in G$, let $f_g \in \mathrm{Aut}_{k-alg}(k^n)$ be the unique k-algebra automorphism such that

$$f_g(e_h) = e_{gh} \text{ for all } h \in G.$$

In other words, f_g is the automorphism

$$f_g \colon \begin{array}{rcl} L_0 & \longrightarrow & L_0 \\ (x_h)_{h \in G} & \longmapsto & (x_{g^{-1}h})_{h \in G}. \end{array}$$

It is easy to check the action of G on k^n given by

$$\begin{array}{rcl} G \times k^n & \longrightarrow & k^n \\ (g, x) & \longmapsto & f_g(x) \end{array}$$

endows k^n with a structure of a Galois G-algebra, that we will denote by L_0.

Lemma V.14.6. *Let k be a field. Then for any finite group G of order n, any structure of a G-Galois algebra on k^n is isomorphic to L_0.*

Proof. Assume that k^n is endowed with a structure of a Galois G-algebra, that we will denote by L. As before, let us index the coordinates of the elements of k^n with the elements of G, and let $(e_g)_{g \in G}$ be the corresponding n idempotents. Since G acts by k-algebra automorphisms, g permutes the idempotents $e_h, h \in G$ by Lemma V.13.4 and its proof. Let us prove that this action is simply transitive. For all $g \in G$, denote by π_g the projection on the gth-coordinate. The elements $\pi_g, g \in G$ are exactly the elements of $X(L)$. By assumption, for all $g \in G$, there exists an unique $g' \in G$ such that $\pi_g = \pi_1 g'$. Applying e_g to this equality yields $1 = \pi_1(g' \cdot e_g)$. Since $\pi_1(e_h) = \delta_h$ for all $h \in G$, we get that $g' \cdot e_g = e_1$, that is $e_g = g'^{-1} \cdot e_1$. This proves that the action of G permutes simply transitively the idempotents. In particular, we have a unique k-algebra automorphism $f : k^n \xrightarrow{\sim} k^n$ such that

$$f(e_g) = g \cdot e_1 \text{ for all } g \in G.$$

It is easy to check that f is G-equivariant. Thus f induces an isomorphism of G-algebras

$$L_0 \simeq_G L,$$

and this concludes the proof. □

Remark V.14.7. If L is a Galois G-algebra over k and K/k is any field extension, then it is straightforward to check that L_K is a Galois G-algebra over K, the action of G on L_K being defined by:

$$g \cdot (u \otimes \lambda) = (g \cdot u) \otimes \lambda \text{ for all } g \in G, u \in L, \lambda \in K.$$

We denote by $G-\mathbf{Gal} : \mathfrak{C}_k \longrightarrow \mathbf{Sets}^*$ the functor of isomorphism classes of Galois G-algebras, where the base point is the isomorphism class of the split Galois G-algebra.

Lemma V.14.8. *Let V be a k-vector space and let K/k be a field extension. Let G be a finite group acting on V linearly. Then we have*

$$(V_K)^G = (V^G)_K,$$

where the action of G is extended to V_K by K-linearity.

Proof. One inclusion is clear. To prove the other one, let $x = v_1 \otimes \lambda_1 + \ldots + v_r \otimes \lambda_r \in V_K$. One can always assume that $\lambda_1, \ldots, \lambda_r$ are linearly independent over k. In this case, the distributivity property of tensor product with respect to direct sums shows that for every $w_1, \ldots, w_r \in V$, we have

$$w_1 \otimes \lambda_1 + \ldots + w_r \otimes \lambda_r = 0 \Rightarrow w_1 = \ldots = w_r = 0.$$

Now assume that $x \in (V_{k'})^G$. For all $g \in G$, we get

$$0 = g \cdot x - x = (g \cdot v_1 - v_1) \otimes \lambda_1 + \ldots + (g \cdot v_r - v_r) \otimes \lambda_r.$$

We then get that $g \cdot v_i - v_i = 0$ for all g and all i, so $v_i \in V^G$ for all i and we are done. □

Theorem V.14.9. *Let k be a field. Let L be a commutative G-algebra of dimension n over k, and assume that $|G| = \dim_k(L)$. Then the following conditions are equivalent:*

(1) *L is a Galois G-algebra.*

(2) *L_{k_s} is G-isomorphic to the Galois G-algebra $(L_0)_{k_s}$.*

(3) *L is étale and $L^G = k$.*

Proof. (1) $\Rightarrow$ (2) Assume that L is a Galois G-algebra. Then L_{k_s} is a Galois G-algebra over k_s. Since L is étale, $L_{k_s} \simeq k_s^n$ as a k_s-algebra. Now transporting the action of G to k_s^n endows k_s^n with a structure of Galois G-algebra, which is isomorphic to the Galois G-algebra $(L_0)_{k_s}$ by Lemma V.14.6. We then get an isomorphism $L_{k_s} \simeq_G (L_0)_{k_s}$.

(2) $\Rightarrow$ (1) If L_{k_s} is isomorphic to the Galois G-algebra $(L_0)_{k_s}$, then in particular L is an étale algebra of rank n. Now we have to check that G acts transitively on $X(L)$. Let $\varphi, \varphi' \in X(L)$. By assumption, L_{k_s} is a Galois G-algebra, so there exists $g \in G$ such that

$$\varphi_{k_s} \cdot g = \varphi'_{k_s}.$$

Applying this equality to $u \otimes 1, u \in L$, we get $\varphi(g \cdot u) = \varphi'(u)$ for all $u \in L$, that is $\varphi \cdot g = \varphi'$, so the action of G on $X(L)$ is transitive.

(2) $\Rightarrow$ (3) If $\psi : L_{k_s} \xrightarrow{\sim} (L_0)_{k_s}$ is an isomorphism of G-algebras, then L is étale. Since G acts by automorphisms of k-algebras, we have $k \subset L^G$. Now let $u \in L^G$. Then $x = \psi(u \otimes 1) \in ((L_0)_{k_s})^G$, since ψ is compatible with the G-actions. Consider $e_1 = (1, 0, \ldots, 0) \in (L_0)_{k_s}$. Since the action of G on $(L_0)_{k_s}$ permutes transitively $e_1, \ldots, e_n$, any element x may be written in a unique way as $x = \sum_{g \in G} x_g (g \cdot e_1)$, for $x_g \in k_s$. For every $g' \in G$, we have

$$g' \cdot x = \sum_{g \in G} x_g (g' g \cdot e_1) = \sum_{g \in G} x_{g'^{-1} g} (g \cdot e_1).$$

Since $g' \cdot x = x$, comparing the first components yields $x_{g'^{-1}} = x_1$ for all $g' \in G$. Hence $x = \lambda 1_{(L_0)_{k_s}}$ for some $\lambda \in k_s$. This implies that $u \otimes 1 = 1 \otimes \lambda$. Now applying $\sigma \in \mathcal{G}_{k_s}$ gives $u \otimes 1 = 1 \otimes (\sigma \cdot \lambda)$. Hence $\lambda = \sigma \cdot \lambda$ for all $\sigma \in \mathcal{G}_{k_s}$, so $\lambda \in k$. We then get $u \otimes 1 = 1 \otimes \lambda = \lambda \otimes 1$, so $u = \lambda \in k$.

(3) $\Rightarrow$ (2) Since L is étale, we have an isomorphism $\psi : L_{k_s} \xrightarrow{\sim} k_s^n$ of k_s-algebras. Now transporting the action of G on k_s^n endows k_s^n with

a structure of G-algebra, and ψ is then G-equivariant with respect to the G-actions. We are going to prove that the action of G permutes transitively the idempotents $e_1, \ldots, e_n$. Assume the contrary, and let $x = \sum_{g \in G} g \cdot e_1 \in k_s^n$. By assumption, x does not have all its coordinates equal, and hence $x \notin k_s 1_{k_s^n}$. Notice that by construction, we have $x \in (k_s^n)^G$. Therefore, to obtain the desired contradiction, it is enough to show that $(k_s^n)^G = k_s 1_{k_s^n}$. Since the isomorphism ψ is G-equivariant, it is enough to prove that $(L_{k_s})^G = k_s(1 \otimes 1)$. In view of the assumption on L, it remains to prove that $(L_{k_s})^G = (L^G)_{k_s}$, which follows from the previous lemma. This concludes the proof. □

We now explain how to construct Galois G-algebras from Galois H-algebras, where H is a subgroup of G.

Let G be a finite group, and let H be a subgroup of G. For any Galois H-algebra M over k, we set

$$\mathrm{Ind}_H^G M = \{f \in Map(G, M) \mid f(hg) = h \cdot f(g) \text{ for all } h \in H, g \in G\}.$$

Notice that G acts on $\mathrm{Ind}_H^G M$ by

$$(g \cdot f)(g') = f(g'g) \text{ for all } g, g' \in G, f \in \mathrm{Ind}_H^G M.$$

Lemma V.14.10. *The k-algebra $\mathrm{Ind}_H^G M$ is a Galois G-algebra.*

Proof. Let us prove first that $\mathrm{Ind}_H^G M$ is an étale k-algebra of dimension $|G|$. Let $Hg_1, \ldots, Hg_r$ the r right cosets of G modulo H. Clearly, any $f \in \mathrm{Ind}_H^G M$ is uniquely determined by its values on $g_1, \ldots, g_r$, and therefore we get an isomorphism

$$\varphi\colon \begin{array}{l} \mathrm{Ind}_H^G M \xrightarrow{\sim} M^r \\ \qquad f \longmapsto (f(g_1), \ldots, f(g_r)). \end{array}$$

Since M is étale, M^r is étale too and so is $\mathrm{Ind}_H^G M$. Now we have

$$\dim_k \mathrm{Ind}_H^G M = r \dim_k M = r|H| = |G|.$$

We now prove that $\mathrm{Ind}_H^G M$ is a G-algebra. It is clear that G acts by k-algebra automorphisms. Let us prove that the action of G is faithful. Let $g \in G, g \neq 1$. We have to show that there exists $f \in \mathrm{Ind}_H^G M$ such that $g \cdot f \neq f$. It is enough to find some $f \in \mathrm{Ind}_H^G M$ satisfying $f(g) \neq f(1)$.

Assume first that $g \notin H$. Then g and 1 represents two different right cosets of G modulo H. In view of the isomorphism above, one can easily construct an element $f \in \mathrm{Ind}_H^G M$ such that $f(g) \neq f(1)$. If now $g \in H$, since H acts faithfully on M, there exists $x \in M$ such that $h \cdot x \neq x$. Now let $f \in \mathrm{Ind}_H^G M$ such that $f(1) = x$. Then we have

$$f(g) = g \cdot f(1) = g \cdot x \neq x = f(1).$$

It remains to prove that the G-algebra $\mathrm{Ind}_H^G M$ is Galois. By Theorem V.14.9, it remains to check that $(\mathrm{Ind}_H^G M)^G = k$. If $f \in (\mathrm{Ind}_H^G M)^G$, then for all $g, g' \in G$, we get $f(g') = (g.f)(g') = f(g'g)$, and therefore $f(g) = f(1)$ for all $g \in G$. But we have $f(1) = f(h) = h \cdot f(1)$ for all $h \in H$, so $f(1) \in M^H = k$. This concludes the proof. □

Definition V.14.11. The Galois G-algebra of $\mathrm{Ind}_H^G M$ is called the **Galois G-algebra induced by M.**

V.14.2 Galois algebras and Galois cohomology

We now relate Galois G-algebras and cohomology. By Theorem V.14.9, the G-algebras which become isomorphic to the G-algebra L_0 over k_s are exactly the Galois G-algebras over k. To use Galois descent, we need to determine $\mathbf{Aut}_{G-alg}(L_0)$.

Lemma V.14.12. *For every field extension K/k, we have an isomorphism of abstract groups*

$$\mathrm{Aut}_{G-alg}((L_0)_K) \simeq G.$$

Proof. With any loss of generality, we may assume that $K = k$. For any $g \in G$, let φ_g be the unique automorphism of k-algebras such that

$$\varphi_g(e_h) = e_{hg^{-1}} \text{ for all } h \in G.$$

It is easy to check that φ_g is an automorphism of G-algebras, and that the map

$$\varphi\colon \begin{array}{l} G \longrightarrow \mathrm{Aut}_{G-alg}(L_0) \\ g \longmapsto \varphi_g \end{array}$$

is a group morphism. Let us prove the bijectivity of φ. If $g \in G$ lies in $\ker(\varphi)$, then we have $e_{hg^{-1}} = e_h$ for all $h \in G$, which is only possible if $g = 1$. Thus φ is injective. We now prove its surjectivity. If $f \in \mathrm{Aut}_{G-alg}(L_0)$, then $f(e_1)$ is an idempotent of L_0, say e_g. Since f is G-equivariant we then get

$$f(e_h) = f(h \cdot e_1) = h \cdot f(e_1) = h \cdot e_g = e_{hg} \text{ for all } h \in G.$$

Therefore $f = \varphi_{g^{-1}} = \varphi(g^{-1})$ and φ is surjective. This concludes the proof. □

In particular, we get a group isomorphism

$$\mathrm{Aut}_{G-alg}((L_0)_{K_s}) \simeq G.$$

One may easily check that this is an isomorphism of $\mathcal{G}_{K_s}$-groups, where the action of $\mathcal{G}_{K_s}$ on G is trivial. By Proposition III.9.7, we then get:

Proposition V.14.13. *Let G be a finite abstract group. We have an isomorphism of functors from $\mathfrak{C}_k$ to* **Sets**

$$G - \mathbf{Gal} \simeq H^1(\_, G),$$

where G is considered as a trivial $\mathcal{G}_{K_s}$-group for every field extension K/k.

Example V.14.14. If $\mathrm{char}(k) \nmid n$ and k contains a primitive n^{th}-root of 1, then $\mu_n(k_s) \simeq \mathbb{Z}/n\mathbb{Z}$ as $\mathcal{G}_{k_s}$-modules, and therefore, we get

$$H^1(k, \mathbb{Z}/n\mathbb{Z}) \simeq k^\times/k^{\times n},$$

using Proposition III.8.27. This isomorphism is just a cohomological translation of Kummer theory of cyclic extensions in the presence of roots of unity.

Let L be a Galois G-algebra over k and $\varphi : \mathcal{G}_{k_s} \longrightarrow G$ a continuous group morphism representing L. Since φ is continuous, $\ker(\varphi)$ is a closed normal subgroup of $\mathcal{G}_{k_s}$. The Galois correspondence then shows the existence of a finite Galois extension M/k such that $\mathrm{Gal}(k_s/M) = \ker(\varphi)$.

Lemma V.14.15. *With the previous notation, for all $\chi \in X(L)$, we have $\chi(L) = M$.*

Proof. Since G acts transitively on $X(L)$, we have $\xi(L) = \chi(L)$ for all $\xi, \chi \in X(L)$. Now applying Proposition V.13.8, we get that $\ker(\varphi) = \mathrm{Gal}(k_s/\chi(L))$, hence $\chi(L) = M$ by Galois correspondence. □

Let L be a Galois G-algebra over k. Set $X = X(L)$ and let us fix $\chi_0 \in X$. Consider the k_s-algebra k_s^X, and let us index the components of $x \in k_s^X$ using elements of X. In particular, $e_\chi, \chi \in X$ will denote the canonical basis of k_s^X as a k-vector space (that is the primitive idempotents).

We now define an action of G by k_s-automorphisms on k_s^X as follows:

$$g \cdot e_\chi = e_{\chi \cdot g^{-1}} \text{ for all } g \in G, \chi \in X.$$

Since G permutes transitively the elements of X, G permutes transitively the e_χ's, $\chi \in X$, so we obtain a structure of Galois G-algebra (which is the only possible one up to isomorphism). We will assume from now on that k_s^X is endowed with this action of G.

Let $f : L_{k_s} \xrightarrow{\sim} k_s^X$ be the canonical isomorphism of k_s-algebras defined in Lemma V.13.7. We claim that f is an isomorphism of Galois G-algebras. Indeed we have

$$f(u \otimes \lambda) = \sum_\chi \chi(u)\lambda e_\chi,$$

and therefore

$$g\cdot(f(u \otimes \lambda)) = \sum_\chi \chi(u)\lambda e_{\chi g^{-1}} = \sum_\chi (\chi\cdot g)(u)\lambda e_\chi.$$

Hence we have

$$f(g\cdot(u \otimes \lambda)) = f(g(u) \otimes \lambda) = \sum_\chi (\chi\cdot g)(u)\lambda e_\chi = g\cdot(f(u \otimes \lambda)),$$

proving the claim. Let $\alpha : \mathcal{G}_{k_s} \longrightarrow \mathrm{Aut}_{G-alg}(k_s^X)$ be the continuous group morphism defined by

$$\alpha_\sigma = f \circ (\sigma\cdot f^{-1}) \text{ for all } \sigma \in \mathcal{G}_{k_s}.$$

We identify $\mathrm{Aut}_{G-alg}(k_s^X)$ to G via the group isomorphism $\eta : G \longrightarrow \mathrm{Aut}_{G-alg}(k_s^X)$ defined by:

$$\eta(g)(e_{\chi_0 h}) = e_{\chi_0\cdot(gh)} \text{ for all } h \in G.$$

Then $\varphi = \eta^{-1} \circ \alpha$ represents the Galois G-algebra L by Galois descent.

We are now going to give a more natural description of this cocycle. Since G acts simply transitively on X on the right, we define a continuous group morphism $\psi : \mathcal{G}_{k_s} \longrightarrow G$ by mapping $\sigma \in \mathcal{G}_{k_s}$ onto the unique element $\psi_\sigma \in G$ satisfying

$$\sigma\cdot\chi_0 = \chi_0\cdot\psi_\sigma.$$

Lemma V.14.16. *Keeping the notation above, we have $\psi = \varphi$.*

Proof. We first show the relation

$$\alpha_\sigma(e_\chi) = e_{\sigma\cdot\chi} \text{ for all } \sigma \in \mathcal{G}_{k_s} \text{ and } \chi \in X.$$

For $\chi \in X$, let $e'_\chi = f^{-1}(e_\chi)$. Then the $e'_\chi, \chi \in X$ form a basis of the

k_s-vector space L_{k_s}, whose dual basis is $\ell_\chi, \chi \in X$. Indeed, by definition of f, we have

$$f(e'_\xi) = (\ell_\chi(e'_\xi))_{\chi\in X} \text{ for all } \xi \in X.$$

Since we have $f(e'_\xi) = e_\xi$ by definition of e'_ξ, we get that $\ell_\chi(e'_\xi) = \delta_{\chi,\xi}$ for all $\chi, \xi \in X$.

Let $\sigma \in \mathcal{G}_{k_s}$ and $\chi \in X$. We have $\sigma \cdot e'_\chi = e'_{\sigma\cdot\chi}$. This results from the relation $\ell_\xi(\sigma{\cdot}e'_\chi) = \sigma(\ell_{\sigma^{-1}\xi}(e'_\chi)) = \delta_{\xi,\sigma\cdot\chi}$, for all $\xi \in X$, the first equality following from Lemma V.13.6. Therefore, we get

$$\alpha_\sigma(e_\chi) = f(\sigma{\cdot}f^{-1}(\sigma^{-1}{\cdot}e_\chi)) = f(\sigma{\cdot}f^{-1}(e_\chi)) = f(\sigma{\cdot}e'_\chi) = f(e'_{\sigma\chi}) = e_{\sigma\chi}.$$

From this, we get

$$\alpha_\sigma(e_{\chi_0 g}) = e_{\sigma(\chi_0 g)} = e_{(\sigma\chi_0)g} \text{ for all } g \in G.$$

By definition of ψ, we have $\sigma\cdot\chi_0 = \chi_0\cdot\psi_\sigma$. We then get

$$\alpha_\sigma(e_{\chi_0\cdot g}) = e_{(\chi_0\cdot\psi_\sigma)g} = e_{\chi_0\cdot(\psi_\sigma g)} = \eta(\psi_\sigma)(e_{\chi_0\cdot g}) \text{ for all } g \in G.$$

Therefore, $\alpha = \eta \circ \psi$, that is $\psi = \varphi$, and this concludes the proof. □

We thus have proved the following theorem:

Theorem V.14.17. *Let L be a Galois G-algebra over k, and let $\chi_0 \in X(L)$. For all $\sigma \in \mathcal{G}_{k_s}$, let $\psi_\sigma \in G$ be the unique element of G satisfying*

$$\sigma\cdot\chi_0 = \chi_0\cdot\psi_\sigma.$$

Then the continuous morphism $\psi : \mathcal{G}_{k_s} \longrightarrow G$ represents the Galois G-algebra L.

Let M/k be a finite Galois extension. For every $\sigma \in \mathcal{G}_{k_s}$, $\sigma_{|M}$ is a k-embedding of M into k_s, so its image is equal to M, since M is a Galois extension. Therefore, one can write $\sigma_{|M} = \iota \circ s_\sigma$ for some $s_\sigma \in \mathcal{G}_M$, where ι is the inclusion $M \subset k_s$.

Lemma V.14.18. *The continuous group morphism*

$$s : \mathcal{G}_{k_s} \longrightarrow \mathcal{G}_M$$

is surjective and represents the Galois $\mathcal{G}_M$-algebra M.

Proof. The surjectivity of s comes from the fact that M/k is a Galois field extension. To prove the rest of the lemma, notice that the inclusion ι is an element of $X(M)$. Now for all $\sigma \in \mathcal{G}_{k_s}$, we have $\sigma{\cdot}\iota = \sigma_{|M} = \iota{\cdot}s_\sigma$. The result then follows from the previous theorem. □

We continue by giving a cohomological interpretation of the induced Galois algebra.

Proposition V.14.19. *Let G be a finite abstract group on which $\mathcal{G}_{k_s}$ acts trivially, and let $H \subset G$ be a subgroup. The map $H^1(k,H) \longrightarrow H^1(k,G)$ induced by the inclusion $H \subset G$ corresponds to $M \longmapsto \mathrm{Ind}_H^G M$.*

Proof. Let M be a Galois H-algebra over k, and let $L = \mathrm{Ind}_H^G(M)$. Let $\chi_0 \in X(M)$, and let ψ be the continuous morphism corresponding to M by Theorem V.14.17. Then the map

$$\chi_0' : \begin{array}{rcl} \mathrm{Ind}_H^G M & \longrightarrow & k_s \\ f & \longmapsto & \chi_0(f(1)) \end{array}$$

is an element of $X(L)$. We are going to show that ψ satisfies

$$\sigma \cdot \chi_0' = \chi_0' \cdot \psi_\sigma \text{ for all } \sigma \in \mathcal{G}_{k_s},$$

which will prove the desired result in view of Theorem V.14.17.

For all $\sigma \in \mathcal{G}_{k_s}$ and all $f \in L$, we have

$$\sigma \cdot \chi_0'(f) = \sigma(\chi_0(f(1))) = (\sigma \cdot \chi_0)(f(1)) = (\chi_0 \cdot \psi_\sigma)(f(1)).$$

Since $f \in L$ and $\psi_\sigma \in H$, we get

$$\sigma \cdot \chi_0'(f) = \chi_0(\psi_\sigma \cdot f(1)) = \chi_0(f(\psi_\sigma)).$$

Now we have

$$\chi_0' \cdot \psi_\sigma(f) = \chi_0'(\psi_\sigma \cdot f) = \chi_0((\psi_\sigma \cdot f)(1)).$$

By definition of the action of G on L, we get

$$\chi_0' \cdot \psi_\sigma(f) = \chi_0(f(\psi_\sigma)) = \chi_0'(\psi_\sigma),$$

and this concludes the proof. $\square$

We are now able to prove that every Galois G-algebra is isomorphic to an induced Galois algebra.

Theorem V.14.20. *Let L be a G-Galois algebra over k, and let $\alpha : \mathcal{G}_{k_s} \longrightarrow G$ be a continuous morphism representing L. Then*

$$L \simeq_G \mathrm{Ind}_H^G M,$$

where $M = \chi(L)$ for any $\chi \in X(L)$ and $H = \mathrm{Im}(\alpha) \simeq \mathcal{G}_M$.

Proof. Let M/k be the finite Galois field extension satisfying

$$\ker(\alpha) = \mathrm{Gal}(\mathcal{G}_{k_s}/M).$$

The first isomorphism theorem shows that $H = \mathrm{Im}(\alpha) \simeq \mathrm{Gal}(M/k)$. Lemma V.14.18 and Proposition V.14.19 then show that $L \simeq_G \mathrm{Ind}_H^G M$. □

Corollary V.14.21. *Let L be a Galois G-algebra L, and let $\alpha : \mathcal{G}_{k_s} \longrightarrow G$ be any continuous morphism representing L. Then L is a field if and only if the corresponding continuous morphism α is surjective, and L is isomorphic to k^n if and only if α is trivial.*

Proof. By the previous theorem, L is a field if and only if $[G : H] = 1$, that is if and only if any continuous morphism representing L is surjective. Similarly, L is isomorphic to L_0 if and only if $[G : H] = n$, that is if and only if $H = \{1\}$, or equivalently if and only if any continuous morphism representing L is trivial (this can also be deduced from Galois descent). □

We now come back to étale algebras. Notice that $H^1(k, S_n)$ has two different interpretations. A cohomology class corresponds on the one hand to an étale k-algebra E, and to a Galois S_n-algebra L over k on the other hand. The following proposition describes the relation between E and L when E is a field.

Proposition V.14.22. *Let E/k be a separable field extension of degree n, let E^{gal}/k be its Galois closure, and let $e : \mathcal{G}_{k_s} \longrightarrow S_n$ be a continuous morphism corresponding to E. Then $\mathcal{G}_{k_s} \longrightarrow \mathrm{Im}(e)$ represents the Galois extension E^{gal}/k. In particular, the S_n-Galois algebra represented by e is $\mathrm{Ind}_{G_E}^{S_n} E^{gal}$, where $G_E = \mathrm{Im}(e)$ is the Galois group of E^{gal}/k.*

Proof. Apply Proposition V.13.8 and Proposition V.14.19. □

Let L be a G-algebra. If H is a normal subgroup of G, then L^H is a G/H-algebra for the action of G/H defined by

$$gH\cdot u = g\cdot u \text{ for all } gH \in G/H \text{ and all } u \in L^H.$$

We then have the following proposition.

Proposition V.14.23. *Let $1 \longrightarrow H \longrightarrow G \xrightarrow{\pi} N \longrightarrow 1$ be an exact sequence of abstract groups. Then for any Galois G-algebra L over k, the k-algebra L^H is a Galois N-algebra. Moreover, the induced map $\pi_* : H^1(k, G) \longrightarrow H^1(k, N)$ corresponds to $L \longmapsto L^H$.*

Proof. Let L be a Galois G-algebra of dimension n over k, and let $f : L_{k_s} \xrightarrow{\sim} k_s^n$ be an isomorphism of G-algebras. Then f induces an isomorphism of N-algebras $f : (L_{k_s})^H \xrightarrow{\sim} (k_s^n)^H$, and therefore an isomorphism $f : (L^H)_{k_s} \xrightarrow{\sim} (k_s^n)^H$ by Lemma V.14.8.

Let $r = [G : H]$. We will index the components of an element of $(k_s)^r$ using the elements of N, and the components of an element of $(k_s)^n$ using the elements of G, and we let act G and N on $(k_s)^n$ and $(k_s)^r$ respectively by left translation on the indices. It is then easy to show that

$$(k_s^n)^H = \{(x_g)_{g\in G} \mid x_{gh} = x_g \text{ for all } g \in G, h \in H\}.$$

Therefore, we get an isomorphism of N-algebras

$$\theta : k_s^r \xrightarrow{\sim} (k_s^n)^H.$$

Composing with f gives rise to an isomorphism of N-algebras

$$f' : (L^H)_{k_s} \xrightarrow{\sim} k_s^r,$$

so L^H is a Galois N-algebra. To prove the last part of the proposition, let $\chi_0 \in X(L)$, and let ψ be the corresponding continuous morphism described in Theorem V.14.17. The map $\chi_0' = (\chi_0)_{|L^H}$ is an element of $X(L^H)$. Moreover, for all $\sigma \in \mathcal{G}_{k_s}$ and all $u \in L^H$, we have

$$\sigma{\cdot}\chi_0'(u) = \sigma(\chi_0(u)) = \chi_0(\psi_\sigma{\cdot}u) = \chi_0(\pi(\psi_\sigma){\cdot}u) = \chi_0'(\pi_*(\psi)_\sigma{\cdot}u).$$

Now use Theorem V.14.17 to conclude. □

We finish this section by giving an interpretation of the map

$$H^1(k, G) \longrightarrow H^1(k, S_n)$$

induced by the inclusion $G \subset S_n$.

Lemma V.14.24. *Let $G \subset S_n$ be a group of order n. Then the map $H^1(k, G) \longrightarrow H^1(k, S_n)$ induced by the inclusion maps the isomorphism class of a Galois G-algebra L onto the isomorphism class of the étale algebra L/k.*

Proof. An isomorphism $f : L_{k_s} \longrightarrow (L_0)_{k_s}$ is also an isomorphism of étale algebras $f : L_{k_s} \longrightarrow k_s^n$. Now use Galois descent to conclude. □

Exercises

1. Let k be a field of characteristic p, where p is a prime number. Let

$$\wp : \begin{array}{l} k \longrightarrow k \\ x \longmapsto x^p - x. \end{array}$$

   (a) Check that $\wp$ is a group morphism.

   (b) Show that $H^1(k, \mathbb{Z}/p\mathbb{Z}) \simeq k/\wp(k)$.

   (c) Given a class $\overline{a} \in k/\wp(k)$, show that the Galois $\mathbb{Z}/p\mathbb{Z}$-algebra corresponding to $\overline{a}$ under the previous isomorphism is isomorphic to $k[X]/(X^p - X - a)$.

2. Let k be a field of characteristic different from 2.

   (a) Show that the first connecting map associated to the exact sequence

$$1 \longrightarrow \mathbb{Z}/2\mathbb{Z} \longrightarrow \mathbb{Z}/4\mathbb{Z} \longrightarrow \mathbb{Z}/2\mathbb{Z} \longrightarrow 1$$

   maps a square class (d) onto $(-1) \cup (d)$.

   (b) Deduce that a quadratic étale k-algebra $k[\sqrt{d}]/k$ can be embedded in a Galois $\mathbb{Z}/4\mathbb{Z}$-algebra L/k such that $L^{\mathbb{Z}/2\mathbb{Z}} = k[\sqrt{d}]$ if and only if d is a sum of two squares of k.

3. Let G be a finite group, let H, H' be two subgroups of G, and let M/k and M'/k be Galois field extensions with Galois groups isomorphic to H and H' respectively. Show that if $\mathrm{Ind}_H^G M$ and $\mathrm{Ind}_{H'}^G M'$ are G-isomorphic, then M and M' are k-isomorphic and $H' = gHg^{-1}$ for some $g \in G$.

4. Let K/k be a field extension. For all $\sigma \in S_n$, we will denote by $f_\sigma : K^n \to K^n$ the K-algebra automorphism defined by

$$f_\sigma(x_1, \ldots, x_n) = (x_{\sigma^{-1}(1)}, \ldots, x_{\sigma^{-1}(n)}), \text{ for all } x_1, \ldots, x_n \in K.$$

   Let E, E' be two étale K-algebras, let $\varphi \in \mathbf{Aut}_{K-alg}(E)$ and let $\varphi' \in \mathbf{Aut}_{K-alg}(E')$. A morphism $f : (E, \varphi) \to (E', \varphi')$ is a morphism of K-algebras such that $f \circ \varphi = \varphi' \circ f$.

   (a) For every pair (E, φ), where $\dim_K(E) = n$, show that there exists a permutation $\sigma \in S_n$, unique up to conjugation, such that

$$(E_{K_s}, \varphi \otimes \mathrm{Id}_{K_s}) \simeq (K_s^n, f_\sigma).$$

We will say that (E, φ) is an étale K-algebra of type σ.

(b) Let $C(\sigma)$ be the centralizer of σ in S_n.

Show that $H^1(\_, C(\sigma))$ classifies isomorphism classes of étale algebras of dimension n of type σ.

5. Let L/k be a Galois field extension with Galois group G. For any subgroup H of G, prove that we have an isomorphism of Galois G-algebras

$$\mathrm{Ind}_H^G L \simeq_G L \otimes_k L^H,$$

where L is considered as a Galois H-algebra over L^H on the left-hand side.

Hint: Write the corresponding cocycles.

VI

Group extensions, Galois embedding problems and Galois cohomology

In this chapter, we give an interpretation of the cohomology group $H^2(G, A)$ in terms of group extensions. We then use this interpretation to give a complete obstruction of a Galois embedding problem in terms of Galois cohomology.

§VI.15 Group extensions

Let E be a finite group, and let A be an abelian normal subgroup of E. We set $G = E/A$, so we have an exact sequence

$$1 \longrightarrow A \longrightarrow E \longrightarrow G \longrightarrow 1.$$

Since A is a normal subgroup of E, E acts on A by conjugation

$$\begin{aligned} E \times A &\longrightarrow A \\ (x, a) &\longmapsto x \cdot a = \mathrm{Int}(x)(a) = xax^{-1}, \end{aligned}$$

so we have a group morphism

$$\mathrm{Int}\colon \begin{array}{l} E \longrightarrow \mathrm{Aut}(A) \\ x \longmapsto \mathrm{Int}(x)_{|A}. \end{array}$$

Morever, if $x \in A$, then $x \cdot a = a$ for all $a \in A$, since A is abelian. Therefore, this action of E on A induces an action of G on A by group automorphisms, defined by

$$\begin{aligned} G \times A &\longrightarrow A \\ (\pi(x), a) &\longmapsto \pi(x) \cdot a = xax^{-1} \end{aligned}$$

where $\pi : E \longrightarrow G$ denotes the canonical projection. We denote by $\varphi : G \longrightarrow \mathrm{Aut}(A)$ the corresponding group morphism. Notice that we have $\varphi \circ \pi = \mathrm{Int}$ by definition. This motivates the following definition.

Definition VI.15.1. Let G, A be abstract finite groups, with A abelian, and assume that G acts by group automorphisms on A, via the group morphism $\varphi : G \longrightarrow \mathrm{Aut}(A)$. In other words, A is a G-module.

A **group extension** of G by A, or more precisely a group extension of (G, A, φ), is an exact sequence

$$1 \longrightarrow A \xrightarrow{\iota} E \xrightarrow{\pi} G \longrightarrow 1$$

such that the action of E by inner automorphisms on A extends the action of G on A. More precisely, we require:

$$x\iota(a)x^{-1} = \iota(\varphi(\pi(x))(a)) \text{ for all } x \in E, a \in A.$$

Remark VI.15.2. If no confusion is possible, the action of G on A will be denoted by $G \times A \longrightarrow (g, a) \mapsto g{\cdot}a$. The previous condition then may be rewritten

$$x\iota(a)x^{-1} = \iota(\pi(x){\cdot}a) \text{ for all } x \in E, a \in A.$$

Things are a bit complicated by the fact that A is not a subgroup of E, but only isomorphic to a subgroup of E via ι. However, we cannot avoid this problem since standard group extensions are constructed on the set $E = A \times G$, as we will see later on.

Example VI.15.3. Examples of group extensions are given by direct products and semi-direct products, but they are not the only ones. For instance, $\mathbb{Z}/8\mathbb{Z}$ is an extension of $\mathbb{Z}/4\mathbb{Z}$ by $\mathbb{Z}/2\mathbb{Z}$, which is **not** a semi-direct product.

Consider an extension of (G, A, φ) given by the exact sequence

$$1 \longrightarrow A \xrightarrow{\iota} E \xrightarrow{\pi} G \longrightarrow 1.$$

This exact sequence allows us to define a bijection between the sets $A \times G$ and E as follows: let us choose a set-theoretic section $s : G \longrightarrow E$, that is a map such that $\pi \circ s = \mathrm{Id}_G$. We also ask for the extra condition $s(1) = 1$. Now if $x \in E$, set $g = \pi(x)$. We have $\pi(xs(g)^{-1}) = gg^{-1} = 1$, so there exists a unique element $a \in A$ such that $\iota(a) = xs(g)^{-1}$. Therefore $x = \iota(a)s(g)$, and a and g are uniquely determined, once s is chosen. The map

$$\begin{aligned} A \times G &\longrightarrow E \\ (a, g) &\longmapsto \iota(a)s(g) \end{aligned}$$

is then bijective. Of course, this is not a group isomorphism if $A \times G$ is endowed with the standard group law.

Let us figure out what happens when we multiply two elements of E. Let $x, x' \in E$ and let $\pi(x) = g, \pi(x') = g'$. We have $\pi(xx') = gg'$, so the previous considerations show that we should be able to write $xx' = \iota(a'')s(gg')$. But we have

$$xx' = \iota(a)s(g)\iota(a')s(g') = \iota(a)s(g)\iota(a')s(g)^{-1}s(g)s(g'),$$

hence

$$xx' = \iota(a)(s(g)\iota(a')s(g)^{-1})(s(g)s(g')s(gg')^{-1})s(gg').$$

By definition of the action of G on A, we have

$$s(g)\iota(a')s(g)^{-1} = \iota(g{\cdot}a'),$$

since s is a section of π. Now it is easy to check that $s(g)s(g')s(gg')^{-1} \in \ker(\pi)$ for all $g, g' \in G$. Therefore, there exists a unique element $\alpha^{(s)}_{g,g'} \in A$ such that

$$\iota(\alpha^{(s)}_{g,g'}) = s(g)s(g')s(gg')^{-1} \text{ for all } g, g' \in G.$$

To sum up, the group law on E is given by the rule

$$(\iota(a)s(g))(\iota(a')s(g')) = \iota(\alpha^{(s)}_{g,g'} a\, g{\cdot}a')s(gg').$$

Straightforward computations show that the map

$$\alpha^{(s)}: \begin{array}{l} G \times G \longrightarrow A \\ (g, g') \longmapsto s(g)s(g')s(gg')^{-1} \end{array}$$

is a 2-cocycle. These considerations lead naturally to the following example.

Example VI.15.4. Let G and A be finite groups, where A is abelian, and assume that G acts by group automorphisms on A. Let $\alpha \in Z^2(G, A)$ be a 2-cocycle, and define an operation on the set $A \times G$ as follows:

$$(a, g)(a', g') = (\alpha_{g,g'}\, ag{\cdot}a', gg'), \text{ for all } a, a' \in A, g, g' \in G.$$

One can show that this defines a group structure on the set $A \times G$ (the nasty thing to check being the associativity, which comes from the cocyclicity condition). The neutral element is $(1, 1)$, and we have an exact sequence (for this particular group law on $A \times G$)

$$1 \longrightarrow A \xrightarrow{\iota_1} A \times G \xrightarrow{\pi_2} G \longrightarrow 1,$$

where $\iota_1(a) = (a, 1)$ and $\pi_2(a, g) = g$. Moreover, one can check that we have

$$(a, g)(a_0, 1)(a, g)^{-1} = (g \cdot a_0, 1),$$

so we get a group extension, denoted by $A \times_\alpha G$. If α is the trivial cocycle, we just recover the external semi-direct product $A \times_\varphi G$.

With this notation, we see that we have a group isomorphism

$$f : \begin{array}{c} A \times_{\alpha^{(s)}} G \xrightarrow{\sim} E \\ (a, g) \longmapsto \iota(a)s(g), \end{array},$$

and that the diagram

$$\begin{array}{ccccccccc} 1 & \longrightarrow & A & \xrightarrow{\iota_1} & A \times_{\alpha^{(s)}} G & \xrightarrow{\pi_2} & G & \longrightarrow & 1 \\ & & \Vert & & \downarrow f & & \Vert & & \\ 1 & \longrightarrow & A & \xrightarrow{\iota} & E & \xrightarrow{\pi} & G & \longrightarrow & 1 \end{array}$$

is commutative

Definition VI.15.5. Two extensions E, E' (for short) of (G, A, φ) are *equivalent* if there exists a group isomorphism $f : E \xrightarrow{\sim} E'$ such that the diagram

$$\begin{array}{ccccccccc} 1 & \longrightarrow & A & \xrightarrow{\iota} & E & \xrightarrow{\pi} & G & \longrightarrow & 1 \\ & & \Vert & & \downarrow f & & \Vert & & \\ 1 & \longrightarrow & A & \xrightarrow{\iota'} & E' & \xrightarrow{\pi'} & G & \longrightarrow & 1 \end{array}$$

commutes.

Thus we have proved that any group extension of (G, A, φ) is equivalent to $A \times_\alpha G$ for some suitable cocycle $\alpha \in Z^2(G, A)$. Now let us see what happens if we choose another set-theoretic section of π. Let $t : G \longrightarrow E$ be another section satisfying $t(1) = 1$, and let $\alpha^{(t)}$ be the corresponding 2-cocycle. For all $g \in G$, we have $s(g)t(g)^{-1} \in \ker(\pi)$ and there exists a unique element $\gamma_g \in A$ such that $s(g) = \iota(\gamma_g)t(g)$. Thus we have

$$\iota(\alpha^{(s)}_{g,g'}) = \iota(\gamma_g)t(g)\iota(\gamma_{g'})t(g')t(gg')^{-1}\iota(\gamma_{gg'})^{-1}.$$

We then get

$$\begin{aligned} \iota(\alpha^{(s)}_{g,g'}) &= \iota(\gamma_g)[\mathrm{Int}(t(g))(\iota(\gamma_{g'}))]t(g)t(g')t(gg')^{-1}\iota(\gamma_{gg'})^{-1} \\ &= \iota(\gamma_g)\iota(g \cdot \gamma_{g'})\iota(\alpha^{(t)}_{g,g'})\iota(\gamma_{gg'})^{-1}, \end{aligned}$$

that is, taking into account that A is abelian and ι is an injective group morphism

$$\alpha^{(s)}_{g,g'} = \alpha^{(t)}_{g,g'}\gamma_g\, g\cdot\gamma_{g'}\,\gamma^{-1}_{gg'}.$$

Moreover, we have $\gamma_1 = 1$ since $s(1) = t(1) = 1$. This means that $\alpha^{(s)}$ and $\alpha^{(t)}$ differ by a 2-coboundary, so they represent the same class in $H^2(G, A)$.

Therefore, we have associated to an extension E a cohomology class in $H^2(G, A)$. Moreover, one may check that two equivalent extensions yield the same cohomology class. Conversely, it is easy to check that if α and α' are two cohomologous cocycles, then the group extensions $A \times_\alpha G$ and $A \times_{\alpha'} G$ are equivalent. Thus, a cohomology class in $H^2(G, A)$ defines an equivalence class of extensions. Morever, the two constructions are easily seen to be mutually inverse. Hence we have (almost) proved:

Theorem VI.15.6. *Let G, A be two finite groups, and assume that A is a G-module via $\varphi : G \longrightarrow \mathrm{Aut}(A)$. Then $H^2(G, A)$ is in one-to-one correspondence with the equivalence classes of group extensions of (G, A, φ).*

Remark VI.15.7. Let us explain once again how this correspondence works. If $[\alpha] \in H^2(G, A)$, the equivalence class of $A \times_\alpha G$ only depends on $[\alpha]$. Conversely, if $1 \longrightarrow A \xrightarrow{\iota} E \xrightarrow{\pi} G \longrightarrow 1$ is a group extension and $s : G \longrightarrow E$ is a set-theoretic section of π satisfying $s(1) = 1$, the cohomology class of the cocycle $\alpha^{(s)} : G \times G \longrightarrow A$ defined by

$$\iota(\alpha_{g,g'}) = s(g)s(g')s(gg')^{-1} \text{ for all } g, g' \in G$$

only depends on the equivalence class of the extension.

Notice from the correspondence that the cocycle corresponding to a given extension $1 \longrightarrow A \longrightarrow E \longrightarrow G \longrightarrow 1$ is trivial if and only if there exists a **group-theoretic** section $s : G \longrightarrow E$, that is, if the exact sequence $1 \longrightarrow A \longrightarrow E \longrightarrow G \longrightarrow 1$ splits. In this case, the extension is equivalent to $1 \longrightarrow A \longrightarrow A \times_\varphi G \longrightarrow G \longrightarrow 1$. Notice also that if two extensions are equivalent, then the corresponding groups E, E' are isomorphic but the converse is not necessarily true.

If G acts trivially on A, that is, if $\varphi(g) = \mathrm{Id}_A$ for all $g \in G$, then for any group extension E of G by A, A identifies with a central subgroup of E

(it is enough to check it on the extensions $A \times_\alpha G$). Conversely, assume that

$$1 \longrightarrow A \xrightarrow{\iota} E \xrightarrow{\pi} G \longrightarrow 1$$

is a group extension such that A (or more precisely $\iota(A)$) is a central subgroup of E. By definition of a group extension, we get

$$x\iota(a)x^{-1} = \iota(a)xx^{-1} = \iota(a) = \iota(\pi(x)\cdot a)$$

for all $x \in E, a \in A$. Using the surjectivity of π and the injectivity of ι, we get that $g\cdot a = a$ for all $a \in A, g \in G$, that is G acts trivially on A.

Definition VI.15.8. A group extension

$$1 \longrightarrow A \xrightarrow{\iota} E \xrightarrow{\pi} G \longrightarrow 1$$

is **central** if $\iota(A)$ is a central subgroup of E.

The previous considerations show that an extension of G by A is central if and only if G acts trivially on A. Therefore, we get the following result.

Corollary VI.15.9. *Let G, A be finite groups, where A is abelian, considered as a trivial G-module. Then $H^2(G, A)$ is in one-to-one correspondence with equivalence classes of central extensions of G by A.*

Example VI.15.10. The group $H^2(S_n, \{\pm 1\})$ is perfectly well-known (the action of S_n on $\{\pm 1\}$ being necessarily the trivial one). Remember that the signature is a morphism $S_n \longrightarrow \{\pm 1\}$, and therefore gives rise to an element $\varepsilon_n \in H^1(S_n, \{\pm 1\})$. We may then consider the cup product $\varepsilon_n \cup \varepsilon_n \in H^2(S_n, \{\pm 1\})$ induced by the $\mathbb{Z}$-bilinear map

$$\begin{aligned} \{\pm 1\} &\longrightarrow \{\pm 1\} \\ ((-1)^m, (-1)^n) &\longmapsto (-1)^{nm}. \end{aligned}$$

If $n = 2, 3$, then $H^2(S_n, \{\pm 1\}) \simeq \mathbb{Z}/2\mathbb{Z}$ and it is generated by $\varepsilon_n \cup \varepsilon_n$. If $n \geq 4$, $H^2(S_n, \{\pm 1\}) \simeq \mathbb{Z}/2\mathbb{Z} \times \mathbb{Z}/2\mathbb{Z}$ and it is generated by $\varepsilon_n \cup \varepsilon_n$ and another class s_n. The extension corresponding to s_n is denoted by $\tilde{S}_n$. This extension can be characterized as follows: any element of $\tilde{S}_n$ whose image in S_n is a transposition (resp. a product of two transpositions with disjoint supports) is an element of order 2 (resp. of order 4). To the presentation of S_n with $n-1$ generators $t_i, 1 \leq i \leq n-1$ subject to the relations

$$t_i^2 = 1, (t_i t_{i+1})^3 = 1, t_i t_j = t_j t_i \text{ if } |j-i| \geq 2$$

corresponds a presentation of $\tilde{S}_n$ with n generators $\tilde{t}_i$ and ω subject to the relations

$$\tilde{t}_i^2 = 1, \omega^2 = 1, \omega\tilde{t}_i = \tilde{t}_i\omega, (\tilde{t}_i\tilde{t}_{i+1})^3 = 1, \tilde{t}_i\tilde{t}_j = \omega\tilde{t}_j\tilde{t}_i \text{ if } |j-i| \geq 2.$$

The maps $\iota : \{\pm 1\} \longrightarrow \tilde{S}_n$ and $\pi : \tilde{S}_n \longrightarrow S_n$ in the exact sequence

$$1 \longrightarrow \{\pm 1\} \xrightarrow{\iota} \tilde{S}_n \xrightarrow{\pi} S_n \longrightarrow 1$$

are characterized by the following relations

$$\begin{aligned} \iota(-1) &= \omega \\ \pi(\tilde{t}_i) &= t_i \text{ for all } i. \end{aligned}$$

The reader will refer to [54] for a proof of these facts. For convenience, we will set $s_n = 0$ if $n = 1, 2, 3$.

Notice that if G is a subgroup of S_n, we can construct central extensions of G by $\{\pm 1\}$ using the restriction map

$$\mathrm{Res} : H^2(S_n, \{\pm 1\}) \longrightarrow H^2(G, \{\pm 1\}).$$

In particular, we can define the extension $\tilde{G}$ corresponding to $\mathrm{Res}(s_n)$. It is then easy to check that $\tilde{G}$ is a subgroup of $\tilde{S}_n$ and that we have a commutative diagram

$$\begin{array}{ccccccccc} 1 & \longrightarrow & \{\pm 1\} & \longrightarrow & \tilde{G} & \longrightarrow & G & \longrightarrow & 1 \\ & & \| & & \downarrow & & \downarrow & & \\ 1 & \longrightarrow & \{\pm 1\} & \longrightarrow & \tilde{S}_n & \longrightarrow & S_n & \longrightarrow & 1 \end{array}$$

where the two vertical maps are just inclusions. In particular, if G is a finite group of order n, we may construct a central extension $\tilde{G}$ of G by $\{\pm 1\}$ after identifying G to a subgroup of S_n. For example, let $n = 4$. One can check that if $G = \mathbb{Z}/4\mathbb{Z}$, then $\tilde{G} = \mathbb{Z}/8\mathbb{Z}$, and that if $G = \mathbb{Z}/2\mathbb{Z} \times \mathbb{Z}/2\mathbb{Z}$, we get $\tilde{G} = Q_8$. If $G = A_n$, we obtain a group $\tilde{A}_n$, which is a central extension of A_n by $\{\pm 1\}$.

§VI.16 Galois embedding problems

Let G be a finite group, and let A be a finite abelian group on which G acts trivially. Let us fix a central extension

$$1 \longrightarrow A \xrightarrow{\iota} G' \xrightarrow{\pi} G \longrightarrow 1\,.$$

Now let k be a field. We consider the following lifting problem: given a continuous morphism $f : \mathcal{G}_{k_s} \longrightarrow G$, does there exists a continuous morphism $f' : \mathcal{G}_{k_s} \longrightarrow G'$ such that the diagram

$$\begin{array}{ccc} \mathcal{G}_{k_s} & \xrightarrow{f'} & G' \\ \| & & \downarrow \pi \\ \mathcal{G}_{k_s} & \xrightarrow{f} & G \end{array}$$

is commutative?

It is straightforward to see that the solvability of the problem only depends on the conjugacy class of f and on the isomorphism class of the extension. Let $\mathcal{G}_{k_s}$ act trivially on G, G' and A. Since A identifies to a central subgroup of G via ι, we have an exact sequence in cohomology

$$H^1(k, G') \xrightarrow{\pi_*} H^1(k, G) \xrightarrow{\delta^1} H^2(k, A) ,$$

and by definition of π_*, the lifting problem is equivalent to ask $[f]$ to be in the image of π_*, that is in the kernel of the connecting map δ^1. Let $s : G \longrightarrow G'$ be a set-theoretic section of π satisfying $s(1) = 1$. Since $s(f_\sigma)$ is a preimage of f_σ under π and $\mathcal{G}_{k_s}$ acts trivially on G, a 2-cocycle $\gamma : \mathcal{G}_{k_s} \times \mathcal{G}_{k_s} \longrightarrow A$ representing $\delta^1([f])$ is uniquely determined by the equalities

$$\iota(\gamma_{\sigma,\tau}) = s(f_\sigma)s(f_\tau)s(f_{\sigma\tau})^{-1} = s(f_\sigma)s(f_\tau)s(f_\sigma f_\tau)^{-1} \text{ for all } \sigma, \tau \in \mathcal{G}_{k_s}.$$

Now a 2-cocycle $\alpha : G \times G \longrightarrow A$ representing the extension G' is uniquely determined by the relations

$$\iota(\alpha_{g,g'}) = s(g)s(g')s(gg')^{-1} \text{ for all } g, g' \in G,$$

by the correspondence between equivalence classes of extensions and $H^2(G, A)$. Hence we get that $\delta^1([f])$ is equal to $f^*([\alpha])$, where $f^*([\alpha])$ is the inverse image of $[\alpha]$ by f, as defined in Example II.3.23. Thus we have proved:

Proposition VI.16.1. *Let $1 \longrightarrow A \longrightarrow G' \longrightarrow G \longrightarrow 1$ be a central extension whose equivalence class corresponds to a class $[\alpha] \in H^2(G, A)$, and let $f : \mathcal{G}_{k_s} \longrightarrow G$ be a continuous group morphism. Then the associated lifting problem has a solution if and only if $f^*([\alpha]) = 0 \in H^2(k, A)$.*

In practice, it is enough to consider the case where f is a **surjective** morphism. Indeed, let $H = \mathrm{Im}(f)$, let $H' = \pi^{-1}(H) \subset G'$ and let

$h' : \mathcal{G}_{k_s} \longrightarrow H'$ be a lifting for $h : \mathcal{G}_{k_s} \longrightarrow H$. Then composing $\tilde{h}$ with the inclusion $H' \subset G'$ gives rise to a lifting of f.

Now assume that $f : \mathcal{G}_{k_s} \longrightarrow G$ is surjective.

Definition VI.16.2. A solution $\tilde{f}$ to the lifting problem for f is said to be **proper** if $\tilde{f}$ is surjective as well, and **improper** otherwise.

The following lemma is left as an easy exercise.

Lemma VI.16.3. *Assume that A has prime order p, and assume that $[\alpha] \neq 0$. Let $f : \mathcal{G}_{k_s} \longrightarrow G$ be a surjective continuous morphism. Then a solution to the associated lifting problem (if any) is proper.*

If we are only looking for proper solutions to the lifting problem, the question may be reformulated as a Galois embedding problem, in view of Corollary V.14.21: given a Galois field extension L/k with Galois group G, does there exists a Galois extension L'/k with Galois group G' such that $L'^A = L$?

This is a classical question in inverse Galois theory. The previous results show that such a field extension L'/k exists if and only if $f^*([\alpha]) = 0$, where $f : \mathcal{G}_{k_s} \longrightarrow G$ is any continuous morphism corresponding to L/k. We will use this result in the next chapter to solve completely a particular Galois embedding problem associated to separable field extensions.

Example VI.16.4. Let p be a prime number, and consider a central extension

$$1 \longrightarrow \mathbb{Z}/p\mathbb{Z} \longrightarrow G' \longrightarrow G \longrightarrow 1.$$

For any field k of characteristic p and for any Galois extension L/k of group G, the associated Galois embedding problem has a solution.

Indeed, we have $H^2(k, \mathbb{Z}/p\mathbb{Z}) = 0$ in this case. This can be seen by applying Galois cohomology to the Artin-Schreier exact sequence

$$0 \longrightarrow \mathbb{Z}/p\mathbb{Z} \longrightarrow \mathbb{G}_a \longrightarrow \mathbb{G}_a \longrightarrow 0,$$

and using the fact that $H^m(k, \mathbb{G}_a) = 0$ for all $m \geq 1$ (see Proposition III.7.31). Proposition VI.16.1 then yields the result.

Exercises

1. Fill the gaps in the proof of Theorem VI.15.6.

2. Prove Lemma VI.16.3.
3. Let G be a finite group acting trivially on a finite abelian group A. Show that $H^2(G, A)$ is killed by $|G|$ and $|A|$. In particular, if G and A have relatively prime orders, any central extension of G by A is trivial.
4. Let $G = A = \mathbb{Z}/3\mathbb{Z}$. For $i = 1, 2$, let $E_i = \mathbb{Z}/9\mathbb{Z}$, and let $\pi_i : E_i \longrightarrow G$ be the canonical projection. Consider the group morphisms

$$\iota_1 : \begin{array}{c} A \longrightarrow E_1 \\ m + 3\mathbb{Z} \longmapsto 3m + 9\mathbb{Z} \end{array}$$

and

$$\iota_2 : \begin{array}{c} A \longrightarrow E_2 \\ m + 3\mathbb{Z} \longmapsto -3m + 9\mathbb{Z}. \end{array}$$

   (a) Show that the central extensions E_1, E_2 of G by A are not isomorphic (although the abstract groups E_1 and E_2 are isomorphic, and even equal).
   (b) Let $[\alpha_i] \in H^2(G, A)$ be the cohomology class corresponding to E_i. Show that $[\alpha_2] = -[\alpha_1]$, and recover the result of the previous question.
   Hint: Use Exercice 3.

5. Let $\tau_1, \ldots, \tau_k$ be k transpositions with disjoint supports in S_n. Show that we have

$$\widetilde{\tau_1 \cdots \tau_k}^2 = \omega^{\frac{k(k-1)}{2}} \in \tilde{S}_n.$$

6. Let G be a cyclic subgroup of S_4 of order 4. Show that $\tilde{G} \simeq \mathbb{Z}/8\mathbb{Z}$.
7. Let G be a cyclic subgroup of S_n of order $n = 2^r \geq 8$. Show that $\operatorname{Res}(s_n) = 0$.
8. Let $n = 2^r \geq 4$, and let $G = D_{2n}$. After a suitable numbering of the elements of G, show that the action of G on itself by left translation identifies G to the subgroup of S_{2n} generated by

$$\sigma = (1 \ldots 2n-1) \text{ and } \tau = (2\, 2n)(3\, 2n-1) \cdots (n\, n+2),$$

and deduce that $\operatorname{Res}(s_{2n}) = 0$.

9. Let $G = Q_8$ be the quaternion group of order 8. Show that $\operatorname{Res}(s_8) = 0$.

Part II
Applications

VII

Galois embedding problems and the trace form

In this chapter, we are going to use the material introduced in the first part of the book to give a full answer to the following Galois embedding problem:

Galois embedding problem: let E/k be a separable field extension of degree n and let E^{gal}/k be its Galois closure (in a fixed separable closure of k). Let G_E be the Galois group of E^{gal}/k, and let $\tilde{G}_E$ be the group extension corresponding to $\mathrm{Res}(s_n) \in H^2(G_E, \{\pm 1\})$, where $s_n \in H^2(S_n, \{\pm 1\})$ is the cohomology class corresponding to the central extension $\tilde{S}_n$ of S_n. Assume that $\mathrm{Res}(s_n) \neq 0$ (otherwise, the problem trivially has a solution). Does there exist a Galois extension $\tilde{E}/k$ with Galois group $\tilde{G}_E$ such that $\tilde{E}^{\{\pm 1\}} = E^{gal}$?

Let us reformulate the problem. Let $e : \mathcal{G}_{k_s} \longrightarrow S_n$ be a continuous morphism representing E. Since E is a field, the compositum of the fields $\chi(E), \chi \in X(E)$, is E^{gal}. By Proposition V.14.22, we then have $\mathrm{Im}(e) = G_E$. Moreover, the continuous morphism $f : \mathcal{G}_{k_s} \longrightarrow G_E$ represents E^{gal}/k by Lemma V.14.18. Using the definitions, it is easy to check that $f^*(\mathrm{Res}(s_n)) = e^*(s_n) \in H^2(k, \{\pm 1\})$. Hence the Galois embedding problem associated to E has solution if and only if $e^*(s_n) = 0$. Therefore, everything boils down to compute $e^*(s_n)$ in more explicit terms.

We will suppose from now on that $\mathrm{char}(k) \neq 2$ (this is not a huge restriction by Example VI.16.4). In this case, we may consider $e^*(s_n)$ as an element of $H^2(k, \mu_2)$.

The crucial idea of this computation is that S_n may be identified to a subgroup of $\mathbf{O}_n(k_s)$, and that this identification may be used to view $\tilde{S}_n$ as a subgroup of $\mathbf{Pin}_n(k_s)$. The computation of $e^*(s_n)$ is then achieved using the computation of the first connecting map associated to the exact sequence

$$1 \longrightarrow \mu_2(k_s) \longrightarrow \mathbf{Pin}_n(k_s) \longrightarrow \mathbf{O}_n(k_s) \longrightarrow 1.$$

With this in mind, let us denote by $\iota : S_n \longrightarrow \mathbf{O}_n(k_s)$ the identification of S_n to a subgroup of $\mathbf{O}_n(k_s)$ by permutation matrices. Notice that ι is a morphism of $\mathcal{G}_{k_s}$-groups. This map induces a map

$$\iota_* : H^1(k, S_n) \longrightarrow H^1(k, \mathbf{O}_n).$$

Using the results of the previous chapters, we see ι_* may be used to attach to each étale k-algebra E of rank n a quadratic form of dimension n, whose isomorphism class only depends on the isomorphism class of E. We will then start with the identification of this quadratic form.

§VII.17 The trace form of an étale algebra

Definition VII.17.1. Let E be an étale algebra of rank n. For all $a \in E$, we denote by $\mathrm{Tr}_{E/k}(a)$ the trace of left multiplication by a in E. The **trace form of** E is the quadratic form $\mathcal{T}_E$ defined by

$$\mathcal{T}_E : \begin{array}{l} E \longrightarrow k \\ x \longmapsto \mathrm{Tr}_{E/k}(x^2). \end{array}$$

Example VII.17.2. If $E = k^n$, then it is easy to check that the canonical basis $e_1, \ldots, e_n$ of k^n is an orthogonal basis for $\mathcal{T}_E$, and that $\mathcal{T}_E(e_i) = 1$ for $1 \leq i \leq n$. Therefore, $\mathcal{T}_E \simeq \langle 1, \ldots, 1 \rangle$.

Lemma VII.17.3. *Let E and E' be two étale k-algebras and let K/k be a field extension. Finally, let $a \in E$ and $a' \in E'$. Then the following properties hold:*

(1) $\mathrm{Tr}_{E\times E'/k}((a, a')) = \mathrm{Tr}_{E/k}(a) + \mathrm{Tr}_{E'/k}(a')$.

(2) $\mathrm{Tr}_{E_K/K}(a \otimes 1) = \mathrm{Tr}_{E/k}(a)$.

(3) *If $\varphi : E \xrightarrow{\sim} E'$ is an isomorphism of k-algebras, then* $\mathrm{Tr}_{E/k}(a) = \mathrm{Tr}_{E'/k}(\varphi(a))$.

(4) *If $\sigma_1, \cdots, \sigma_n$ are the distinct elements of $X(E)$, then* $\mathrm{Tr}_{E/k}(a) = \sigma_1(a) + \ldots + \sigma_n(a)$.

Proof. For $a \in E$, we will denote by ℓ_a the endomorphism of left multiplication by a in E.

(1) Let $\mathbf{e} = (e_1, \ldots, e_n)$ and $\mathbf{e}' = (e'_1, \ldots, e'_m)$ be k-bases of E and E' respectively. Then $\mathbf{e}'' = ((e_1, 0), \ldots, (e_n, 0), (0, e'_1), \ldots, (0, e'_m))$ is

a k-basis of $E\times E'$. Now if $M = \mathrm{Mat}(\ell_a, \mathbf{e})$ and $M' = \mathrm{Mat}(\ell_{a'}, \mathbf{e}')$, then we have

$$\mathrm{Mat}(\ell_{(a,a')}, \mathbf{e}'') = \begin{pmatrix} M & 0 \\ 0 & M' \end{pmatrix}.$$

The result follows immediately.

(2) If $\mathbf{e} = (e_1, \ldots, e_n)$ is a k-basis of E, then $\mathbf{e}' = (e_1 \otimes 1, \ldots, e_n \otimes 1)$ is a K-basis of E_K. Clearly, we have

$$\mathrm{Mat}(\ell_{a\otimes 1}, \mathbf{e}') = \mathrm{Mat}(\ell_a, \mathbf{e}),$$

hence the result.

(3) If $\mathbf{e} = (e_1, \ldots, e_n)$ is a k-basis of E, then the family

$$\varphi(\mathbf{e}) = (\varphi(e_1), \ldots, \varphi(e_n))$$

is a k-basis of E'. It is easy to check that we have

$$\mathrm{Mat}(\ell_{\varphi(a)}, \varphi(\mathbf{e})) = \mathrm{Mat}(\ell_a, \mathbf{e}),$$

and the result follows.

(4) Assume first that $E = k^n$. In this case, the matrix of left multiplication by $a = (a_1, \cdots, a_n)$ in the canonical basis is the diagonal matrix

$$\begin{pmatrix} a_1 & & \\ & \ddots & \\ & & a_n \end{pmatrix}.$$

Therefore, $\mathrm{Tr}_{E/k}(a) = a_1 + \ldots + a_n$ in this case. Since the elements of $X(E)$ are the n projections, we are done. Now assume that E is an arbitrary étale k-algebra of rank n. By definition, we have a k_s-algebra isomorphism

$$f : E_{k_s} \xrightarrow{\sim} k_s^n.$$

Using (2) and (3), we get

$$\mathrm{Tr}_{E/k}(a) = \mathrm{Tr}_{E_{k_s}/k_s}(a \otimes 1) = \mathrm{Tr}_{k_s^n/k_s}(f(a \otimes 1)).$$

Moreover, if $\sigma_1, \ldots, \sigma_n$ are the distinct elements of $X(E)$, then the morphisms $(\sigma_1)_{k_s} \circ f^{-1}, \ldots, (\sigma_n)_{k_s} \circ f^{-1}$ are the n elements of $X(k_s^n)$. Using the previous case, we get

$$\mathrm{Tr}_{E/k}(a) = \sum_{i=1}^{n} ((\sigma_i)_{k_s} \circ f^{-1})(f(a \otimes 1)) = \sigma_1(a) + \ldots + \sigma_n(a).$$

This concludes the proof.

□

The next result is an easy consequence of the previous lemma, and its proof is left to the reader.

Lemma VII.17.4. *Let E and E' be two étale algebras of rank n over k. Then the following properties hold:*

(1) $\mathcal{T}_{E\times E'} \simeq \mathcal{T}_E \perp \mathcal{T}_{E'}$.

(2) *For any field extension K/k, we have $\mathcal{T}_{E_K} \simeq (\mathcal{T}_E)_K$.*

(3) *If $E \simeq E'$, then $\mathcal{T}_E \simeq \mathcal{T}_{E'}$.*

Corollary VII.17.5. *For any étale k-algebra E, the trace form is a regular quadratic form whose isomorphism class only depends on the isomorphism class of E.*

Proof. It only remains to prove that $\mathcal{T}_E$ is regular. But this property is invariant by scalar extension. Since we have

$$(\mathcal{T}_E)_{k_s} \simeq \mathcal{T}_{E_{k_s}} \simeq \mathcal{T}_{k_s^n} \simeq \langle 1, \ldots, 1 \rangle,$$

we are done. □

Recall now that for any field K, we have an injective morphism $S_n \longrightarrow \mathbf{GL}_n(K)$, which sends a permutation $s \in S_n$ to the matrix of the automorphism of φ_s of K^n which sends e_i to $e_{s(i)}$. In fact, this automorphism preserves the quadratic form $\langle 1, \ldots, 1 \rangle$, so we get a morphism $S_n \longrightarrow \mathbf{O}_n(K)$. In particular, we get an injective group morphism

$$\iota : S_n \longrightarrow \mathbf{O}_n(k_s),$$

and it is easy to check that it is a morphism of $\mathcal{G}_{k_s}$-groups.

Lemma VII.17.6. *The map $H^1(k, S_n) \longrightarrow H^1(k, \mathbf{O}_n)$ induced by the injection $S_n \longrightarrow \mathbf{O}_n(k_s)$ is the map $E \mapsto \mathcal{T}_E$.*

Proof. Let $f : E_{k_s} \xrightarrow{\sim} k_s^n$ be an isomorphism of k_s-algebras, so that the map

$$\alpha : \begin{array}{l} \mathcal{G}_{k_s} \longrightarrow S_n \\ \quad \sigma \longmapsto f \circ \sigma \cdot f^{-1} \end{array}$$

is a cocycle representing E. We saw that f induces an isomorphism $\mathcal{T}_{E_{k_s}} \simeq \mathcal{T}_{k_s^n}$, that is an isomorphism $(\mathcal{T}_E)_{k_s} \simeq \langle 1, \ldots, 1 \rangle$. Hence, by Galois descent, the map $\iota \circ \alpha$ is also a cocycle representing $\mathcal{T}_E$. □

Definition VII.17.7. The **discriminant** of E is the square-class

$$d_E = \det(\mathcal{T}_E) \in k^\times/k^{\times 2}.$$

Notice that the discriminant commutes with scalar extensions.

Lemma VII.17.8. *Let $\varepsilon : S_n \longrightarrow \{\pm 1\} = \mu_2(k_s)$ be the signature morphism. Then $\varepsilon_* : H^1(k, S_n) \longrightarrow H^1(k, \mu_2)$ is the map $E \longmapsto d_E$.*

Proof. Since ε is the composite map

$$S_n \longrightarrow \mathbf{O}_n(k_s) \xrightarrow{\det} \mu_2(k_s)\ ,$$

the previous lemma and Proposition IV.11.2 give the result. □

Lemma VII.17.9. *Let E/k be a separable field extension of degree n, let α be a primitive element of E and let $\alpha_1 = \alpha, \dots, \alpha_n$ be its conjugates. Then we have*

$$d_E = \prod_{i<j} (\alpha_i - \alpha_j)^2 \in k^\times/k^{\times 2}.$$

Proof. A k-basis of E is $1, \alpha, \dots, \alpha^{n-1}$, so the representative matrix of $\mathcal{T}_E$ in this basis is given by $B = (\mathrm{Tr}_{E/k}(\alpha^{i+j-2}))_{i,j}$. Since by Lemma VII.17.3 (4) we have $\mathrm{Tr}_{E/k}(\alpha^m) = \alpha_1^m + \dots + \alpha_n^m$, it is easy to see that we have $B = M^t M$, where $M = (\alpha_i^{j-1})_{i,j}$. Now apply the Vandermonde determinant formula to get the desired equality. □

Corollary VII.17.10. *Let E/k be a* **Galois** *extension of degree n, and let G be its Galois group. Let us identify G to a subgroup of S_n by letting G act on itself by left translations. Then d_E is a square if and only if $G \subset A_n$.*

Proof. Let $\sigma_1 = \mathrm{Id}_E, \sigma_2, \dots, \sigma_n$ be the elements of G. Let us choose a primitive element α of E/k, and let $\alpha_i = \sigma_i(\alpha), i = 1, \dots, n$. The previous lemma shows that d_E is a square in k if and only if the element

$$x = \prod_{i<j} (\alpha_i - \alpha_j) = \prod_{i<j} (\sigma_i(\alpha) - \sigma_j(\alpha)) \in E$$

lies in k, which means that it is fixed by all the elements of G. But we have

$$\sigma \cdot x = \prod_{i<j} (\sigma \circ \sigma_i(\alpha) - \sigma \circ \sigma_j(\alpha)) = \varepsilon(s_\sigma) x,$$

where $s_\sigma \in S_n$ is the permutation induced on the elements of G by left multiplication by σ. Thus $x \in k$ if and only if G is a subgroup of A_n. □

Remark VII.17.11. One can easily show that a group G of order n, identified to a subgroup of S_n by acting on itself by left translations, is a subgroup of A_n if and only if it has odd order or its Sylow 2-subgroup is not cyclic.

We would like to end this section by explaining how the trace form may be useful to count the number of real roots of a separable polynomial of $\mathbb{Q}[X]$.

Proposition VII.17.12. *Let $f \in \mathbb{Q}[X]$ be a non-zero separable polynomial, and let $E = \mathbb{Q}[X]/(f)$. Let r_1 be the number of real roots over f, and let r_2 be the number of pairwise non-conjugate complex roots of f. Then we have*

$$(\mathcal{T}_E)_{\mathbb{R}} \simeq r_1 \times \langle 1 \rangle \perp r_2 \times \langle 1, -1 \rangle.$$

In particular, the signature of the trace form $\mathcal{T}_E$ is the number of real roots of f.

Proof. Without loss of generality, one can assume that f is monic. By Lemma VII.17.4, we have $(\mathcal{T}_E)_{\mathbb{R}} \simeq \mathcal{T}_{E_{\mathbb{R}}}$. Now we have $E_{\mathbb{R}} \simeq \mathbb{R}[X]/(f)$. Let $f_1, \ldots, f_m \in \mathbb{R}[X]$ be the distinct monic irreducible factors of f. By the Chinese Remainder Theorem, we have

$$E \otimes_{\mathbb{Q}} \mathbb{R} \simeq \mathbb{R}[X]/(f_1) \times \cdots \times \mathbb{R}[X]/(f_m).$$

If f_i has degree 1, then $\mathbb{R}[X]/(f_i) \simeq \mathbb{R}$. If f_i has degree 2, then $\mathbb{R}[X]/(f_i) \simeq \mathbb{C}$. Therefore, we have an isomorphism of $\mathbb{R}$-algebras

$$E \otimes_{\mathbb{Q}} \mathbb{R} \simeq \mathbb{R}^{r_1} \times \mathbb{C}^{r_2},$$

where r_1 is the number of real roots of f and r_2 is the number of pairwise non-conjugate complex roots of f. Now it is easy to check that $\mathcal{T}_{\mathbb{R}} \simeq \langle 1 \rangle$ and $\mathcal{T}_{\mathbb{C}} \simeq \langle 1, -1 \rangle$. By Lemma VII.17.4 (4), we get

$$(\mathcal{T}_E)_{\mathbb{R}} \simeq r_1 \times \langle 1 \rangle \perp r_2 \times \langle 1, -1 \rangle.$$

This concludes the proof. □

§VII.18 Computation of $e^*(s_n)$

This subsection is devoted to the proof of the following theorem, originally proved in [61] in a different way.

Theorem VII.18.1 (Serre's formula). *Let E/k be an étale algebra represented by $e : \mathcal{G}_{k_s} \longrightarrow S_n$. Then we have*

$$e^*(s_n) = w_2(\mathcal{T}_E) + (2) \cup (d_E).$$

We will assume that $n \geq 4$ and leave the remaining cases to the reader.

Recall that $\iota : S_n \longrightarrow \mathbf{O}_n(k_s)$ denotes the orthogonal representation of S_n by permutation matrices. Let $X \subset \mathbf{Pin}_n(k_s)$ be the preimage of $\iota(S_n)$ under the map $\alpha_{k_s} : \mathbf{Pin}_n(k_s) \longrightarrow \mathbf{O}_n(k_s)$, so that we have a commutative diagram

$$\begin{array}{ccccccccc} 1 & \longrightarrow & \mu_2(k_s) & \longrightarrow & X & \xrightarrow{\pi} & S_n & \longrightarrow & 1 \\ & & \| & & \downarrow & & \downarrow \iota & & \\ 1 & \longrightarrow & \mu_2(k_s) & \longrightarrow & \mathbf{Pin}_n(k_s) & \xrightarrow{\alpha_{k_s}} & \mathbf{O}_n(k_s) & \longrightarrow & 1 \end{array}$$

where π is just the composite of α_{k_s} with $\iota^{-1} : \iota(S_n) \longrightarrow S_n$. Recall for later use that the maps $\mu_2(k_s) \longrightarrow \mathbf{Pin}(q)(k_s)$ and $\mu_2(k_s) \longrightarrow X$ are just inclusions.

For any $\sigma \in \mathcal{G}_{k_s}$, write $\dfrac{\sigma(\sqrt{2})}{\sqrt{2}} = (-1)^{\chi(\sigma)}$. In particular, the map

$$\begin{aligned} \mathcal{G}_{k_s} &\longrightarrow \mu_2(k_s) \\ \sigma &\longmapsto (-1)^{\chi(\sigma)} \end{aligned}$$

is a cocycle representing $(2) \in H^1(k, \mu_2)$. We also write $\varepsilon(s) = (-1)^{\nu(s)}$ for any $s \in S_n$.

The next lemma describes X as a $\mathcal{G}_{k_s}$-group.

Lemma VII.18.2. *The extension of abstract groups*

$$1 \longrightarrow \{\pm 1\} \longrightarrow X \longrightarrow S_n \longrightarrow 1$$

is equivalent to

$$1 \longrightarrow \{\pm 1\} \longrightarrow \tilde{S}_n \longrightarrow S_n \longrightarrow 1.$$

Moreover, for all $\sigma \in \mathcal{G}_{k_s}$ and all $x \in X$, we have

$$\sigma \cdot x = (-1)^{\chi(\sigma)\nu(\pi(x))} x.$$

In particular, X is **not** isomorphic to $\tilde{S}_n$ as a $\mathcal{G}_{k_s}$-group.

Proof. Let $e_1, \ldots, e_n$ be the canonical basis of k_s^n, and let $q_0 = \langle 1, \ldots, 1 \rangle$, so that this basis is an orthogonal basis for q_0. Let $(ij) \in S_n$ be a transposition, so its image in $\mathbf{O}_n(k_s)$ is the map f_{ij} which switches e_i and e_j and fixes e_m if $m \neq i, j$.

Let us check that we have $f_{ij} = \tau_{e_i - e_j}$. It is enough to prove that these maps coincide on a basis. First observe that

$$q_0(e_i - e_j) = q_0(e_i) - 2b_{q_0}(e_i, e_j) + q_0(e_j) = 2,$$

so we have $\tau_{e_i-e_j}(x) = x - b_{q_0}(x, e_i - e_j)(e_i - e_j)$. If $m \neq i, j$, we have $f_{ij}(e_m) = e_m$. Since e_m is orthogonal to e_i and to e_j, hence to $e_i - e_j$, we have $\tau_{e_i-e_j}(e_m) = e_m = f_{ij}(e_m)$. Moreover, since $b_{q_0}(e_i, e_j) = 0$ and $b_{q_0}(e_i, e_i) = 1$, we have

$$\begin{array}{lllllll} \tau_{e_i-e_j}(e_i) & = & e_i - (e_i - e_j) & = & e_j & = & f_{ij}(e_i); \\ \tau_{e_i-e_j}(e_j) & = & e_j + (e_i - e_j) & = & e_i & = & f_{ij}(e_j). \end{array}$$

Hence the preimages of f_{ij} in $\mathbf{Pin}_n(k_s)$ are $\pm\frac{1}{\sqrt{2}}(e_i - e_j)$ by Corollary IV.10.19. Now we have

$$\left(\frac{1}{\sqrt{2}}(e_i - e_j)\right)^2 = \frac{1}{2}q_0(e_i - e_j) = 1,$$

so these elements have order 2. Now let $(ij)(kl) \in S_n$ be a product of two transpositions with disjoint supports. The image of this element under ι is $\tau_{e_i-e_j} \circ \tau_{e_k-e_l}$, and the corresponding preimages in $\mathbf{Pin}_n(k_s)$ are $\pm\frac{1}{2}(e_i - e_j)(e_k - e_l)$. Since $e_i - e_j$ and $e_k - e_l$ are orthogonal, they anticommute in the Clifford algebra, so

$$\left(\frac{1}{2}(e_i - e_j)(e_k - e_l)\right)^2 = -\frac{1}{4}(e_i - e_j)^2(e_k - e_l)^2 = -1,$$

and thus these elements have order 4. Hence we proved that if $x \in X$ is mapped under π onto a transposition (respectively a product of two transpositions with disjoint supports), then x has order 2 (respectively has order 4). These properties characterize $\tilde{S}_n$.

Now let us describe the action of $\mathcal{G}_{k_s}$ on X. Let $x \in X$ and write $\pi(x) = (i_1 j_1) \cdots (i_r j_r)$ as a product of transpositions with disjoint supports. Then $x = \pm\frac{1}{(\sqrt{2})^r}(e_{i_1} - e_{j_1}) \cdots (e_{i_r} - e_{j_r})$. It is then clear that $\sigma \cdot x = (-1)^{\chi(\sigma)} x$ if r is odd and that $\sigma \cdot x = x$ if r is even, which is equivalent to the last statement of the lemma. □

Proof of Theorem VII.18.1. The commutative diagram

$$\begin{array}{ccccccccc}
1 & \longrightarrow & \mu_2(k_s) & \longrightarrow & X & \xrightarrow{\pi} & S_n & \longrightarrow & 1 \\
 & & \| & & \downarrow & & \downarrow{\scriptstyle \iota} & & \\
1 & \longrightarrow & \mu_2(k_s) & \longrightarrow & \mathbf{Pin}_n(k_s) & \xrightarrow{\alpha_{k_s}} & \mathbf{O}_n(k_s) & \longrightarrow & 1
\end{array}$$

induces a commutative diagram in cohomology

$$\begin{array}{ccc}
H^1(k, S_n) & \xrightarrow{\delta^1} & H^2(k, \mu_2) \\
\downarrow{\scriptstyle \iota_*} & & \| \\
H^1(k, \mathbf{O}_n) & \longrightarrow & H^2(k, \mu_2)
\end{array}$$

by Theorem III.7.39 (4). By Lemma VII.17.6, the first vertical map sends an étale algebra E to its trace form $\mathcal{T}_E$, and the second horizontal map sends a quadratic form q to $w_2(q)$ by Corollary IV.11.7. Hence we have $\delta^1(E) = w_2(\mathcal{T}_E)$. Now we have to compute $\delta^1(E)$ in a different way. Let $e : \mathcal{G}_{k_s} \longrightarrow S_n$ be a continuous morphism representing E.

Let $t : S_n \longrightarrow X$ be a section of π satisfying $t(\mathrm{Id}) = 1$. By Theorem VI.15.6, the cocycle $\alpha : X \times X \longrightarrow \{\pm 1\}$ defined by

$$\alpha_{s,s'} = t(s)t(s')t(ss')^{-1} \text{ for all } s, s' \in S_n$$

represents the extension $1 \longrightarrow \{\pm 1\} \longrightarrow X \longrightarrow S_n \longrightarrow 1$. Since this extension is equivalent to the extension $1 \longrightarrow \{\pm 1\} \longrightarrow \tilde{S}_n \longrightarrow S_n \longrightarrow 1$ by Lemma VII.18.2, α is cohomologous to s_n by the same theorem. Let $x_\sigma = t(e_\sigma)$, so that $\pi(x_\sigma) = e_\sigma$ for all $\sigma \in \mathcal{G}_{k_s}$. By construction, $\delta^1(E)$ is represented by the cocycle $\beta : \mathcal{G}_{k_s} \longrightarrow \mu_2(k_s)$ defined by

$$\beta_{\sigma,\tau} = x_\sigma \sigma{\cdot}x_\tau \, x_{\sigma\tau}^{-1} \text{ for all } \sigma, \tau \in \mathcal{G}_{k_s}.$$

By Lemma VII.18.2, we get

$$\beta_{\sigma,\tau} = (-1)^{\chi(\sigma)\nu(e_\tau)} x_\sigma x_\tau x_{\sigma\tau}^{-1} \text{ for all } \sigma, \tau \in \mathcal{G}_{k_s}.$$

Now we have

$$x_\sigma x_\tau x_{\sigma\tau}^{-1} = t(e_\sigma)t(e_\tau)t(e_{\sigma\tau})^{-1} = \alpha_{e_\sigma, e_\tau} = e^*(\alpha)_{\sigma,\tau}.$$

Therefore,

$$\beta_{\sigma,\tau} = (-1)^{\chi(\sigma)\nu(e_\tau)} e^*(\alpha)_{\sigma,\tau} = (-1)^{\chi(\sigma)\nu(e_\tau)} e^*(\alpha)_{\sigma,\tau}.$$

But the cocycle $\mathcal{G}_{k_s} \longrightarrow \mu_2(k_s), \sigma \longmapsto (-1)^{\nu(e_\sigma)}$ is just $\varepsilon \circ e$, that is

$\varepsilon_*(e)$, which represents $(d_E) \in H^1(k, \mu_2)$ by Lemma VII.17.8. Hence the cocycle

$$\begin{aligned} \mathcal{G}_{k_s} \times \mathcal{G}_{k_s} &\longrightarrow \mu_2(k_s) \\ (\sigma, \tau) &\longmapsto (-1)^{\chi(\sigma)\nu(e_\sigma)} \end{aligned}$$

represents $(2) \cup (d_E)$. Now since α and s_n are cohomologous, $e^*(\alpha)$ is cohomologous to $e^*(s_n)$. Therefore, we get

$$\delta^1(E) = e^*(s_n) + (2) \cup (d_E),$$

hence the result. □

Remark VII.18.3. A generalization of Serre's formula has been found by Fröhlich [24] in the context of orthogonal representations. This has been proved very useful to solve Galois embedding problems associated to central extensions of the type

$$1 \longrightarrow \{\pm 1\} \longrightarrow X \longrightarrow G \longrightarrow 1.$$

For more details and numerous references on the applications of this formula, the reader will refer to [16] and [24].

§VII.19 Applications to inverse Galois theory

In this section, we would like to apply the previous result to the Galois inverse problem.

Galois inverse problem: let k be a field. Describe all the finite groups which may occur as the Galois group of a field extension of k.

This problem is not very thrilling if k is algebraically closed, real closed or even finite. However, this problem is particularly interesting (and far from being solved) if $k = \mathbb{Q}$. It is conjectured that every finite group G may occur as a Galois group over a given number field. The most striking result in this direction is a deep theorem of Shafarevich which asserts that every finite solvable group occurs as a Galois group of a given finite field extension of $\mathbb{Q}$ (see [62]). The case of non-solvable finite groups is still open. We would like to apply the result of the previous sections in the case of central extensions with kernel $\{\pm 1\}$.

Let us start with the case of cyclic extensions of degree 4.

Example VII.19.1. Let E/k be a Galois extension of group $G_E = \mathbb{Z}/4\mathbb{Z}$ and discriminant d, so $\tilde{G}_E = \mathbb{Z}/8\mathbb{Z}$. Then one can show that $d = a^2 + b^2$ for some $a, b \in k^\times$, and that there exists $q \in k^\times$ such that

$E = k(\sqrt{q(d + a\sqrt{d})})$ (see [20]). It is a good exercise on quadratic forms to check that $\mathcal{T}_E \simeq \langle 1, d, q, q\rangle$ (and in particular $d_E = d$).

Hence $w_2(\mathcal{T}_E) = (q) \cup (q) = (q) \cup (-1)$, and E/k can be embedded in a Galois extension of group $\mathbb{Z}/8\mathbb{Z}$ if and only if $(q) \cup (-1) = (2) \cup (d)$.

Unfortunately, this does not generalize for cyclic groups of higher order. Indeed, one can show that $\mathrm{Res}(s_{2^m}) \in H^2(\mathbb{Z}/2^m\mathbb{Z}, \{\pm 1\})$ is trivial for $m \geq 3$, and so the group extension obtained is **not** $\mathbb{Z}/2^{m+1}\mathbb{Z}$ (see [20], for example).

Now let E/k be a Galois extension of group $G_E = (\mathbb{Z}/2\mathbb{Z})^2$, so $\tilde{G}_E = Q_8$. We have $E = k(\sqrt{a}, \sqrt{b})$, where $a, b, \in k^\times, ab^{-1} \notin k^{\times 2}$, and the trace form is $\mathcal{T}_E \simeq \langle 1, a, b, ab\rangle$, so here $d_E = 1$. It is easy to check that we have

$$w_2(\mathcal{T}_E) = (a) \cup (b) + (-1) \cup (ab),$$

so E/k can be embedded in a Galois extension of group Q_8 if and only if $(a) \cup (b) + (-1) \cup (ab) = 0$. Notice that $(a) \cup (b) + (-1) \cup (ab)$ is also the Hasse-Witt invariant of the quadratic form $\langle a, b, (ab)^{-1}\rangle$. Since quadratic forms of dimension 3 are classified by dimension, determinant and Hasse-Witt invariant (see [53, Chapter 2,Theorem 13.5]), one can reformulate the previous result as follows:

Proposition VII.19.2. *Let $E = k(\sqrt{a}, \sqrt{b})/k$ be a biquadratic field extension. Then E/k can be embedded in a Galois extension $\tilde{E}/k$ of group Q_8 if and only if we have an isomorphism of quadratic forms*

$$\langle a, b, (ab)^{-1}\rangle \simeq \langle 1, 1, 1\rangle.$$

This result is due to Witt (see [70]). Moreover, from an isomorphism between these quadratic forms, Witt gives an explicit construction of $\tilde{E}$. We now describe it briefly. Set $q = \langle a, b, (ab)^{-1}\rangle$ and $q_0 = \langle 1, 1, 1\rangle$. Let $f : k^3 \xrightarrow{\sim} k^3$ be an isomorphism of k-vector spaces satisfying $q \circ f = q_0$.

If $P = (p_{ij})$ denotes the representative matrix of f in the canonical basis of k^3, we have

$$P^t \begin{pmatrix} a & & \\ & b & \\ & & (ab)^{-1} \end{pmatrix} P = I_3.$$

Therefore, we have $\det(P)^2 = 1$, that is $\det(f) = \pm 1$. Composing f on the right by a permutation endomorphism, we then may assume that

$\det(f) = 1$. In this case, for any $r \in k^\times$ the field extension $\tilde{E}/k$ defined by

$$\tilde{E} = k\left(\sqrt{r\left(1 + p_{11}\sqrt{a} + p_{22}\sqrt{b} + \frac{p_{33}}{\sqrt{a}\sqrt{b}}\right)}\right)$$

is a Galois extension of Galois group Q_8. Moreover, every such extension may be obtained in this way (cf. [70]).

Let us continue with more general considerations. If k is an arbitrary field, then $e^*(s_n) = 0$ if $d_E = 1$ and $w_2(T_E) = 0$. We would like to give examples of situations when this occurs. First we need a little lemma.

Lemma VII.19.3. *Let k be a field. Let $r_1, r_2 \geq 0$ be two integers, and let q_{r_1,r_2} be the quadratic form*

$$r_1 \times \langle 1 \rangle \perp r_2 \times \langle 1, -1 \rangle.$$

(1) *If $r_2 \equiv 0 \mod 4$, then $\det(q_{r_1,r_2}) = 1$ and $w_2(q_{r_1,r_2}) = 0$.*

(2) *If -1 is not a sum of two squares of k, the converse also holds.*

Proof. Clearly, $\det(q_{r_1,r_2}) = (-1)^{r_2}$. Moreover, we have

$$w_2(q_{r_1,r_2}) = w_2(r_2 \times \langle -1 \rangle) = \frac{r_2(r_2-1)}{2}(-1) \cup (-1).$$

If r_2 is a multiple of 4, then r_2 and $\frac{r_2(r_2-1)}{2}$ are both even, hence the first part of the lemma. Now assume that -1 is not a sum of two squares of k; in particular $-1 \notin k^{\times 2}$. If r_2 is odd, then $d_E = -1 \in k^\times/k^{\times 2}$, so $d_E \neq 1$. Finally, if $r_2 \equiv 2 \mod 4$, $\frac{r_2(r_2-1)}{2}$ is odd, so $w_2(q_{r_1,r_2}) = (-1) \cup (-1)$. Since -1 is not a sum of two squares of k, this is non-zero by Proposition III.9.15 (2). This completes the proof of the lemma. □

The next result follows from a direct application of this lemma.

Corollary VII.19.4. *Let E/k be a separable field extension of finite degree. Assume that we have an isomorphism of quadratic forms*

$$\mathcal{T}_E \simeq q_{r_1,r_2} \text{ for some } r_1, r_2 \geq 0.$$

If $r_2 \equiv 0 \mod 4$, then $d_E = 1$ and $e_*(s_n) = 0$. In particular, this holds if $\mathcal{T}_E \simeq n \times \langle 1 \rangle$.

The following proposition describes exactly in which cases the Galois embedding problem has a solution for $k = \mathbb{Q}$, provided that $G_E \subset A_n$. It was originally proven in [61].

Proposition VII.19.5. *Let $E/\mathbb{Q}$ be a finite extension of degree n. Denote by r_1 (resp. r_2) the number of real embeddings (resp. the number of pairwise non-conjugate complex embeddings) of E into $\mathbb{C}$; in particular $n = r_1 + 2r_2$. Then the following conditions are equivalent:*

(1) $d_E = 1$ *and* $e^*(s_n) = 0$.

(2) $r_2 \equiv 0 \mod 4$ *and* $\mathcal{T}_E \simeq q_{r_1,r_2}$.

Proof. The implication (2) $\Rightarrow$ (1) follows from the previous corollary. Now assume that $d_E = 1$ and $e^*(s_n) = 0$. Therefore, $d_E = 1$ and $w_2(\mathcal{T}_E) = 0$. In particular, we have $d_E = 1 \in \mathbb{R}^\times/\mathbb{R}^{\times 2}$. Since we have $\det((\mathcal{T}_E)_\mathbb{R}) = \det(\mathcal{T}_E) \in \mathbb{R}^\times/\mathbb{R}^{\times 2}$, we get

$$\det((\mathcal{T}_E)_\mathbb{R}) = 1 \in \mathbb{R}^\times/\mathbb{R}^{\times 2}.$$

We also have $\mathrm{Res}_{\mathbb{R}/\mathbb{Q}}(w_2(\mathcal{T}_E)) = 0$. Since the Hasse invariant commutes with scalar extension by Remark IV.11.9, we get

$$w_2((\mathcal{T}_E)_\mathbb{R}) = 0.$$

By Proposition VII.17.12, we have $(\mathcal{T}_E)_\mathbb{R} \simeq q_{r_1,r_2}$, and the two equalities above then yield

$$\det(q_{r_1,r_2}) = 1 \in \mathbb{R}^\times/\mathbb{R}^{\times 2} \text{ and } w_2(q_{r_1,r_2}) = 0 \in H^2(\mathbb{R}, \mu_2).$$

Since -1 is not a sum of squares in $\mathbb{R}$, Lemma VII.19.3 implies that $r_2 \equiv 0 \mod 4$. This implies now that the quadratic form q_{r_1,r_2} over $\mathbb{Q}$ has trivial determinant and Hasse invariant. Notice also that $\dim(q_{r_1,r_2}) = r_1 + 2r_2 = n$. Hence we have the following equalities:

$$\begin{array}{lclcl} \dim(\mathcal{T}_E) & = & n & = & \dim(q_{r_1,r_2}) \\ \det(\mathcal{T}_E) & = & 1 & = & \det(q_{r_1,r_2}) \\ w_2(\mathcal{T}_E) & = & 0 & = & w_2(q_{r_1,r_2}) \\ \mathrm{sign}(\mathcal{T}_E) & = & r_1 & = & \mathrm{sign}(q_{r_1,r_2}), \end{array}$$

the last one coming from Proposition VII.17.12, and from the fact that the number of real embeddings of E equals the number or real roots of a polynomial defining E. By Theorem IV.12.5, this implies that $\mathcal{T}_E \simeq q_{r_1,r_2}$. □

This result is particularly interesting for the following reason. If the Galois embedding problem associated to E/k has a solution and $\mathcal{T}_E \simeq q_{r_1,r_2}$, then an element $\gamma \in E^{gal}$ such that $E^{gal}(\sqrt{\gamma})$ is a Galois extension with Galois group $\tilde{G}_E$ may be computed by explicit methods. We refer to reader to [16] and the associated references for more details.

Even if the trace form $\mathcal{T}_E$ may be computed and diagonalized by algorithmic methods, it is not always straightforward to decide whether or not $(2) \cup (d_E) + w_2(\mathcal{T}_E) = 0$, or if $\mathcal{T}_E$ is isomorphic to q_{r_1,r_2}, especially over general fields. Therefore, it is preferable to find a nice diagonalization of this quadratic form, for example a diagonalization containing a lot of hyperbolic planes $\langle 1, -1 \rangle$. This can be done if a polynomial f defining E has sufficiently many zero coefficients. For example in [65], Vila computed the trace form of the étale algebra defined by a polynomial of the form $f(X) = X^n + a(bX + c)^k, k$ odd, and used her results to construct Galois extensions of $\mathbb{Q}$ with group $\tilde{A}_n$ for infinitely many values of n.

If $f(X) = X^n + aX + b$, the trace form has been computed by Serre in [61]. He obtains in particular the following result.

Proposition VII.19.6. *Assume that n is even and that* $\mathrm{char}(k) \nmid n$. *Assume that the k-algebra $E = k[X]/(X^n + aX + b)$ is étale. Then we have*

$$\mathcal{T}_E \simeq \langle n, -(-1)^{n/2} n d_E \rangle \perp \frac{n-2}{2} \times \langle 1, -1 \rangle.$$

Proof. Let $E' = \{x \in E \mid \mathrm{Tr}_{E/k}(x) = 0\}$. Since $\mathrm{Tr}_{E/k}(1) = n \neq 0$, it follows easily that we have $E = k \perp E'$. Therefore, we have

$$\mathcal{T}_E \simeq \langle n \rangle \perp (\mathcal{T}_E)_{|E'}.$$

Let α be the image of X in E, so $1, \alpha, \ldots, \alpha^{n-1}$ is a k-basis of E. Let us denote by $\alpha_1 = \alpha, \cdots, \alpha_n$ the images of α under the various k-algebra morphisms $\sigma_i : E \longrightarrow k_s$. Notice that α_i is a root of $f(X) = X^n + aX + b$ in k_{alg}. By Lemma VII.17.3 (4), we have

$$\mathrm{Tr}_{E/k}(\alpha^j) = \alpha_1^j + \ldots + \alpha_n^j \text{ for all } j \geq 0.$$

Since $\alpha_1, \cdots, \alpha_n$ are also the distinct roots of f, using Newton identities, it is easy to check that we have

$$\mathrm{Tr}_{E/k}(\alpha^j) = 0 \text{ for } 1 \leq j \leq n-2.$$

Hence the linear subspace of E' generated by $\alpha, \cdots, \alpha^{(n-2)/2}$ is totally isotropic. Therefore, we get

$$(\mathcal{T}_E)_{|E'} \simeq \langle c \rangle \perp \frac{n-2}{2} \times \langle 1, -1 \rangle.$$

Thus, we have

$$\mathcal{T}_E \simeq \langle n \rangle \perp \langle c \rangle \perp \frac{n-2}{2} \times \langle 1, -1 \rangle.$$

Comparing determinants, we see that $c = -(-1)^{n/2} n d_E \in k^\times/k^{\times 2}$. This concludes the proof. □

Corollary VII.19.7. *Assume that* $n = 2(2m+1)^2$ *for some* $m \geq 1$, *and that* $\mathrm{char}(k) \nmid n$. *Assume that the* k*-algebra* $E = k[X]/(X^n + aX + b)$ *is étale. Then* $w_2(\mathcal{T}_E) = (2) \cup (d_E)$. *In particular, if* E/k *is a field, the associated Galois embedding problem has a solution.*

Proof. Assume that $n = 2(2m+1)^2$ for some $m \geq 1$. In this case, $n = 8r + 2$ (with $r = m^2 + m$) and the previous proposition implies that we have

$$\mathcal{T}_E \simeq \langle 2, 2d_E \rangle \perp 4r \times \langle 1, -1 \rangle.$$

By Lemma IV.11.5, we have

$$w_2(\mathcal{T}_E) = w_2(\langle 2, 2d_E \rangle) + (d_E) \cup ((-1)^{4r}) + w_2(4r \times \langle 1, -1 \rangle).$$

This yields

$$w_2(\mathcal{T}_E) = (2) \cup (2d_E) + \frac{4r(4r-1)}{2}(-1) \cup (-1) = (2) \cup (2d_E).$$

Since 2 is a sum of two squares, we have

$$(2) \cup (2) = (2) \cup (-1) = 0,$$

by Proposition III.9.15. We now conclude the proof using the bilinearity of the cup-product. □

Example VII.19.8. Let $n \geq 1$, and let $f(X) = X^n - X - 1 \in \mathbb{Q}[X]$. Set $E = \mathbb{Q}[X]/(f)$. By a result of Selmer [56], f is irreducible over $\mathbb{Q}$ and $\mathrm{Gal}(E^{gal}/\mathbb{Q}) \simeq S_n$. Therefore, if $n = 2(2m+1)^2$, the previous corollary implies that there exists a Galois extension $\tilde{E}/\mathbb{Q}$ with Galois group $\tilde{S}_n$ containing E.

Further results in inverse Galois theory and Galois embedding problems may be found in [60].

Exercises

1. Let E/k be a Galois extension of group G. If $n = |G|$ is odd, show that $\mathcal{T}_E \simeq n \times \langle 1 \rangle$.

2. Let E and E' be two étale k-algebras. Show that $\mathcal{T}_{E\otimes_k E'} \simeq \mathcal{T}_E \otimes \mathcal{T}_{E'}$.

3. Let E/k be a separable field extension.

   (a) Assume that E is the compositum of pairwise linearly disjoint field extensions $E_1, \ldots, E_r$ of even degree. Show that $[\mathcal{T}_E] \in I^r(k)$.

   (b) Assume that E/k is a Galois extension of group G. Let r be the 2-rank of G, that is the largest integer $r \geq 0$ such that G contains a subgroup isomorphic to $(\mathbb{Z}/2\mathbb{Z})^r$. Show that $[\mathcal{T}_E] \in I^r(k)$.

   Hint: If K/k is a separable field extension, and if $[q] \in I^r(K)$, then $[\mathrm{Tr}_{K/k}(q)] \in I^r(k)$ (see [53, Corollary 14.9, p. 93]). Moreover, if $k \subset K \subset E$, then $\mathrm{Tr}_{E/k} = \mathrm{Tr}_{K/k} \circ \mathrm{Tr}_{E/K}$.

   (c) Assume that E/k is a Galois extension of group G, and let r be the 2-rank of G. If $r \geq 3$, deduce carefully from the previous questions that $e^*(s_n) = 0$, that is the associated Galois embedding problem has a solution.

4. Prove Serre's formula for $n = 1, 2, 3$.

5. Let $\hat{S}_n$ be the central extension of S_n by $\{\pm 1\}$ corresponding to $\varepsilon_n \cup \varepsilon_n \in H^2(S_n, \{\pm 1\})$. Let E/k be a separable field extension. Let $G_E \subset S_n$ be the Galois group of E^{gal}/k, and let $\hat{G}_E$ be the central extension of G_E corresponding to $\mathrm{Res}(\varepsilon_n \cup \varepsilon_n)$. Compute the obstruction to the associated embedding problem.

VIII

Galois cohomology of central simple algebras

In this chapter, we study the cohomology of central simple algebras with or without involution. The first half is devoted to the proof of the isomorphism

$$\mathrm{Br}(k) \simeq H^2(k, \mathbb{G}_m).$$

In the second half, we will define several quadratic forms attached to an algebra with involution. As an illustration of Galois cohomology techniques, we will compute their Hasse invariants. These quadratic forms will be useful later when dealing with rationality problems of algebraic groups.

§VIII.20 Central simple algebras

Let us recollect some results on central simple k-algebras. Let k be a field. If A is a central simple k-algebra, we denote by A^{op} the opposite k-algebra . As a set,

$$A^{op} = \{a^{op} \mid a \in A\},$$

and we define the various operations on A^{op} as follows: for all $a, b \in A, \lambda \in k$, we have

$$a^{op} + b^{op} = (a+b)^{op}, \lambda a^{op} = (\lambda a)^{op}, a^{op}b^{op} = (ba)^{op}.$$

Then A^{op} is also a central simple k-algebra. The tensor product of two central simple k-algebras is again a central simple k-algebra. In particular, if A is a central simple k-algebra, so is $\mathrm{M}_r(A) \simeq \mathrm{M}_r(k) \otimes_k A$ for all $r \geq 1$. We may then define an equivalence relation on the set of central simple k-algebras as follows. We say that two central simple

k-algebras A and B are **Brauer-equivalent** if there exist two integers $r, s \geq 1$ such that

$$\mathrm{M}_r(A) \simeq \mathrm{M}_s(B).$$

The Brauer-equivalence class of A is denoted by $[A]$, and the set of Brauer-equivalence classes is denoted by $\mathrm{Br}(k)$. One can show that the operations

$$[A] + [B] = [A \otimes_k B], -[A] = [A^{op}]$$

endow $\mathrm{Br}(k)$ with the structure of an abelian group, whose neutral element is the class of k. The group $\mathrm{Br}(k)$ is called the **Brauer group** of k. If K is a field and L/K is a field extension, then for every central simple K-algebra A, the class $[A_L]$ only depends on $[A]$. Therefore, we have a group morphism

$$\mathrm{Res}_{L/K} : \mathrm{Br}(K) \longrightarrow \mathrm{Br}(L).$$

We then get a functor $\mathrm{Br}(\_) : \mathfrak{C}_k \longrightarrow$ **AbGrps**.

Finally, a field extension L/k is called a **splitting field** of A if $A_L \simeq \mathrm{M}_n(L)$ for some $n \geq 1$. We also say that L/k splits A or that A is split by L/k. Every central simple k-algebra can be split by a finite Galois field extension L/k. A **maximal commutative subfield** of A is a subalgebra L of A which is also a field, and such that $\deg_k(A) = [L : k]$ (that is $\dim_k(A) = [L : k]^2$). A maximal commutative subfield L of A is a splitting field of A. More precisely, we have an isomorphism of L-algebras

$$f: \begin{array}{l} A_L \xrightarrow{\sim} \mathrm{End}_L(A) \\ a \otimes \lambda \longmapsto (z \longmapsto az\lambda), \end{array}$$

where A is viewed as a right L-vector space. One can show that for any central simple k-algebra A, there exists a central simple k-algebra A' which is Brauer-equivalent to A, and which contains a Galois extension L/k as a maximal commutative subfield.

If $u \in A^\times$, we denote by $\mathrm{Int}(u)$ the automorphism

$$\mathrm{Int}(u): \begin{array}{l} A \longrightarrow A \\ x \longmapsto uxu^{-1}. \end{array}$$

Such an automorphism is called **inner**. By Skolem-Noether's theorem, every k-automorphism of A is inner. We refer to [19] for proofs of these facts.

If A is a central simple k-algebra, we denote by $\mathbf{PGL}_1(A)$ the algebraic group-scheme of automorphisms of A. In particular, we have $\mathbf{PGL}_1(\mathrm{M}_n(k)) = \mathbf{PGL}_n$. If K/k is a field extension, as for the case of matrix algebras, we have

$$\mathbf{PGL}_1(A)(K) \simeq A_K^\times/K^\times,$$

since every automorphism of A_K is inner.

For any commutative k-algebra R, recall that the algebraic group-scheme $\mathbf{GL}_1(A)$ is defined by

$$\mathbf{GL}_1(A)(R) = A_R^\times$$

for any commutative k-algebra R. In particular, $\mathbf{GL}_1(\mathrm{M}_n(k)) = \mathbf{GL}_n$.

If Ω/k be a Galois extension, and A is a central simple k-algebra which is split by Ω, Galois descent shows that A is obtained by twisting $\mathrm{M}_n(\Omega)$ by an appropriate cocycle. We are going to show that the same is true at the level of invertible elements and automorphisms. In fact, this happens to be true for any pairs of k-algebras which become isomorphic over Ω.

Let A and B be two arbitrary finite dimensional k-algebras such that $A_\Omega \simeq B_\Omega$, and let γ be a cocycle with values in $\mathbf{Aut}_{alg}(A)(\Omega)$ representing the isomorphism class of B. Since any automorphism of the Ω-algebra A_Ω induces a group automorphism of $A_\Omega^\times$, γ induces by restriction a cocycle of $\mathcal{G}_\Omega$ with values in the automorphism group of $A_\Omega^\times$, that we will denote by β.

Lemma VIII.20.1. *Keeping the notation above, we have isomorphisms of $\mathcal{G}_\Omega$-groups*

$$\mathbf{GL}_1(A)(\Omega)_\beta \simeq \mathbf{GL}_1(B)(\Omega) \text{ and } \mathbf{Aut}_{alg}(A)(\Omega)_\gamma \simeq \mathbf{Aut}_{alg}(B)(\Omega).$$

Proof. Twisting by two cohomologous cocycles giving rise to two isomorphic $\mathcal{G}_\Omega$-groups, one may assume that we have

$$\gamma_\sigma = \varphi\,\sigma\cdot\varphi^{-1} \text{ for all } \sigma \in \mathcal{G}_\Omega,$$

where $\varphi : B_\Omega \xrightarrow{\sim} A_\Omega$ is an isomorphism of Ω-algebras. The map φ induces by restriction a group isomorphism $\varphi : B_\Omega^\times \xrightarrow{\sim} (A_\Omega^\times)_\beta$, since $(A_\Omega^\times)_\beta$ and $A_\Omega^\times$ are equal as abstract groups. Therefore, it remains to check that φ is $\mathcal{G}_\Omega$-equivariant. Let $\sigma \in \mathcal{G}_\Omega$, and let $b \in B_\Omega^\times$. By definition of the $\mathcal{G}_\Omega$-action on $(A_\Omega^\times)_\beta$, we have

$$\sigma * \varphi(b) = \beta_\sigma(\sigma\cdot\varphi(b)) = \gamma_\sigma(\sigma\cdot\varphi(b)) = \gamma_\sigma((\sigma\cdot\varphi)(\sigma\cdot b)).$$

Thus we have $\sigma * \varphi(b) = \varphi(\sigma \cdot b)$, meaning that φ is $\mathcal{G}_\Omega$-equivariant. This concludes the proof of the first part.

Now we have a group isomorphism

$$\psi : \begin{array}{rcl} \mathrm{Aut}(B_\Omega) & \xrightarrow{\sim} & \mathrm{Aut}(A_\Omega) \\ f & \longmapsto & \varphi \circ f \circ \varphi^{-1}. \end{array}$$

We are going to show that ψ is $\mathcal{G}_\Omega$-equivariant. For, let $\sigma \in \mathcal{G}_\Omega$, and let $f \in \mathrm{Aut}(A_\Omega)$. By definition, we have

$$\sigma * \psi(f) = \gamma_\sigma \circ (\sigma \cdot \psi(f)) \circ \gamma_\sigma^{-1}.$$

Since we have

$$\sigma \cdot \psi(f) = \sigma \cdot (\varphi \circ f \circ \varphi^{-1}) = (\sigma \cdot \varphi) \circ (\sigma \cdot f) \circ (\sigma \cdot \varphi)^{-1},$$

we get

$$\begin{array}{rcl} \sigma * \psi(f) & = & \varphi \circ (\sigma \cdot \varphi)^{-1} \circ (\sigma \cdot \varphi) \circ (\sigma \cdot f) \circ (\sigma \cdot \varphi)^{-1} \circ (\sigma \cdot \varphi) \circ \varphi^{-1} \\ & = & \varphi \circ \sigma \cdot f \circ \varphi^{-1} \\ & = & \psi(\sigma \cdot f). \end{array}$$

Hence ψ is an isomorphism of $\mathcal{G}_\Omega$-groups. This concludes the proof. □

The rest of this section is devoted to establishing an isomorphism of functors from $\mathfrak{C}_k$ to **AbGrps**

$$\mathrm{Br}(\_) \simeq H^2(\_, \mathbb{G}_m).$$

Our first goal is to define a natural transformation

$$\theta : \mathrm{Br}(\_) \longrightarrow H^2(\_, \mathbb{G}_m).$$

Let $n \geq 1$ be an integer. For every field extension K/k, let

$$\delta^1_{n,K} : H^1(K, \mathbf{PGL}_n) \longrightarrow H^2(K, \mathbb{G}_m)$$

be the first connecting map associated to the exact sequence

$$1 \longrightarrow K_s^\times \longrightarrow \mathbf{GL}_n(K_s) \longrightarrow \mathbf{PGL}_n(K_s) \longrightarrow 1.$$

By the last part of Theorem III.7.39, these maps give rise to a natural transformation of functors

$$\delta^1_n : H^1(\_, \mathbf{PGL}_n) \longrightarrow H^2(\_, \mathbb{G}_m).$$

We denote by $\theta_n : \mathbf{CSA}_n \longrightarrow H^2(\_, \mathbb{G}_m)$ the natural transformation obtained by composing δ^1_n with the bijection $\mathbf{CSA}_n \simeq H^1(\_, \mathbf{PGL}_n)$ obtained by Galois descent.

Lemma VIII.20.2. *Let K/k be a field extension. Let A and A' be central simple K-algebras of degree n and m respectively. Then the following properties hold:*

(1) *We have $\theta_{nm,K}(A \otimes_k A') = \theta_{n,K}(A) + \theta_{m,K}(A')$.*

(2) *If A and A' are Brauer-equivalent, then $\theta_{n,K}(A) = \theta_{m,K}(A')$.*

Proof. Let $\varphi : A_{K_s} \xrightarrow{\sim} \mathrm{M}_n(K_s)$ be an isomorphism of K_s-algebras, and let

$$\alpha\colon \begin{array}{l} \mathcal{G}_{K_s} \longrightarrow \mathbf{PGL}_n(K_s) \\ \sigma \longmapsto \varphi\sigma\cdot\varphi^{-1}. \end{array}$$

Write $\alpha_\sigma = \mathrm{Int}(M_\sigma)$, for some $M_\sigma \in \mathbf{GL}_n(K_s)$. Then by definition, $\theta_{n,K}(A)$ is the cohomology class represented by the cocycle $\beta \in Z^2(\mathcal{G}_{K_s}, K_S^\times)$ defined by

$$M_\sigma\sigma\cdot M_\tau M_{\sigma\tau}^{-1} = \beta_{\sigma,\tau} I_n \text{ for all } \sigma,\tau \in \mathcal{G}_{K_s}.$$

Similarly, let $\varphi' : A'_{K_s} \xrightarrow{\sim} \mathrm{M}_m(K_s)$ be an isomorphism of K_s-algebras, let

$$\alpha'\colon \begin{array}{l} \mathcal{G}_{K_s} \longrightarrow \mathbf{PGL}_m(K_s) \\ \sigma \longmapsto \varphi'\sigma\cdot\varphi'^{-1}, \end{array}$$

and write $\alpha'_\sigma = \mathrm{Int}(M'_\sigma)$, for some $M'_\sigma \in \mathbf{GL}_m(K_s)$, so that $\theta_{n,K}(A')$ is the cohomology class represented by the cocycle $\beta' \in Z^2(\mathcal{G}_{K_s}, K_S^\times)$ defined by

$$M'_\sigma\sigma\cdot M'_\tau M'^{-1}_{\sigma\tau} = \beta'_{\sigma,\tau} I_n \text{ for all } \sigma,\tau \in \mathcal{G}_{K_s}.$$

Let $u : \mathrm{M}_n(K_s) \otimes_{K_s} \mathrm{M}_m(K_s) \xrightarrow{\sim} \mathrm{M}_{nm}(K_s)$ be the isomorphism of K_s-algebras given by the Kronecker product of matrices. The map

$$u \circ (\varphi \otimes \varphi') : (A \otimes_K A') \otimes_K K_s \longrightarrow \mathrm{M}_{nm}(K_s)$$

is then an isomorphism of K_s-algebras. Straightforward computations show that we have

$$\sigma\cdot(u \circ (\varphi \otimes \varphi')) = u \circ (\sigma\cdot\varphi \otimes \sigma\cdot\varphi') \text{ for all } \sigma \in \mathcal{G}_{K_s}.$$

Hence a cocycle representing $A \otimes_K B$ is given by

$$\alpha''\colon \begin{array}{l} \mathcal{G}_{K_s} \longrightarrow \mathbf{PGL}_{nm}(K_s) \\ \sigma \longmapsto u \circ (\alpha_\sigma \otimes \alpha'_\sigma) \circ u^{-1}. \end{array}$$

Now we have

$$u \circ (\mathrm{Int}(M_\sigma) \otimes \mathrm{Int}(M'_\sigma)) = u \circ (\mathrm{Int}(M_\sigma \otimes M'_\sigma)) = \mathrm{Int}(u(M_\sigma \otimes M'_\sigma)) \circ u,$$

and thus $\alpha''_\sigma = \mathrm{Int}(M''_\sigma)$, where $M''_\sigma = u(M_\sigma \otimes M'_\sigma)$ for all $\sigma \in \mathcal{G}_{K_s}$. We easily have

$$\begin{aligned} M''_\sigma \sigma \cdot M''_\tau M''^{-1}_{\sigma\tau} &= u(\beta_{\sigma,\tau} I_n \otimes \beta'_{\sigma,\tau} I_n) \\ &= \beta_{\sigma,\tau}\beta'_{\sigma,\tau} u(I_n \otimes I_m) \\ &= \beta_{\sigma,\tau}\beta'_{\sigma,\tau} I_{nm}. \end{aligned}$$

It follows that $\theta_{nm}(A \otimes_K A')$ is represented by the cocycle $\beta\beta'$, that is

$$\theta_{nm,K}(A \otimes_K A') = [\beta\beta'] = [\beta] + [\beta'] = \theta_{n,K}(A) + \theta_{m,K}(A').$$

Assume now that A and A' are Brauer-equivalent. Let $r, s \geq 1$ be two integers such that $\mathrm{M}_r(A) \simeq \mathrm{M}_s(A')$. Notice that we have $rn = sm$, as we can see by comparing dimensions. To prove the desired result, it is enough to check that

$$\theta_{n,K}(A) = \theta_{nr,K}(\mathrm{M}_r(A)) \text{ for all } r \geq 1.$$

Indeed, we will have

$$\theta_{n,K}(A) = \theta_{nr,K}(\mathrm{M}_r(A)) = \theta_{sm,K}(\mathrm{M}_r(A)) = \theta_{sm,K}(\mathrm{M}_s(A')),$$

that is $\theta_{n,K}(A) = \theta_{m,K}(A')$. Notice that we implicitely used here the fact that $\theta_{n,K}$ maps K-isomorphic algebras onto the same cohomology class, which is obvious from its definition. Now $\theta_{r,K}(\mathrm{M}_r(K)) = 0$, since $\mathrm{M}_r(K)$ corresponds to the class of the trivial cocycle in $H^1(K, \mathbf{PGL}_r)$. By the first point, we have

$$\theta_{nr,K}(\mathrm{M}_r(A)) = \theta_{nr,K}(\mathrm{M}_r(K) \otimes_K A) = \theta_{r,K}(\mathrm{M}_r(K)) + \theta_{n,K}(A),$$

and therefore $\theta_{nr,K}(\mathrm{M}_r(A)) = \theta_{n,K}(A)$. This concludes the proof. □

Let $[A] \in \mathrm{Br}(K)$, and let $n = \deg_k(A)$. We set $\theta_K([A]) = \theta_{n,K}(A)$. Lemma VIII.20.2 implies immediately that the map

$$\theta_K : \mathrm{Br}(K) \longrightarrow H^2(K, \mathbb{G}_m)$$

is a well-defined group morphism. The various maps θ_K then induce a natural transformation of functors

$$\theta : \mathrm{Br}(\_) \longrightarrow H^2(\_, \mathbb{G}_m).$$

Indeed, if L/K is a field extension and A is a central simple K-algebra of degree n, then A_L has also degree n over L, and we have

$$\theta_L([A_L]) = \theta_{n,L}(A_L) = \mathrm{Res}_{L/K}(\theta_{n,K}(A)) = \mathrm{Res}_{L/K}(\theta_K([A])).$$

Lemma VIII.20.3. *The natural transformation of functors*

$$\theta : \mathrm{Br}(\_) \longrightarrow H^2(\_, \mathbb{G}_m)$$

is injective.

Proof. Let K/k be a field extension and let $[A] \in \mathrm{Br}(K)$ such that $\theta_K([A]) = 0$. Let $n = \deg_K(A)$. We then have

$$\theta_K([A]) = \theta_{n,K}(A) = \delta^1_{n,K}([\alpha]) = 0,$$

where $[\alpha] \in H^1(K, \mathbf{PGL}_n)$ is the cohomology class corresponding to A. Applying cohomology to the exact sequence

$$1 \longrightarrow K_s^\times \longrightarrow \mathbf{GL}_n(K_s) \longrightarrow \mathbf{PGL}_n(K_s) \longrightarrow 1$$

and Hilbert 90 show that $\delta^1_{n,K}$ has trivial kernel. Hence, we get $[\alpha] = 1$ and thus $A \simeq \mathrm{M}_n(K)$. Therefore, $[A] = 0$ and θ_K is injective. □

Lemma VIII.20.4. *Let L/K be a finite Galois extension of degree n. The connecting map $H^1(\mathcal{G}_L, \mathbf{PGL}_n(L)) \longrightarrow H^2(\mathcal{G}_L, L^\times)$ associated to the exact sequence of $\mathcal{G}_L$-groups*

$$1 \longrightarrow L^\times \longrightarrow \mathbf{GL}_n(L) \longrightarrow \mathbf{PGL}_n(L) \longrightarrow 1$$

is bijective.

Proof. Let $[c] \in H^2(\mathcal{G}_L, L^\times)$. Consider the product $V = \prod_{g \in \mathcal{G}_L} L$ of n copies of L, indexed by the elements of $\mathcal{G}_L$, and let $(e_g)_{g \in \mathcal{G}_L}$ be the corresponding canonical basis of this product, considered as a right L-vector space. For $g \in \mathcal{G}_L$, let $a_g \in \mathrm{GL}(V)$ defined by

$$a_g(e_{g'}) = e_{gg'} c_{g,g'} \text{ for all } g' \in \mathcal{G}_L.$$

For all $g, g', g'' \in G$, we have

$$a_g(g \cdot a_{g'}(e_{g''})) = a_g(e_{g'g''} g \cdot c_{g',g''}) = e_{gg'g''} c_{g,g'g''}\, g \cdot c_{g',g''}.$$

Since c is a 2-cocycle, we have $g \cdot c_{g',g''} c_{g,g'g''} = c_{gg',g''} c_{g,g'}$, and therefore

$$a_g(g \cdot a_{g'}(e_{g''})) = e_{gg'g''} c_{gg',g''} c_{g,g'} = a_{gg'}(e_{g''}) c_{g,g'}.$$

We then have $a_g \circ g \cdot a_{g'} \circ a_{gg'}^{-1} = c_{g,g'} \mathrm{Id}_V$.

Let $M_g \in \mathrm{GL}_n(L)$ be the matrix of a_g in the basis $(e_g)_{g \in \mathcal{G}_L}$. It follows from the previous computation that the map

$$\alpha : \begin{array}{l} \mathcal{G}_L \times \mathcal{G}_L \longrightarrow L^\times \\ \qquad\quad g \longmapsto \mathrm{Int}(M_g) \end{array}$$

is a cocycle with values in $\mathbf{PGL}_n(L)$ whose cohomology class maps onto $[c]$ under the connecting map $\delta^1_L : H^1(\mathcal{G}_L, \mathbf{PGL}_n(L)) \longrightarrow H^2(\mathcal{G}_L, L^\times)$.

Let us prove the injectivity of this connecting map. Notice first that, as before, δ^1_L has trivial kernel. Now let $\gamma \in Z^1(\mathcal{G}_L, \mathbf{PGL}_n(L))$ be a cocycle. The cocycle $\beta \in Z^1(\mathcal{G}_L, \mathbf{Aut}(\mathbf{GL}_n(L)))$ deduced from γ as explained just before Proposition II.5.6 is easily seen to be

$$\begin{aligned} \mathcal{G}_L &\longrightarrow \mathbf{Aut}(\mathbf{GL}_n(L)) \\ \sigma &\longmapsto \gamma_{\sigma_{|\mathbf{GL}_n(L)}}. \end{aligned}$$

By the first part of Proposition II.5.6, we have an exact sequence

$$1 \longrightarrow L^\times \longrightarrow \mathbf{GL}_n(L)_\beta \longrightarrow \mathbf{PGL}_n(L)_\gamma \longrightarrow 1,$$

and by Lemma VIII.20.1, this is nothing but that the exact sequence

$$1 \longrightarrow L^\times \longrightarrow \mathbf{GL}_1(A)(L) \longrightarrow \mathbf{PGL}_1(A)(L) \longrightarrow 1$$

where A is the central simple K-algebra corresponding to $[\gamma]$. We denote by $\delta^1_{A,L}$ the corresponding connecting map. Since $H^1(\mathcal{G}_L, \mathbf{GL}_1(A)(L)) = 1$ by Hilbert 90, applying cohomology to the exact sequence above shows that $\delta^1_{A,L}$ has trivial kernel. By Lemma II.5.5, δ^1_L is injective, and this concludes the proof. □

Remark VIII.20.5. By Galois descent, for any $c \in Z^2(\mathcal{G}_L, L^\times)$, the cohomology class $[c] \in H^2(\mathcal{G}_L, L^\times)$ corresponds to the isomorphism class of the central simple K-algebra

$$A_c = \{M \in \mathrm{M}_n(L) \mid M_g\, \sigma{\cdot}M = MM_g \text{ for all } g \in \mathcal{G}_L\}.$$

We now have the following result:

Theorem VIII.20.6. *The natural transformation of functors*

$$\theta : \mathrm{Br}(\_) \longrightarrow H^2(\_, \mathbb{G}_m)$$

is an isomorphism. Moreover, for every field extension K/k and every central simple K-algebra A of degree n, the connecting map

$$\delta^1_{n,K} : H^1(K, \mathbf{PGL}_n) \longrightarrow H^2(K, \mathbb{G}_m)$$

corresponds to the map

$$\begin{aligned} \mathbf{CSA}_n(K) &\longrightarrow \mathrm{Br}(K) \\ A &\longmapsto [A]. \end{aligned}$$

Proof. Let K/k be a field extension. By Lemma VIII.20.3, θ_K is injective. Let us prove the surjectivity of θ_K. Let $[\xi] \in H^2(K, \mathbb{G}_m)$. By Theorem III.7.30, there exists a finite Galois extension L/K and $[c] \in H^2(\mathcal{G}_L, L^\times)$ such that $[\xi] = \rho_L([c])$, where

$$\rho_L : H^2(\mathcal{G}_L, L^\times) \longrightarrow H^2(K, \mathbb{G}_m)$$

is the map induced in cohomology by the compatible maps

$$\mathcal{G}_\Omega \longrightarrow \mathcal{G}_L \text{ and } L^\times \longrightarrow K_s^\times.$$

Abusing notation, we will also denote by ρ_L the map

$$H^1(\mathcal{G}_L, \mathbf{PGL}_n(L)) \longrightarrow H^1(K, \mathbf{PGL}_n).$$

By Lemma VIII.20.4, there exists $[\alpha] \in H^1(\mathcal{G}_L, \mathbf{PGL}_n(L))$ which is mapped onto $[c]$ by the connecting map

$$H^1(\mathcal{G}_L, \mathbf{PGL}_n(L)) \longrightarrow H^2(\mathcal{G}_L, L^\times).$$

It is easy to check that the diagram

$$\begin{array}{ccc} H^1(\mathcal{G}_L, \mathbf{PGL}_n(L)) & \longrightarrow & H^2(\mathcal{G}_L, L^\times) \\ \downarrow & & \downarrow \\ H^1(K, \mathbf{PGL}_n) & \longrightarrow & H^2(K, \mathbb{G}_m) \end{array}$$

is commutative. Therefore, $\rho_L([\alpha])$ is mapped onto $\rho_L([c]) = [\xi]$ by $\delta^1_{n,K}$. Notice that the isomorphism class of A_c then corresponds to the cohomology class $\rho_L([\alpha]) \in H^1(K, \mathbf{PGL}_n)$ by Remark VIII.20.5 and Remark III.8.18. Thus we get $\theta_K([A_c]) = [\xi]$ and θ_K is surjective.

We now prove the last statement of the theorem. Let A be a central simple K-algebra of degree n, and let $[\alpha] \in H^1(K, \mathbf{PGL}_n)$ be the corresponding cohomology class. We have to prove that $\theta_K^{-1}(\delta^1_{n,K}([\alpha])) = [A]$, that is $\delta^1_{n,K}([\alpha]) = \theta_K([A])$. But this comes from the definition of θ_K. This concludes the proof. □

Remark VIII.20.7. This result admits a slight generalization. For any Galois field extension Ω/K, we have a group isomorphism

$$\mathrm{Br}(\Omega/K) \simeq H^2(\mathcal{G}_\Omega, \Omega^\times),$$

where $\mathrm{Br}(\Omega/K) = \ker(\mathrm{Br}(K) \longrightarrow \mathrm{Br}(\Omega))$, and the connecting map

$$H^1(\mathcal{G}_\Omega, \mathbf{PGL}_n(\Omega)) \longrightarrow H^2(\mathcal{G}_\Omega, \Omega^\times)$$

corresponds to $A \longmapsto [A]$.

The arguments needed to prove this result are almost identical to the ones used to establish the previous theorem. Since we will not really need this generalization, we leave the details to the reader.

Corollary VIII.20.8. *Let $n \geq 2$, and let k be a field whose characteristic does not divide n. We have a group isomorphism*

$$H^2(k, \mu_n) \simeq \mathrm{Br}_n(k),$$

where $\mathrm{Br}_n(k)$ is the n-torsion part of $\mathrm{Br}(k)$.

Proof. The exact sequence

$$1 \longrightarrow \mu_n \longrightarrow k_s^\times \xrightarrow{\times^n} k_s \longrightarrow 1$$

induces an exact sequence in cohomology (using Hilbert 90)

$$1 \longrightarrow H^2(k, \mu_n) \longrightarrow H^2(k, \mathbb{G}_m) \longrightarrow H^2(k, \mathbb{G}_m),$$

where the last map is induced by raising to the nth-power. Hence the abelian group $H^2(k, \mu_n)$ identifies to the n-torsion part of $H^2(k, \mathbb{G}_m)$, and therefore to the n-torsion part of $\mathrm{Br}(k)$ by the previous theorem. □

Remark VIII.20.9. The isomorphism above is described as follows: if $[A] \in \mathrm{Br}_n(k) \subset \mathrm{Br}(k)$, then the corresponding cohomology class $[\alpha] \in H^2(k, \mu_n)$ is the unique element of $H^2(k, \mu_n)$ which satisfies

$$\iota_*([\alpha]) = \theta_k([A]),$$

where $\iota_* : H^2(k, \mu_n) \longrightarrow H^2(k, \mathbb{G}_m)$ is the map induced by the inclusion $\mu_n(k_s) \subset k_s^\times$.

Corollary VIII.20.10. *Let k be a field of characteristic different from 2. The isomorphism*

$$\mathrm{Br}_2(k) \simeq H^2(k, \mu_2)$$

maps the Brauer class of a quaternion algebra (a, b) onto $(a) \cup (b)$.

Proof. The commutative diagram

$$\begin{array}{ccccccccc}
1 & \longrightarrow & \mu_2 & \longrightarrow & \mathbf{SL}_2(k_s) & \longrightarrow & \mathbf{PGL}_2(k_s) & \longrightarrow & 1 \\
 & & \downarrow & & \downarrow & & \| & & \\
1 & \longrightarrow & k_s^\times & \longrightarrow & \mathbf{GL}_2(k_s) & \longrightarrow & \mathbf{PGL}_2(k_s) & \longrightarrow & 1
\end{array}$$

induces a commutative diagram

$$\begin{array}{ccc} H^1(k, \mathbf{PGL}_2) & \longrightarrow & H^2(k, \mu_2) \\ \| & & \downarrow \\ H^1(k, \mathbf{PGL}_2) & \xrightarrow{\delta^1} & H^2(k, \mathbb{G}_m) \end{array}$$

by Theorem III.7.39 (4). Let $[\xi] \in H^1(k, \mathbf{PGL}_2)$ be the cohomology class corresponding to the isomorphism class of Q. By Proposition III.9.19 and the commutativity of the diagram above, we have

$$\theta_k([Q]) = \delta^1([\xi]) = \iota_*((a) \cup (b)),$$

the first equality coming from the definition of θ. The result follows. □

§VIII.21 Algebras with involutions

In this section, k is a field, and A is a central simple k-algebra. To simplify the exposition, we will assume that $\mathrm{char}(k) \neq 2$.

VIII.21.1 Basic concepts

Definition VIII.21.1. An **involution** $\sigma : A \longrightarrow A$ is a ring anti-automorphism of order dividing 2.

In other words, an involution is a map $\sigma : A \longrightarrow A$ satisfying for all $x, y \in A$:

(1) $\sigma(x + y) = \sigma(x) + \sigma(y)$.

(2) $\sigma(1) = 1$.

(3) $\sigma(xy) = \sigma(y)\sigma(x)$.

(4) $\sigma(\sigma(x)) = x$.

For example, the transposition is an involution on $\mathrm{M}_n(k)$.

An element $x \in A$ will be called **symmetric** if $\sigma(x) = x$, and **skew-symmetric** if $\sigma(x) = -x$. We denote by $\mathrm{Sym}(A, \sigma)$ the set of symmetric elements, and by $\mathrm{Skew}(A, \sigma)$ the set of skew-symmetric elements. We also set

$$\mathrm{Sym}(A, \sigma)^\times = \mathrm{Sym}(A, \sigma) \cap A^\times \text{ and } \mathrm{Skew}(A, \sigma)^\times = \mathrm{Skew}(A, \sigma) \cap A^\times.$$

It easy to check that for every $\lambda \in k$, $\sigma(\lambda)$ lies in the center of A, that is k. Hence $\sigma_{|_k}$ is an automorphism of order dividing 2 of k.

We set

$$k_0 = \{\lambda \in k | \sigma(\lambda) = \lambda\}.$$

We say that σ is an **involution of the first kind** if $\sigma_{|_k} = \mathrm{Id}_k$, that is if $k = k_0$, and an **involution of the second kind (or unitary)** otherwise. In this latter case, k/k_0 is a quadratic field extension, and $\sigma_{|_k}$ is the unique non-trivial k_0-automorphism of k/k_0.

Remark VIII.21.2. Let A be a central simple k-algebra. If A carries an involution of the first kind, then $[A] \in \mathrm{Br}_2(k)$. Indeed, if σ is such an involution, the map

$$f: \begin{array}{l} A \longrightarrow A^{op} \\ a \longmapsto (\sigma(a))^{op} \end{array}$$

is easily checked to be an isomorphism of k-algebras. In particular, we have $[A] = [A^{op}] = -[A]$, that is $2[A] = 0$.

From now on, we will only consider involutions of the first kind.

Proposition VIII.21.3. *Let σ, σ' be two involutions of the first kind on A. Then there exists $u \in A^\times$ such that $\sigma' = \mathrm{Int}(u) \circ \sigma$ and $\sigma(u) = \pm u$. Moreover, u is uniquely determined up to multiplication by an element of $k^\times$.*

Proof. Notice that $\sigma' \circ \sigma^{-1}$ is a k-algebra automorphism of A. By Skolem-Noether's theorem, there exists $u \in A^\times$ such that $\sigma' \circ \sigma^{-1} = \mathrm{Int}(u)$, that is

$$\sigma' = \mathrm{Int}(u) \circ \sigma.$$

Easy computations show that we have $\sigma \circ \mathrm{Int}(u) = \mathrm{Int}(\sigma^{-1}(u)) \circ \sigma$. Thus, we have

$$\sigma'^2 = \mathrm{Int}(u) \circ \mathrm{Int}(\sigma^{-1}(u)) \circ \sigma^2 = \mathrm{Int}(u\sigma^{-1}(u)) \circ \sigma^2.$$

Since σ and σ' have order dividing 2, we get $\mathrm{Id}_A = \mathrm{Int}(u\sigma^{-1}(u))$, so $u\sigma^{-1}(u)$ lies in the center of A. Hence there exists $\lambda \in k$ such that

$$u = \lambda\sigma(u).$$

We then have

$$u = \lambda\sigma(\lambda\sigma(u)) = \lambda u \sigma(\lambda) = \lambda\sigma(\lambda)u,$$

since $\sigma(\lambda)$ lies in the center of A. Since $u \in A^\times$, we get $\lambda\sigma(\lambda) = \lambda^2 = 1$. Therefore $\lambda = \pm 1$, and it follows that $\sigma(u) = \pm u$.

Finally, if $\sigma' = \text{Int}(u_1)\circ\sigma = \text{Int}(u_2)\circ\sigma$, where u_1, u_2 are both symmetric or skew-symmetric, we have

$$\text{Int}(u_2 u_1^{-1}) \circ \sigma = \sigma.$$

It implies that $\text{Int}(u_2u_1^{-1}) = \text{Id}_A$. Thus, as before, there exists $\lambda \in k^\times$ such that

$$u_2 = \lambda u_1.$$

This concludes the proof. □

Remark VIII.21.4. This proposition shows in particular that involutions of first kind on $\text{M}_n(k)$ have the form $\sigma_B = \text{Int}(B)\circ^t$, where B is an invertible symmetric or skew-symmetric matrix.

Lemma VIII.21.5. *Let σ be an involution on A, and let $u \in A^\times$ such that $\sigma(u) = \pm u$. Set $\sigma' = \text{Int}(u) \circ \sigma$.*

(1) *If $\sigma(u) = u$, we have*

$$\text{Sym}(A, \sigma') = u\text{Sym}(A, \sigma) = \text{Sym}(A, \sigma)u^{-1}$$

and

$$\text{Skew}(A, \sigma') = u\text{Skew}(A, \sigma) = \text{Skew}(A, \sigma)u^{-1}.$$

(2) *If $\sigma(u) = -u$, we have*

$$\text{Sym}(A, \sigma') = u\text{Skew}(A, \sigma) = \text{Skew}(A, \sigma)u^{-1}$$

and

$$\text{Skew}(A, \sigma') = u\text{Sym}(A, \sigma) = \text{Sym}(A, \sigma)u^{-1}.$$

Proof. Assume that $\sigma(u) = \varepsilon u$. For all $x \in A$, we have

$$\begin{aligned} \sigma'(x) &= u\sigma(x)u^{-1} \\ &= \varepsilon(\varepsilon u)\sigma(x)u^{-1} \\ &= \varepsilon\sigma(u)\sigma(x)u^{-1} \\ &= \varepsilon\sigma(xu)u^{-1}. \end{aligned}$$

Similarly, we have $\sigma'(x) = \varepsilon u\sigma(u^{-1}x)$. Therefore, we have

$$\sigma'(x) = \varepsilon' x \iff \sigma(xu) = \varepsilon\varepsilon' xu \iff \sigma(u^{-1}x) = \varepsilon\varepsilon' u^{-1}x.$$

The lemma follows easily. □

We now introduce the concept of isomorphic algebras with involutions.

Definition VIII.21.6. We say that two central simple k-algebras with involutions (A, σ) and (A', σ') are **isomorphic** if there exists an isomorphism of k-algebras $f : A \xrightarrow{\sim} A'$ such that

$$\sigma' \circ f = f \circ \sigma.$$

In this case, it is easy to check that σ and σ' are involutions of the same kind. Moreover, f then induces isomorphisms of k-vector spaces

$$\mathrm{Sym}(A, \sigma) \simeq \mathrm{Sym}(A', \sigma') \text{ and } \mathrm{Skew}(A, \sigma) \simeq \mathrm{Skew}(A', \sigma').$$

Two involutions σ, σ' on A are **conjugate** if (A, σ) and (A, σ') are isomorphic. Since every automorphism of A is inner, it follows that σ and σ' are conjugate if and only if there exists $a \in A^\times$ such that

$$\sigma' = \mathrm{Int}(a\sigma(a)) \circ \sigma.$$

Let σ be an involution of the first kind on A, let L/k be a splitting field of A, and let

$$f : A_L \xrightarrow{\sim} \mathrm{M}_n(L)$$

be an isomorphism of L-algebras. Let us denote by σ_L the involution $\sigma \otimes \mathrm{Id}_L$. Then $f \circ \sigma_L \circ f^{-1}$ is an involution of the first kind on $\mathrm{M}_n(L)$. In view of Remark VIII.21.4, $f \circ \sigma_L \circ f^{-1} = \sigma_B$ for some invertible symmetric or skew-symmetric matrix. In other words, for every splitting field L/k, there exists $B \in \mathrm{GL}_n(L)$ and $B^t = \pm B$ such that

$$(A_L, \sigma_L) \simeq (\mathrm{M}_n(L), \sigma_B).$$

The next result allows us to decide in which case B is symmetric or skew-symmetric.

Proposition VIII.21.7. *Assume that σ is an involution of the first kind on a central simple k-algebra A of degree n. Then we have*

$$\dim_k(\mathrm{Sym}(A, \sigma)) = \frac{n(n+1)}{2} \text{ or } \frac{n(n-1)}{2}.$$

The second case only holds if n is even.

Moreover, we have $\dim_k(\mathrm{Sym}(A, \sigma)) = \dfrac{n(n+\varepsilon)}{2}$ if and only if for every splitting field L/k of A, (A_L, σ_L) is isomorphic to $(\mathrm{M}_n(L), \sigma_B)$ for some $B \in \mathrm{GL}_n(L)$ satisfying $B^t = \varepsilon B$.

Proof. For every field extension L/k, it is easy to check that we have

$$\mathrm{Sym}(A_L, \sigma_L) = \mathrm{Sym}(A, \sigma)_L.$$

Assume that L/k is a splitting field of A, so (A_L, σ_L) is isomorphic to $(\mathrm{M}_n(L), \sigma_B)$ for some invertible symmetric or skew-symmetric matrix B. We then have

$$\mathrm{Sym}(A, \sigma)_L = \mathrm{Sym}(A_L, \sigma_L) \simeq \mathrm{Sym}(\mathrm{M}_n(L), \sigma_B).$$

Therefore, we get

$$\dim_k(\mathrm{Sym}(A, \sigma)) = \dim_L(\mathrm{Sym}(A, \sigma)_L) = \dim_L(\mathrm{Sym}(\mathrm{M}_n(L), \sigma_B)).$$

Assume first that B is symmetric. By Lemma VIII.21.5, we have

$$\mathrm{Sym}(\mathrm{M}_n(L), \sigma_B) = B\,\mathrm{Sym}(\mathrm{M}_n(L), t),$$

and therefore we get

$$\dim_k(\mathrm{Sym}(A, \sigma)) = \dim_L(\mathrm{Sym}(\mathrm{M}_n(L), t)) = \frac{n(n+1)}{2}.$$

If B is skew-symmetric, by Lemma VIII.21.5, we have

$$\mathrm{Sym}(\mathrm{M}_n(L), \sigma_B) = B\,\mathrm{Skew}(\mathrm{M}_n(L), t),$$

and therefore we get

$$\dim_k(\mathrm{Sym}(A, \sigma)) = \dim_L(\mathrm{Skew}(\mathrm{M}_n(L), t)) = \frac{n(n-1)}{2}.$$

Moreover in this case, we have

$$\det(B) = \det(B^t) = \det(-B) = (-1)^n \det(B).$$

Since B is invertible, we get $1 = (-1)^n$, which implies that n is even. This concludes the proof. □

Definition VIII.21.8. We say that an involution σ of the first kind on a central simple k-algebra A of degree n is **orthogonal** (or of **type** 1) if for any splitting field L of A, the involution σ_L is isomorphic to σ_B for some symmetric invertible matrix $B \in \mathrm{M}_n(L)$. It is equivalent to say that $\dim_k(\mathrm{Sym}(A, \sigma)) = \dfrac{n(n+1)}{2}$. We say that σ is **symplectic** (or of **type** -1) if, for any splitting field L of A, the involution σ_L is isomorphic to σ_B for some skew-symmetric invertible matrix $B \in \mathrm{M}_n(L)$. It is equivalent to say that $\dim_k \mathrm{Sym}((A, \sigma)) = \dfrac{n(n-1)}{2}$. If A carries a symplectic involution, then n is necessarily even.

Examples VIII.21.9.

(1) The involution σ_B on $\mathrm{M}_n(k)$ is orthogonal if B is symmetric and symplectic if B is skew-symmetric.

(2) Let $Q = (a, b)_k$. The map

$$\gamma\colon \begin{array}{rcl} Q & \longrightarrow & Q \\ x + yi + zj + tij & \longmapsto & x - yi - zj - tij \end{array}$$

is a symplectic involution on Q.

Let K be a field. If $\varepsilon = \pm 1$ and $n \geq 1$, we denote by $\mathbf{CSA}_n^{\varepsilon}(K)$ the set of isomorphism classes of algebras with involution (A, σ), where $\deg_K(A) = n$ and σ has type ε. If L/K is a field extension and $(A, \sigma) \in \mathbf{CSA}_n^{\varepsilon}(K)$, then $(A_L, \sigma_L) \in \mathbf{CSA}_n^{\varepsilon}(L)$.

We then get a functor $\mathbf{CSA}_n^{\varepsilon} : \mathfrak{C}_k \longrightarrow \mathbf{Sets}$.

VIII.21.2 Hyperbolic involutions

In this section, we define the concept of a hyperbolic involution, which generalizes in some sense the notion of a hyperbolic quadratic form or hyperbolic alternating form. Let us start with some easy considerations.

Lemma VIII.21.10. *Let σ be an involution on $\mathrm{M}_n(k)$.*

(1) *If σ is symplectic, then there exists $E \in \mathrm{M}_n(k)$ satisfying $E^2 = E$ and $\sigma(E) = I_n - E$.*

(2) *If $\sigma = \sigma_B$ is orthogonal, then there exists $E \in \mathrm{M}_n(k)$ satisfying $E^2 = E$ and $\sigma(E) = I_n - E$ if and only if the quadratic form associated to B is hyperbolic.*

In both cases, such an element E satisfies $\mathrm{rank}(E) = \dfrac{n}{2}$. In particular n is even.

Proof. Write $\sigma = \sigma_B$ for some $B \in \mathrm{GL}_n(k)$ satisfying $B^t = \pm B$, and let

$$b\colon \begin{array}{rcl} k^n \times k^n & \longrightarrow & k \\ (X, Y) & \longmapsto & X^t B Y \end{array}$$

be the bilinear form associated to B.

Assume first that an idempotent E of $\mathrm{M}_n(k)$ such that $\sigma(E) = I_n - E$

exists. Since $E^2 = E$, the endomorphisms corresponding to E and $I_n - E$ are projectors. Thus we have

$$k^n = \mathrm{Im}(E) \oplus \mathrm{Im}(I_n - E).$$

But $\sigma(E) = I_n - E$ and E have the same rank (since the rank is preserved by transposition and multiplication by an invertible matrix). Therefore $n = 2\mathrm{rank}(E)$.

Assume now that $B^t = -B$, so that σ is symplectic. In particular, $n = 2m$. It is known in this case that there exists a basis $E_1, F_1, \ldots, E_m, F_m$ of k^n such that

$$b(E_i, F_i) = 1, b(F_i, E_j) = b(F_i, F_j) = 0 \text{ for all } i \neq j.$$

It follows that there exists $P \in \mathrm{GL}_n(k)$ such that

$$B = P^t \begin{pmatrix} 0 & I_m \\ -I_m & 0 \end{pmatrix} P.$$

Let $E = P^t \begin{pmatrix} I_m & 0 \\ 0 & 0 \end{pmatrix} (P^t)^{-1}$. It is easy to check that we have

$$E^2 = E \text{ and } \sigma_B(E) = I_n - E.$$

Assume that $B^t = B$, so that σ is orthogonal and b is symmetric. Suppose that the quadratic form

$$q\colon \begin{array}{l} k^n \longrightarrow k \\ X \longmapsto X^t B X \end{array}$$

is hyperbolic. Then it is known that in this case that $n = 2m$, and that there exists a basis $E_1, F_1, \ldots, E_m, F_m$ of k^n such that

$$b(E_i, F_i) = 1, b(F_i, E_j) = b(F_i, F_j) = 0 \text{ for all } i \neq j.$$

Similar computations to those done in the previous case show that there exists $E \in \mathrm{M}_n(k)$ satisfying

$$E^2 = E \text{ and } \sigma_B(E) = I_n - E.$$

Conversely, assume that such an element $E \in \mathrm{M}_n(k)$ exists. We have to prove that q has a totally isotropic subspace of dimension m. Let W be the image of E^t. Then $\dim_k(W) = \mathrm{rank}(E) = m$. We claim that W is totally isotropic. To check it, it is enough to show that $EBE^t = 0$.

By assumption, we have $E^2 = E$ and $BE^tB^{-1} = I_n - E$. Therefore we have

$$EBE^tB^{-1} = E - E^2 = 0,$$

and therefore $EBE^t = 0$. Hence q is hyperbolic. □

This result motivates the following definition:

Definition VIII.21.11. Let (A, σ) be a central simple k-algebra with an involution of the first kind. We say that σ is **hyperbolic** if there exists an **idempotent** $e \in A$ such that $\sigma(e) = 1 - e$.

Remark VIII.21.12. Notice that if A is a division algebra, then A does not carry any hyperbolic involution. Indeed, the only idempotents e of A are 0 and 1 in this case, which clearly do not satisfy the relation $\sigma(e) = 1 - e$ for any involution σ of the first kind. In fact, if A carries a hyperbolic involution, one can show that $A = \mathrm{M}_2(B)$ for some central simple k-algebra B. See [30] for more details.

Example VIII.21.13. Let $\sigma = \sigma_B$ be an involution of the first kind on $\mathrm{M}_n(k)$. By Lemma VIII.21.10, if σ is symplectic, then σ is hyperbolic, and if σ is orthogonal, σ is hyperbolic if and only the quadratic form

$$q\colon \begin{array}{l} k^n \longrightarrow k \\ X \longmapsto X^tBX \end{array}$$

is hyperbolic.

It is known that hyperbolic quadratic forms are isomorphic. This easily implies that all hyperbolic involutions on $\mathrm{M}_n(k)$ of same type are conjugate. We would like to generalize this fact to hyperbolic involutions on an arbitrary central simple k-algebra. We start with a lemma:

Lemma VIII.21.14. *Let A be a central simple k-algebra. If A carries a hyperbolic involution, then A has even degree. Moreover, if σ, σ' are two hyperbolic involutions of the same type on A and $e, e' \in A$ are idempotents such that $\sigma(e) = 1 - e$ and $\sigma'(e) = 1 - e'$, then there exists $a \in A^\times$ such that $e' = aea^{-1}$.*

Proof. Assume first that $A = \mathrm{M}_n(k)$, and let σ be a hyperbolic involution on A. In this case, we know from Lemma VIII.21.10 that $n = 2m$, where m is the rank of any idempotent E satisfying $\sigma(E) = I_n - E$. Now let σ' be another hyperbolic involution of the same type, and let E' be an idempotent satisfying $\sigma'(E') = I_n - E'$. We then have

$\mathrm{rank}(E') = \mathrm{rank}(E) = m$. Hence there exist two invertible matrices P and P' such that

$$PEP^{-1} = P'E'P'^{-1} = \begin{pmatrix} I_m & 0 \\ 0 & 0 \end{pmatrix}.$$

We then have $(P'^{-1}P)E(P'^{-1}P)^{-1} = E'$, and the result is proved in this case.

Let us go back to the general case. If k is finite, A is isomorphic to a matrix algebra, and the result is already known. Hence one may assume that k is infinite. Let $\varphi : A_{k_s} \xrightarrow{\sim} \mathrm{M}_n(k_s)$ be an isomorphism of k_s-algebras, let $\sigma_s = \varphi \circ \sigma_{k_s} \circ \varphi^{-1}$ and let $E = \varphi(e \otimes 1)$. Then $E^2 = E$ and $\sigma_s(E) = I_n - E$. Hence σ_s is hyperbolic, and therefore n is even by the previous point. Since $n = \deg_k(A)$, this proves the first part.

Now let $\sigma'_s = \varphi \circ \sigma'_{k_s} \circ \varphi^{-1}$ and let $E' = \varphi(e' \otimes 1)$. Since σ'_s is also hyperbolic, the first point shows that there exists an invertible matrix $M \in \mathrm{M}_n(k_s)$ such that $MEM^{-1} = E'$. Let $a = \varphi^{-1}(M)$. Then $a \in A_{k_s}^\times$ and $e' \otimes 1 = a(e \otimes 1)a^{-1}$. Let W be the affine k-variety defined by

$$W(R) = \{a \in A_R \mid a(e \otimes 1) = (e' \otimes 1)a\}$$

for any commutative k-algebra R. This variety is not empty and isomorphic to an affine space, since it is the variety associated to a finite dimensional k-vector space. Therefore W is an affine rational variety. The subset $U \subset W$ defined by

$$U(R) = W(R) \cap A_R^\times$$

is an open subset of W, and the previous considerations show that $U(k_s)$ is not empty. Since k is infinite, it follows that $U(k) \neq \emptyset$, which is exactly what we wanted to prove. For those who are not familiar with this kind of geometric argument, one may also argue more explicitely as follows. Let us keep the notation of Example III.7.19 (2). Let $w_1, \ldots, w_r$ be a k-basis of W, let $e_1, \ldots, e_m$ be a k-basis of A (where $m = n^2$), and write

$$w_j = \sum_{i=1}^m a_{ij} e_i \text{ for all } j = 1, \ldots, r.$$

Then $a = \sum_{j=1}^r w_j \otimes \mu_j = \sum_{i=1}^m (e_i \otimes 1)(\sum_{j=1}^r a_{ij}\mu_j) \in U(R)$ if and only if $Q(\mu_1, \ldots, \mu_r) \neq 0$, where $Q \in k[Y_1, \ldots, Y_r]$ is the polynomial

$$Q = P_A\left(\sum_{j=1}^{r} a_{1j}Y_j, \ldots, \sum_{j=1}^{r} a_{mj}Y_j\right).$$

Since $U(k_s)$ is not empty, the polynomial Q is non-zero. Since k is infinite, this implies that Q does not vanish on k^r, and therefore $U(k)$ is not empty. □

We are now ready to prove the following result:

Theorem VIII.21.15. *Let A be a central simple k-algebra. Then two hyperbolic involutions of same type on A are conjugate.*

Proof. We are grateful to J.-P.Tignol for providing us the following arguments. Let σ, σ' be two hyperbolic involutions of same type on A, and let e, e' be the corresponding idempotents of A. By the previous lemma, there exists $a \in A^\times$ such that $e' = aea^{-1}$. Let $\sigma'' = \mathrm{Int}(a)^{-1} \circ \sigma' \circ \mathrm{Int}(a)$. It is enough to show that σ and σ'' are conjugate. We have

$$\sigma''(e) = \sigma''(a^{-1}e'a) = \sigma'' \circ \mathrm{Int}(a)^{-1}(e') = \mathrm{Int}(a)^{-1}(\sigma'(e')).$$

Thus, we get

$$\sigma''(e) = \mathrm{Int}(a)^{-1}(1 - e') = 1 - e = \sigma(e).$$

By Proposition VIII.21.3, there exists $u \in \mathrm{Sym}(A, \sigma)^\times$ such that

$$\sigma'' = \mathrm{Int}(u) \circ \sigma.$$

Since $\sigma(e) = \sigma''(e) = u\sigma(e)u^{-1} = \sigma(u^{-1}eu)$, we have $u^{-1}eu = e$, that is $ue = eu$. Set

$$v = eue + (1 - e).$$

It easily follows from the relations $e^2 = e$ and $ue = eu$ that we have $v\sigma(v) = u$. Therefore, we get

$$\sigma'' = \mathrm{Int}(v\sigma(v)) \circ \sigma,$$

and σ and σ'' are conjugate. This concludes the proof. □

VIII.21.3 Similitudes

In this paragraph, we define several group-schemes attached to a central simple k-algebra with involution (A, σ), where σ is an involution of the first kind.

If R is a commutative k-algebra, we set $\sigma_R = \sigma \otimes_k \mathrm{Id}_R$.

Definition VIII.21.16. The group-scheme of **similitudes** of (A, σ), denoted by $\mathbf{Sim}(A, \sigma)$, is defined by

$$\mathbf{Sim}(A, \sigma)(R) = \{g \in A_R^\times \mid g\sigma_R(g) \in R^\times\}.$$

If $g \in \mathbf{Sim}(A, \sigma)(R)$, the element $\mu(g) = g\sigma(g) \in R^\times$ is called the **multiplier** of the similitude g. If σ is orthogonal, we denote it by $\mathbf{GO}(A, \sigma)$, and if σ is symplectic, we denote it by $\mathbf{GSp}(A, \sigma)$.

The group-scheme of **isometries** of (A, σ), denoted by $\mathbf{Iso}(A, \sigma)$, is defined by

$$\mathbf{Iso}(A, \sigma)(R) = \{g \in A_R^\times \mid g\sigma_R(g) = 1\}.$$

If σ is orthogonal, we denote it by $\mathbf{O}(A, \sigma)$, and if σ is symplectic, we denote it by $\mathbf{Sp}(A, \sigma)$.

We also define the group-scheme $\mathbf{Aut}(A, \sigma)$ of automorphisms of (A, σ) by

$$\mathbf{Aut}(A, \sigma)(R) = \{f \in \mathrm{Aut}(A_R) \mid f \circ \sigma = \sigma \circ f\}.$$

If σ is orthogonal, we denote it by $\mathbf{PGO}(A, \sigma)$, and if σ is symplectic, we denote it by $\mathbf{PGSp}(A, \sigma)$.

Notice that we have a morphism of group-schemes

$$\mathrm{Int} : \mathbf{Sim}(A, \sigma) \longrightarrow \mathbf{Aut}(A, \sigma),$$

defined by

$$\mathrm{Int}_R : \begin{array}{rcl} \mathbf{Sim}(A, \sigma)(R) & \longrightarrow & \mathbf{Aut}(A, \sigma)(R) \\ g & \longmapsto & \mathrm{Int}(g). \end{array}$$

Indeed, if $g\sigma_R(g) \in R^\times$, we have $\mathrm{Int}(g\sigma_R(g)) = \mathrm{Id}_{A_R}$, and thus we have

$$\mathrm{Int}(g) \circ \sigma_R \circ \mathrm{Int}(g)^{-1} = \mathrm{Int}(g) \circ \mathrm{Int}(\sigma_R(g)) \circ \sigma = \mathrm{Int}(g\sigma_R(g)) \circ \sigma_R = \sigma_R.$$

Moreover, $\mathrm{Int}(g) = \mathrm{Id}_{A_R}$ if and only if $g \in R^\times$, so we get an exact sequence

$$1 \longrightarrow R^\times \longrightarrow \mathbf{Sim}(A, \sigma)(R) \longrightarrow \mathbf{Aut}(A, \sigma)(R).$$

If K is a field, every automorphism of A_K is inner. If $f = \mathrm{Int}(g)$ is such an automorphism, the previous computations show that f is an automorphism of (A_K, σ_K) if and only if $\mathrm{Int}(g\sigma(g)) = \mathrm{Id}_{A_K}$, that is if and only if $g\sigma(g) \in K^\times$. Therefore, Int_K is surjective for every field K, so we get an exact sequence

$$1 \longrightarrow K^{\times} \longrightarrow \mathbf{Sim}(A,\sigma)(K) \longrightarrow \mathbf{Aut}(A,\sigma)(K) \longrightarrow 1.$$

If we restrict the morphism Int to $\mathbf{Iso}(A,\sigma)$, we easily see that we have an exact sequence

$$1 \longrightarrow \mu_2(R) \longrightarrow \mathbf{Iso}(A,\sigma)(R) \longrightarrow \mathbf{Aut}(A,\sigma)(R).$$

If K is a field and K_s denotes a separable closure of K, the map Int is surjective on the K_s-points. Indeed, if $f = \mathrm{Int}(g) \in \mathbf{Aut}(A,\sigma)(K_s)$, then g is a similitude and $g' = \dfrac{1}{\sqrt{\mu(g)}} g$ is an isometry of (A_{K_s}, σ_{K_s}) satisfying $f = \mathrm{Int}(g')$. We then get an exact sequence of $\mathcal{G}_{K_s}$-groups

$$1 \longrightarrow \mu_2(K_s) \longrightarrow \mathbf{Iso}(A,\sigma)(K_s) \longrightarrow \mathbf{Aut}(A,\sigma)(K_s) \longrightarrow 1.$$

VIII.21.4 Cohomology of algebras with involution

We now give a description of the first Galois cohomology set of the group-schemes in the previous section. We start with a lemma.

Lemma VIII.21.17. *Let (A,σ) and (A',σ') be two central simple k-algebras with involution of same type. Then we have*

$$(A_{k_s}, \sigma_{k_s}) \simeq (A'_{k_s}, \sigma'_{k_s}).$$

Proof. Let $f : A_{k_s} \xrightarrow{\sim} \mathrm{M}_n(k_s)$ be an isomorphism of central simple k_s-algebras, and let $\sigma_0 = f \circ \sigma_{k_s} \circ f^{-1}$. We know that we have $\sigma_0 = \sigma_B$ for some invertible matrix $B \in \mathrm{M}_n(k_s)$ satisfying $B^t = \pm B$. By definition, we have

$$(A_{k_s}, \sigma_{k_s}) \simeq (\mathrm{M}_n(k_s), \sigma_B).$$

Set $B_0 = I_n$ if $B^t = B$ and $B_0 = \begin{pmatrix} 0 & 1 & & & \\ -1 & 0 & & & \\ & & \ddots & & \\ & & & 0 & 1 \\ & & & -1 & 0 \end{pmatrix}$ if $B^t = -B$.

The map $k_s^n \times k_s^n \longrightarrow k_s, (X,Y) \longmapsto X^tBY$ is a regular symmetric bilinear form if $B^t = B$ and a regular skew-symmetric form if $B^t = -B$. The theory of symmetric forms and skew-symmetric bilinear forms shows that in both cases, there exists $P \in \mathrm{GL}(k_s)$ such that $B = P^t B_0 P$. We then have

$$\sigma_B = \mathrm{Int}(P^t) \circ \sigma_{B_0} \circ \mathrm{Int}(P^t)^{-1}.$$

In particular, we get

$$(A_{k_s}, \sigma_{k_s}) \simeq (\mathrm{M}_n(k_s), \sigma_B) \simeq (\mathrm{M}_n(k_s), \sigma_{B_0}).$$

The lemma follows easily. □

Reasoning as in the proof of Proposition III.9.7, we get the following result:

Proposition VIII.21.18. *Let k be a field, and let (A, σ) be a central simple k-algebra with an involution of the first kind of type ε. Then we have an isomorphism of functors $\mathfrak{C}_k \longrightarrow$* **Sets*****

$$H^1(\_, \mathbf{PSim}(A, \sigma)) \simeq \mathbf{CSA}_n^{\varepsilon},$$

where the base point of $\mathbf{CSA}_n^{\varepsilon}$ is chosen to be the isomorphism class of (A, σ).

Now we would like to give a description of the functor $H^1(\_, \mathbf{Iso}(A, \sigma))$. We define a functor $\mathbf{Sym}(A, \sigma)^{\times} : \mathfrak{C}_k \longrightarrow \mathbf{Sets}^*$ by setting

$$\mathbf{Sym}(A, \sigma)^{\times}(K) = \mathrm{Sym}(A_K, \sigma_K)^{\times},$$

for every field extension K/k, the base point being 1_{A_K}.

Notice now that $\mathbf{GL}_1(A)$ acts on $\mathbf{Sym}(A, \sigma)^{\times}$ as follows:

$$\begin{aligned} \mathbf{GL}_1(A)(K) \times \mathbf{Sym}(A, \sigma)^{\times}(K) &\longrightarrow \mathbf{Sym}(A, \sigma)^{\times}(K) \\ (a, u) &\longmapsto a u \sigma_K(a). \end{aligned}$$

We denote by $\mathbf{Sym}(A, \sigma)^{\times}/_{\sim}$ the corresponding functor of equivalence classes, and by $u/_{\sim}$ the equivalence class of an invertible symmetric element u. Now by definition of the action of $\mathbf{GL}_1(A)$, we have

$$\mathbf{Stab}_{\mathbf{GL}_1(A)}(1_A)(K) = \mathbf{Iso}(A, \sigma)(K)$$

for every field extension K/k. If $u \in \mathrm{Sym}(A, \sigma)^{\times}(K)$ and L/K is a field extension, we will denote by u_L the element $u \otimes 1 \in \mathrm{Sym}(A, \sigma)^{\times}(L)$.

We claim that for all $u \in \mathrm{Sym}(A_K, \sigma_K)^{\times}$, we have $u_{K_s} \sim_{K_s} 1$.

Indeed, by Lemma VIII.21.17, the involutions $\mathrm{Int}(u_{K_s}) \circ \sigma_{K_s}$ and σ_{K_s} on A_{K_s} are conjugate. Therefore, there exists $a \in A_{K_s}^{\times}$ such that

$$\mathrm{Int}(a(u_{K_s})\sigma_{K_s}(a)) \circ \sigma_{K_s} = \sigma_{K_s}.$$

We deduce that there exists $\lambda \in K_s^\times$ such that $au_{K_s}\sigma_{K_s}(a) = \lambda$. Let $\sqrt{\lambda} \in K_s^\times$ be a square-root of λ. Setting $a' = \dfrac{a}{\sqrt{\lambda}}$, we see that we have

$$a'u_{K_s}\sigma_{K_s}(a') = 1,$$

that is $u_{K_s} \sim 1$.

It is easy to check that all the hypotheses of the Galois descent lemma are fulfilled. Since $H^1(\_, \mathbf{GL}_1(A)) = 1$ by Hilbert 90, the Galois descent lemma yields

Proposition VIII.21.19. *We have an isomorphism of functors from* $\mathfrak{C}_k$ *to* **Sets***

$$H^1(\_, \mathbf{Iso}(A,\sigma)) \simeq \mathbf{Sym}(A,\sigma)^\times/_\sim.$$

We end this paragraph by identifying some maps in cohomology.

Since an automorphism of an algebra with involution (A,σ) is an automorphism of A, we have an inclusion $\mathbf{Aut}(A,\sigma) \subset \mathbf{PGL}_1(A)$. The proof of the following lemma is left to the reader as an exercise.

Lemma VIII.21.20. *For every field extension K/k, the map*

$$H^1(K, \mathbf{Aut}(A,\sigma)) \longrightarrow H^1(K, \mathbf{PGL}_1(A))$$

induced by the inclusion maps the isomorphism class of (A',σ') onto the isomorphism class of A'.

Lemma VIII.21.21. *For every field extension K/k, the map*

$$\mathrm{Int}_{*,K} : H^1(K, \mathbf{Iso}(A,\sigma)) \longrightarrow H^1(K, \mathbf{Aut}(A,\sigma))$$

induced by the group morphism $\mathrm{Int} : \mathbf{Iso}(A,\sigma) \longrightarrow \mathbf{Aut}(A,\sigma)$ *maps* $u/_\sim$ *onto the isomorphism class of* $(A_K, \mathrm{Int}(u)\circ\sigma_K)$.

Proof. Let $a \in A_{K_s}$ such that $au_{K_s}\sigma_{K_s}(a) = 1$. Galois descent shows that a cocycle α representing $u/_\sim$ is given by

$$\alpha\colon \begin{array}{l} \mathcal{G}_{K_s} \longrightarrow \mathbf{Iso}(A,\sigma)(K_s) \\ \rho \longmapsto a\,\rho{\cdot}a^{-1}. \end{array}$$

Notice that $\mathrm{Int}(a)$ induces an isomorphism of algebras with involution between $(A_{K_s}, \mathrm{Int}(u_{K_s})\circ\sigma_{K_s})$ and (A_{K_s}, σ_{K_s}). Indeed, we have

$$\begin{aligned} \mathrm{Int}(a)\circ\mathrm{Int}(u_{K_s})\circ\sigma_{K_s}\circ\mathrm{Int}(a)^{-1} &= \mathrm{Int}(au_{K_s})\circ\sigma_{K_s}\circ\mathrm{Int}(a)^{-1} \\ &= \mathrm{Int}(au_{K_s})\circ\mathrm{Int}(\sigma_{K_s}(a))\circ\sigma_{K_s} \\ &= \mathrm{Int}(au_{K_s}\sigma_{K_s}(a))\circ\sigma_{K_s} \\ &= \sigma_{K_s}. \end{aligned}$$

By Galois descent, a cocycle β representing $(A_K, \mathrm{Int}(u) \circ \sigma_K)$ is then given by

$$\beta\colon \begin{array}{l} \mathcal{G}_{K_s} \longrightarrow \mathbf{Aut}(A,\sigma)(K_s) \\ \rho \longmapsto \mathrm{Int}(a)\,\rho{\cdot}\mathrm{Int}(a)^{-1}. \end{array}$$

For all $x \in A_{K_s}$, we have

$$(\rho{\cdot}\mathrm{Int}(a))(x) = \rho{\cdot}(\mathrm{Int}(a)(\rho^{-1}{\cdot}x)) = \rho{\cdot}(a(\rho^{-1}{\cdot}x)a^{-1}).$$

Since ρ acts on A_{K_s} by algebra automorphisms, we get

$$(\rho{\cdot}\mathrm{Int}(a))(x) = (\rho{\cdot}a)x(\rho{\cdot}a)^{-1},$$

hence $\rho{\cdot}\mathrm{Int}(a) = \mathrm{Int}(\rho{\cdot}a)$. We then get

$$\beta_\rho = \mathrm{Int}(a\,\rho{\cdot}a^{-1}) = \mathrm{Int}(\alpha_\rho) \text{ for all } \rho \in \mathcal{G}_{K_s}.$$

This means that β represents $\mathrm{Int}_{*,K}(u/_\sim)$. This concludes the proof. □

VIII.21.5 Trace forms

In this paragraph, we associate some quadratic forms to central simple algebras with an involution of the first kind. We first define a characteristic polynomial for certain types of algebras.

Definition VIII.21.22. Let R be a commutative ring with unit. Let $\mathcal{A}$ be an associative R-algebra with unit. Assume that there exist a commutative faithfully flat R-algebra S, a projective S-module P of finite rank n and an isomorphism of S-algebras

$$\varphi : \mathcal{A} \otimes_R S \xrightarrow{\sim} \mathrm{End}_S(P).$$

For all $a \in \mathcal{A}$, the polynomial $\chi_{\mathcal{A},a} = \det(t\mathrm{Id} - \varphi(a \otimes 1_S))$ has coefficients in R and does not depend on the choice of (S, P, φ) (see [31] for a proof). It is called the **reduced characteristic polynomial** of a. Let us write

$$\chi_{\mathcal{A},a} = t^n - a_{n-1}t^{n-1} + \ldots + (-1)^n a_0.$$

The elements a_{n-1} and a_0 are respectively called the **reduced trace** and **reduced norm** of a, and are denoted by $\mathrm{Trd}_{\mathcal{A}}(a)$ and $\mathrm{Nrd}_{\mathcal{A}}(a)$. In other words, we have

$$\mathrm{Trd}_{\mathcal{A}}(a) = \mathrm{tr}(\varphi(a \otimes 1_S)) \text{ and } \mathrm{Nrd}_{\mathcal{A}}(a) = \det(\varphi(a \otimes 1_S)).$$

Example VIII.21.23. Let A be a central simple k-algebra, let R be a commutative k-algebra, and let $\mathcal{A} = A_R$. Let L/k be any splitting field of A, and let $\rho : A_L \xrightarrow{\sim} \mathrm{M}_n(L)$ be an isomorphism of L-algebras. Then the R-algebra $S = R \otimes_k L$ is faithfully flat. Moreover, the S-module $P = S^n$ is free (hence projective) of rank n, and we have the following isomorphisms of L-algebras

$$A_R \otimes_R S \simeq A_L \otimes_k S \simeq \mathrm{M}_n(L) \otimes_k S \simeq \mathrm{End}_S(P),$$

the second one being induced by ρ_S.

We then may define a characteristic polynomial for A_R, and we have in particular reduced trace and reduced norm maps.

Lemma VIII.21.24. *Let A and B be two central simple k-algebras. Then the following properties hold:*

(1) *The map $\mathrm{Trd}_A : A \longrightarrow k$ is k-linear. Moreover for all $a, a' \in A$, we have*

$$\mathrm{Trd}_A(aa') = \mathrm{Trd}_A(a'a) \text{ and } \mathrm{Nrd}_A(aa') = \mathrm{Nrd}_A(a)\mathrm{Nrd}_A(a').$$

(2) *For all $n \geq 1$, the maps $\mathrm{Trd}_{\mathrm{M}_n(k)}$ and $\mathrm{Nrd}_{\mathrm{M}_n(k)}$ coincide with the usual trace and determinant maps.*

(3) *For every field extension K/k, and every $a \in A$, we have*

$$\mathrm{Trd}_{A_K}(a \otimes 1) = \mathrm{Trd}_A(a), \mathrm{Nrd}_{A_K}(a \otimes 1) = \mathrm{Nrd}_A(a).$$

(4) *If $f : A \xrightarrow{\sim} B$ is an isomorphism of k-algebras, then for all $a \in A$, we have*

$$\mathrm{Trd}_B(f(a)) = \mathrm{Trd}_A(a) \text{ and } \mathrm{Nrd}_B(f(a)) = \mathrm{Nrd}_A(a).$$

(5) *For all $a \in A$, we have*

$$\mathrm{Trd}_{A^{op}}(a^{op}) = \mathrm{Trd}_A(a) \text{ and } \mathrm{Nrd}_{A^{op}}(a^{op}) = \mathrm{Nrd}_A(a).$$

(6) *We have $\mathrm{Nrd}_A(a) \neq 0$ if and only if $a \in A^\times$.*

(7) *For all $a \in A, b \in B$, we have*

$$\mathrm{Trd}_{A \otimes_k B}(a \otimes b) = \mathrm{Trd}_A(a)\mathrm{Trd}_B(b)$$

and

$$\mathrm{Nrd}_{A \otimes_k B}(a \otimes b) = \mathrm{Nrd}_A(a)^{\deg_k(B)}\mathrm{Nrd}_B(b)^{\deg_k(A)}.$$

Proof. Properties (1), (2), (3) and (7) come from the definitions and the properties of the trace and determinant maps. For example, let K/k be a field extension and let $a \in A$. Let L/k be a splitting field of A containing K (for example $L = K_s$), and let

$$\varphi : A_L \xrightarrow{\sim} \mathrm{M}_n(L)$$

be an isomorphism of L-algebras. Composing φ with the canonical isomorphism

$$A_K \otimes_K L \simeq A_L$$

gives rise to an isomorphism of L-algebras $\psi : A_K \otimes_K L \xrightarrow{\sim} \mathrm{M}_n(L)$ which satisfies

$$\psi((a \otimes 1) \otimes 1) = \varphi(a \otimes 1).$$

Therefore, we get

$$\mathrm{Trd}_{A_K}(a \otimes 1) = \mathrm{tr}(\psi((a \otimes 1) \otimes 1)) = \mathrm{tr}(\varphi(a \otimes 1)) = \mathrm{Trd}_A(a).$$

We will prove (4), (5) and (6) in detail. For the rest of the proof, let us fix a splitting field L of A, an isomorphism $\varphi : A_L \xrightarrow{\sim} \mathrm{M}_n(L)$ of L-algebras, and an element $a \in A$.

Let us prove (4). Let $f : A \xrightarrow{\sim} B$ be an isomorphism of k-algebras. Then the map

$$\varphi \circ (f^{-1} \otimes \mathrm{Id}_L) : B \longrightarrow \mathrm{M}_n(L)$$

is an isomorphism of L-algebras. We then have

$$\mathrm{Trd}_B(f(a)) = \mathrm{tr}(\varphi \circ (f^{-1} \otimes \mathrm{Id}_L)(f(a) \otimes 1)) = \mathrm{tr}(\varphi(a \otimes 1)) = \mathrm{Trd}_A(a),$$

and similarly for $\mathrm{Nrd}_B(f(a))$. This proves (4).

We now prove (5). One can show that the map

$$\psi : \begin{array}{l} (A^{op})_L \longrightarrow \mathrm{M}_n(L) \\ a^{op} \otimes \lambda \longmapsto \varphi(a \otimes \lambda)^t \end{array}$$

is a well-defined isomorphism of L-algebras. Therefore, we have

$$\mathrm{Trd}_{A^{op}}(a^{op}) = \mathrm{tr}(\varphi(a \otimes 1)^t) = \mathrm{tr}(\varphi(a \otimes 1)) = \mathrm{Trd}_A(a),$$

and similarly for the second equality.

It remains to prove (6). If $a \in A^\times$, then $a \otimes 1 \in A_L^\times$, and thus $\varphi(a \otimes 1)$ is an invertible matrix. Therefore

$$\mathrm{Nrd}_A(a) = \det(\varphi(a \otimes 1)) \neq 0.$$

Now assume that a is not invertible. This means that the k-linear map

$$\ell_a : \begin{array}{l} A \longrightarrow A \\ a' \longmapsto aa' \end{array}$$

is not surjective. Since A is finite dimensional over k, it is not injective either, so there exists $a' \in A, a' \neq 0$ such that $aa' = 0$. Hence a is either zero or a zero divisor, and therefore so is $\varphi(a \otimes 1) \in \mathrm{M}_n(L)$. In particular, $\varphi(a \otimes 1)$ is not invertible. It follows that $\mathrm{Nrd}_A(a) = 0$ by the properties of the determinant. This concludes the proof. □

Remark VIII.21.25. All these properties are true if we replace A by a R-algebra $\mathcal{A}$ satisfying the assumptions of Definition VIII.21.22. For example, assume that we have an isomorphism of R-algebras $f : \mathcal{A} \xrightarrow{\sim} \mathcal{B}$, and let (S, P, φ) be a triple as in Definition VIII.21.22. We then have an isomorphism

$$\varphi \circ (f^{-1} \otimes \mathrm{Id}_S) : \mathcal{B} \xrightarrow{\sim} \mathrm{End}_S(P),$$

and therefore

$$\mathrm{Trd}_{\mathcal{B}}(f(a)) = \mathrm{tr}(\varphi \circ (f^{-1} \otimes \mathrm{Id}_S)(f(a) \otimes 1)) = \mathrm{tr}(\varphi(a \otimes 1)) = \mathrm{Trd}_{\mathcal{A}}(a).$$

We are ready to define our first quadratic form.

Definition VIII.21.26. Let A be a central simple k-algebra. The map

$$\mathcal{T}_A : \begin{array}{l} A \longrightarrow k \\ a \longmapsto \mathrm{Trd}_A(a^2) \end{array}$$

is a quadratic form, called the **trace form** of A.

Proposition VIII.21.27. *Let A be a central simple k-algebra, and let K/k be a field extension. Then the following properties hold:*

(1) *The isomorphism class of $\mathcal{T}_A$ only depends on the isomorphism class of A.*

(2) $\mathcal{T}_{A_K} \simeq (\mathcal{T}_A)_K$.

(3) $\mathcal{T}_{\mathrm{M}_n(K)} \simeq n \times \langle 1 \rangle \perp n \times \langle 1, -1 \rangle$.

(4) *The quadratic form $\mathcal{T}_A$ is regular.*

Proof. Let us prove (1). Let $\varphi : A \xrightarrow{\sim} B$ be an isomorphism of k-algebras. For all $a \in A$, we have

$$\mathcal{T}_B(\varphi(a)) = \mathrm{Trd}_B(\varphi(a)^2) = \mathrm{Trd}_B(\varphi(a^2)) = \mathrm{Trd}_A(a^2) = \mathcal{T}_A(a).$$

Hence φ induces an isomorphism between the quadratic spaces $(A, \mathcal{T}_A)$ and $(B, \mathcal{T}_B)$.

Point (2) is immediate using Lemma VIII.21.24 (3). Let us prove (3). The polar form b_A associated to the trace form is

$$b_A : \begin{array}{l} A \times A \longrightarrow k \\ (a, a') \longmapsto \mathrm{Trd}_A(aa'). \end{array}$$

Indeed, b_A is bilinear and symmetric by Lemma VIII.21.24 (1), and satisfies

$$b_A(a, a) = \mathrm{Trd}_A(a^2) \text{ for all } a \in A.$$

In particular, we get

$$b_{\mathrm{M}_n(K)}(M, M') = \mathrm{tr}(MM') \text{ for all } M, M' \in \mathrm{M}_n(K).$$

Using this formula, it is easy to check that the elements

$$E_{ii}, \frac{E_{ij} + E_{ji}}{2}, \frac{E_{ij} - E_{ji}}{2}, 1 \leq i < j \leq n$$

form an orthogonal basis of A with respect to the trace form. The result follows by computing the value of $\mathcal{T}_{\mathrm{M}_n(K)}$ at each element of this basis. It remains to prove (4). To check that $\mathcal{T}_A$ is regular, it is enough to do it after extending scalars to k_{alg}. By (2), we have

$$(\mathcal{T}_A)_{k_{alg}} \simeq \mathcal{T}_{A_{k_{alg}}}.$$

Since $A_{k_{alg}} \simeq \mathrm{M}_n(k_{alg})$, (1) yields

$$(\mathcal{T}_A)_{k_{alg}} \simeq \mathcal{T}_{\mathrm{M}_n(k_{alg})}.$$

Now apply (3) to conclude. □

We now associate some quadratic forms to algebras with involutions.

Definition VIII.21.28. Let (A, σ) be a central simple k-algebra with an involution of the first kind. The map

$$\mathcal{T}_\sigma : \begin{array}{l} A \longrightarrow k \\ a \longmapsto \mathrm{Trd}_A(\sigma(a)a) \end{array}$$

is a quadratic form called the **trace form** of (A, σ) . We denote respectively by $\mathcal{T}_\sigma^+$ and $\mathcal{T}_\sigma^-$ the restriction of $\mathcal{T}_\sigma$ to $\mathrm{Sym}(A, \sigma)$ and $\mathrm{Skew}(A, \sigma)$. They are called the **restricted trace forms** of (A, σ).

We now establish some auxiliary results on these various trace forms.

Lemma VIII.21.29. *Let (A,σ) be a central simple k-algebra with an involution of the first kind. The following properties hold:*

(1) *For all $a \in A$, we have $\mathrm{Trd}_A(\sigma(a)) = \mathrm{Trd}_A(a)$ and $\mathrm{Nrd}_A(\sigma(a)) = \mathrm{Nrd}_A(a)$.*

(2) *For all $a \in \mathrm{Skew}(A,\sigma), s \in \mathrm{Sym}(A,\sigma)$, we have $\mathrm{Trd}_A(as) = 0$.*

Proof. Observe first that the map

$$\varphi\colon \begin{array}{l} A \longrightarrow A^{op} \\ a \longmapsto \sigma(a)^{op} \end{array}$$

is an isomorphism of k-algebras. Therefore, for all $x \in A$, we have

$$\mathrm{Trd}_A(x) = \mathrm{Trd}_{A^{op}}(\sigma(x)^{op}) = \mathrm{Trd}_A(\sigma(x))$$

by Lemma VIII.21.24 (4) and (5). We then have

$$\begin{aligned} \mathrm{Trd}_A(as) &= \mathrm{Trd}_A(\sigma(as)) \\ &= \mathrm{Trd}_A(\sigma(s)\sigma(a)) \\ &= \mathrm{Trd}_A(-sa) \\ &= -\mathrm{Trd}_A(sa) \\ &= -\mathrm{Trd}_A(as). \end{aligned}$$

We then get $2\mathrm{Trd}_A(as) = 0$, and therefore $\mathrm{Trd}_A(as) = 0$. □

Proposition VIII.21.30. *Let (A,σ) be a central simple k-algebra with an involution of the first kind.*

(1) *The isomorphism classes of $\mathcal{T}_\sigma, \mathcal{T}_\sigma^+$ and $\mathcal{T}_\sigma^-$ only depend on the isomorphism class of (A,σ).*

(2) *We have $\mathcal{T}_A \simeq \mathcal{T}_\sigma^+ \perp -\mathcal{T}_\sigma^-$ and $\mathcal{T}_\sigma \simeq \mathcal{T}_\sigma^+ \perp \mathcal{T}_\sigma^-$.*

(3) *For every field extension K/k, we have*

$$\mathcal{T}_{\sigma_K} \simeq (\mathcal{T}_\sigma)_K, \mathcal{T}_{\sigma_K}^+ \simeq (\mathcal{T}_\sigma^+)_K \text{ and } \mathcal{T}_{\sigma_K}^- \simeq (\mathcal{T}_\sigma^-)_K.$$

(4) *The quadratic forms $\mathcal{T}_\sigma, \mathcal{T}_\sigma^+$ and $\mathcal{T}_\sigma^-$ are regular.*

Proof. Let us prove (1). Let $\varphi : (A,\sigma) \xrightarrow{\sim} (A',\sigma')$ be an isomorphism of algebras with involution. For all $a \in A$, we have

$$\begin{aligned} \mathcal{T}_{\sigma'}(\varphi(a)) &= \mathrm{Trd}_{A'}(\sigma'(\varphi(a))\varphi(a)) \\ &= \mathrm{Trd}_{A'}(\varphi(\sigma(a))\varphi(a)) \\ &= \mathrm{Trd}_{A'}(\varphi(\sigma(a)a)) \\ &= \mathrm{Trd}_A(\sigma(a)a) \\ &= \mathcal{T}_\sigma(a). \end{aligned}$$

Hence φ induces an isomorphism between the quadratic spaces $(A, \mathcal{T}_\sigma)$ and $(A', \mathcal{T}_{\sigma'})$. We have already noticed that φ induces isomorphisms of k-vector spaces

$$\mathrm{Sym}(A,\sigma) \simeq \mathrm{Sym}(A',\sigma') \text{ and } \mathrm{Skew}(A,\sigma) \simeq \mathrm{Skew}(A',\sigma').$$

The rest follows from the equalities

$$\mathrm{Trd}_{A'}(\varphi(a)^2) = \mathrm{Trd}_{A'}(\varphi(a^2)) = \mathrm{Trd}_A(a^2) \text{ for all } a \in A.$$

To prove (2), notice first that we have

$$A = \mathrm{Sym}(A,\sigma) \oplus \mathrm{Skew}(A,\sigma).$$

Recall that the polar form of $\mathcal{T}_A$ is

$$b_A : \begin{array}{l} A \times A \longrightarrow k \\ (a, a') \longmapsto \mathrm{Trd}_A(aa'). \end{array}$$

The previous lemma shows that the direct sum above is orthogonal with respect to the trace form. Moreover, for all $s \in \mathrm{Sym}(A,\sigma), a \in \mathrm{Skew}(A,\sigma)$, we have

$$\mathcal{T}_A(s) = \mathrm{Trd}_A(s^2) = \mathcal{T}_\sigma^+(s)$$

and

$$\mathcal{T}_A(a) = \mathrm{Trd}_A(a^2) = -\mathrm{Trd}_A(-a^2) = -\mathcal{T}_\sigma^-(a).$$

We then get $\mathcal{T}_A \simeq \mathcal{T}_\sigma^+ \perp -\mathcal{T}_\sigma^-$.

Notice now that the map

$$b_\sigma : \begin{array}{l} A \times A \longrightarrow k \\ (a, a') \longmapsto \mathrm{Trd}_A(\sigma(a)a') \end{array}$$

is bilinear and symmetric. Indeed, for all $a, a' \in A$, we have

$$\mathrm{Trd}_A(\sigma(a)a') = \mathrm{Trd}_A(\sigma(\sigma(a)a')) = \mathrm{Trd}_A(\sigma(a')a),$$

the first equality coming from Lemma VIII.21.29 (1). Since $b_\sigma(a,a) = \mathrm{Trd}_A(\sigma(a)a)$, it turns out that b_σ is the polar form of $\mathcal{T}_\sigma$. It follows easily that $\mathrm{Sym}(A,\sigma)$ and $\mathrm{Skew}(A,\sigma)$ are orthogonal with respect to $\mathcal{T}_\sigma$, and we get $\mathcal{T}_\sigma \simeq \mathcal{T}_\sigma^+ \perp \mathcal{T}_\sigma^-$.

Point (3) follows from Lemma VIII.21.24. To prove (4), notice that by definition of orthogonal sum and regularity, $\mathcal{T}_\sigma^+$ and $\mathcal{T}_\sigma^-$ will be regular if and only if $\mathcal{T}_\sigma^+ \perp -\mathcal{T}_\sigma^-$ is. By (2), this last orthogonal sum is isomorphic to $\mathcal{T}_A$, which is regular by Proposition VIII.21.27 (4). Therefore, $\mathcal{T}_\sigma^+$ and $\mathcal{T}_\sigma^-$ are regular, and thus so is $\mathcal{T}_\sigma \simeq \mathcal{T}_\sigma^+ \perp \mathcal{T}_\sigma^-$. □

The following lemma is useful to compute the various trace forms.

Lemma VIII.21.31. *Let A and A' be two central simple algebras. Then we have*

$$\mathcal{T}_{A\otimes B} \simeq \mathcal{T}_A \otimes \mathcal{T}_B.$$

Moreover, if σ and σ' are involutions of type ε and ε' on A and A' respectively, then $\sigma\otimes\sigma'$ is an involution of type $\varepsilon\varepsilon'$ on $A\otimes_k B$, and we have the following isomorphisms:

(1) $\mathcal{T}_{\sigma\otimes\sigma'} \simeq \mathcal{T}_\sigma \otimes \mathcal{T}_{\sigma'}$.

(2) $\mathcal{T}^+_{\sigma\otimes\sigma'} \simeq \mathcal{T}^+_\sigma \otimes \mathcal{T}^+_{\sigma'} \perp \mathcal{T}^-_\sigma \otimes \mathcal{T}^-_{\sigma'}$.

(3) $\mathcal{T}^-_{\sigma\otimes\sigma'} \simeq \mathcal{T}^+_\sigma \otimes \mathcal{T}^-_{\sigma'} \perp \mathcal{T}^-_\sigma \otimes \mathcal{T}^+_{\sigma'}$.

Proof. Let b_A, b_B and $b_{A\otimes_k B}$ the polar forms of $\mathcal{T}_A, \mathcal{T}_B$ and $\mathcal{T}_{A\otimes_k B}$ respectively. For all $a, a'\in A, b, b'\in B$, we have

$$\begin{aligned} b_{A\otimes_k B}(a\otimes b, a'\otimes b) &= \mathrm{Trd}_{A\otimes_k B}((a\otimes b)(a'\otimes b')) \\ &= \mathrm{Trd}_{A\otimes_k B}(aa'\otimes bb') \\ &= \mathrm{Trd}_A(aa')\mathrm{Trd}_B(bb') \\ &= b_A(a,a')b_B(b,b'). \end{aligned}$$

The first isomorphism follows. The isomorphism $\mathcal{T}_{\sigma\otimes\sigma'} \simeq \mathcal{T}_\sigma\otimes\mathcal{T}_{\sigma'}$ may be proved in a similar way.

The reader will check that $\mathrm{Sym}(A\otimes_k B, \sigma\otimes\sigma')$ is the direct sum of the linear subspaces $\mathrm{Sym}(A,\sigma)\otimes_k \mathrm{Sym}(A',\sigma')$ and $\mathrm{Skew}(A,\sigma)\otimes_k\mathrm{Skew}(A',\sigma')$, and that $\mathrm{Skew}(A\otimes_k B, \sigma\otimes\sigma')$ is the direct sum of the two linear subspaces $\mathrm{Sym}(A,\sigma)\otimes_k\mathrm{Skew}(A',\sigma')$ and $\mathrm{Skew}(A,\sigma)\otimes_k\mathrm{Sym}(A',\sigma')$. It follows in particular by dimension count that if σ and σ' are involutions of type ε and ε' on A and A' respectively, then $\sigma\otimes\sigma'$ is an involution of type $\varepsilon\varepsilon'$ on $A\otimes_k A'$. Moreover, Lemma VIII.21.29 (2) implies easily that these direct sums are orthogonal. The rest of the lemma follows. □

Let us give some examples of computation of these trace forms.

Examples VIII.21.32.

(1) Let $A = \mathrm{M}_n(k)$, and let ^{t} be the transposition. It is easy to check that the matrices

$$E_{ii}, \frac{E_{ij}+E_{ji}}{2}, i<j$$

form an orthogonal basis of $\mathrm{Sym}(A, {}^t)$. Straightforward computations show that we have

$$\mathcal{T}_t^+ \simeq n \times \langle 1 \rangle \perp \frac{n(n-1)}{2} \times \langle 2 \rangle.$$

Similarly, the matrices

$$\frac{E_{ij} - E_{ji}}{2}, i < j$$

form an orthogonal basis of $\mathrm{Skew}(A, {}^t)$, and we have

$$\mathcal{T}_t^- \simeq \frac{n(n-1)}{2} \times \langle 2 \rangle.$$

(2) Let $Q = (a, b)$ be a quaternion k-algebra and let us consider the symplectic involution

$$\gamma\colon \begin{array}{c} Q \longrightarrow Q \\ x + yi + zj + tij \longmapsto x - yi - zj - tij. \end{array}$$

Then 1 is a basis of $\mathrm{Sym}(Q, \gamma)$, and i, j, ij is an orthogonal basis of $\mathrm{Skew}(Q, \gamma)$, so we have

$$\mathcal{T}_\gamma^+ \simeq \langle 2 \rangle \text{ and } \mathcal{T}_\gamma^- \simeq \langle -2a, -2b, 2ab \rangle.$$

Therefore, we also get

$$\mathcal{T}_\gamma \simeq 2\langle\langle a, b \rangle\rangle.$$

(3) Let $A = \mathrm{M}_m(k) \otimes_k Q$, and let $\sigma = {}^t \otimes \gamma$. Lemma VIII.21.31 and the previous examples imply that we have

$$\mathcal{T}_\sigma^+ \simeq m \times \langle 2 \rangle \perp \frac{m(m-1)}{2} \times \langle 1, -a, -b, ab \rangle$$

and

$$\mathcal{T}_\sigma^- \simeq m \times \langle -2a, -2b, 2ab \rangle \perp \frac{m(m-1)}{2} \times \langle 1, -a, -b, ab \rangle.$$

(4) Let $A = \mathrm{M}_{2m}(k)$ and let $\sigma = \sigma_{B_0}$, where B_0 is the matrix

$$B_0 = \begin{pmatrix} 0 & 1 & & & \\ -1 & 0 & & & \\ & & \ddots & & \\ & & & 0 & 1 \\ & & & -1 & 0 \end{pmatrix}.$$

Notice that we have an isomorphism $\varphi : (1,-1) \xrightarrow{\sim} \mathrm{M}_2(k)$ of k-algebras such that

$$\varphi(i) = \begin{pmatrix} 1 & 0 \\ 0 & -1 \end{pmatrix} \text{ and } \varphi(j) = \begin{pmatrix} 0 & 1 \\ -1 & 0 \end{pmatrix}.$$

It is easy to check this isomorphism induces an isomorphism of algebras of involutions

$$((1,-1),\gamma) \simeq (\mathrm{M}_2(k), \sigma_H),$$

where $H = \begin{pmatrix} 0 & 1 \\ -1 & 0 \end{pmatrix}$. We then get isomorphisms of algebras with involution

$$\begin{aligned} (\mathrm{M}_{2m}(k), \sigma) &\simeq (\mathrm{M}_m(k) \otimes_k \mathrm{M}_2(k), t \otimes \sigma_H) \\ &\simeq (\mathrm{M}_m(k) \otimes_k (1,-1), t \otimes \gamma). \end{aligned}$$

Therefore, the computations done in the previous case give

$$\mathcal{T}_\sigma^+ \simeq m \times \langle 2 \rangle \perp m(m-1) \times \langle 1, -1 \rangle$$

and

$$\begin{aligned} \mathcal{T}_\sigma^- &\simeq m \times \langle -2, 2, -2 \rangle \perp m(m-1) \times \langle 1, -1 \rangle \\ &\simeq m \times \langle -2 \rangle \perp m^2 \times \langle 1, -1 \rangle. \end{aligned}$$

We now would like to give a cohomological interpretation of the restricted trace form.

Lemma VIII.21.33. *Let (A,σ) be a central simple k-algebra with an involution of the first kind , let R be a commutative k-algebra and let f be an automorphism of (A_R, σ_R). Then the restriction of f to* $\mathrm{Sym}(A_R, \sigma_R)$ *is an automorphism of $(\mathcal{T}_\sigma^+)_R$.*

Proof. By Remark VIII.21.25, for all $a \in A_R$ and any automorphism f of A_R, we have

$$\mathrm{Trd}_{A_R}(f(a)) = \mathrm{Trd}_{A_R}(a).$$

The lemma follows immediately. □

The previous result shows that we have a morphism of algebraic groups

$$\rho : \mathbf{Aut}(A,\sigma) \longrightarrow \mathbf{O}(\mathcal{T}_\sigma^+),$$

which is obtained by restricting an automorphism of a central simple algebra with involution to the corresponding set of σ-symmetric elements.

Lemma VIII.21.34. *Let (A,σ) be a central simple k-algebra with an involution of the first kind. For every field extension K/k, the induced map*

$$\rho_{*,K} : H^1(K, \mathbf{Aut}(A,\sigma)) \longrightarrow H^1(K, \mathbf{O}(\mathcal{T}_\sigma^+))$$

takes the isomorphism class of (A',σ') onto the isomorphism class of $\mathcal{T}_{\sigma'}^+$.

Proof. Let K/k be a field extension, let (A',σ') be a central simple K-algebra with an involution σ' of same type as σ, and let

$$\varphi : A'_{K_s} \xrightarrow{\sim} A_{K_s}$$

be an isomorphism of K_s-algebras of $(A',\sigma')_{K_s}$ onto $(A,\sigma)_{K_s}$. The map

$$\alpha : \begin{array}{l} \mathcal{G}_{K_s} \longrightarrow \mathbf{Aut}(A,\sigma)(K_s) \\ \sigma \longmapsto \varphi\sigma{\cdot}\varphi^{-1} \end{array}$$

is a cocycle representing (A',σ'). Now φ induces by restriction an isomorphism between the restricted trace forms of $(A',\sigma')_{K_s}$ and $(A,\sigma)_{K_s}$, that is an isomorphism between $(\mathcal{T}_{\sigma'}^+)_{K_s}$ and $(\mathcal{T}_\sigma^+)_{K_s}$ by Proposition VIII.21.30 (3). It follows that the map $\rho\circ\alpha$ is a cocycle representing $\mathcal{T}_{\sigma'}^+$. The result follows. □

As an application of Galois cohomology, we are going to compute the Hasse invariant of $\mathcal{T}_\sigma^+$ when σ is a symplectic involution. The computation of the Hasse invariants of $\mathcal{T}_A$ and $\mathcal{T}_\sigma$ will be done in the exercices. Before starting the proof, we would like to recall two facts on algebraic group-schemes, which we will use without further reference.

(1) Any morphism of algebraic group-schemes $\rho : G \longrightarrow H$ induces by restriction a morphism $G^0 \longrightarrow H^0$, where G^0 and H^0 denote the connected component of the neutral element of G and H respectively. Indeed, such a morphism ρ is continuous (for the Zariski topology) and therefore maps connected sets to connected sets.

(2) If G and H are **connected** algebraic group-schemes, then any morphism of algebraic group-schemes $\rho : G \longrightarrow H$ induces a morphism $\tilde{\rho} : \tilde{G} \longrightarrow \tilde{H}$ between the universal covers of G and H

(see [7, Proposition 2.24 (i),p.262]) such that the diagram

$$\begin{array}{ccc} \tilde{G} & \xrightarrow{\pi_G} & G \\ \downarrow{\tilde{\rho}} & & \downarrow{\rho} \\ \tilde{H} & \xrightarrow{\pi_H} & H \end{array}$$

is commutative. We then get in particular an induced morphism $\tau : \ker(\pi_G) \longrightarrow \ker(\pi_H)$.

Since $\mathbf{PGSp}(A,\sigma)$ is connected (see[30], for example), the group-scheme morphism $\rho : \mathbf{PGSp}_(A,\sigma) \longrightarrow \mathbf{O}(\mathcal{T}_\sigma^+)$ restricts to a group-scheme morphism (still denoted by ρ)

$$\rho : \mathbf{PGSp}(A,\sigma) \longrightarrow \mathbf{O}^+(\mathcal{T}_\sigma^+).$$

We then get the following result:

Corollary VIII.21.35. *Let (A,σ) be a central simple k-algebra with a symplectic involution. For any other symplectic involution σ' on A, we have*

$$\det(\mathcal{T}_{\sigma'}^+) = \det(\mathcal{T}_\sigma^+).$$

Proof. This follows from the previous lemma and Corollary IV.11.3. □

We are now ready to prove the following result, due to Quéguiner[46].

Theorem VIII.21.36. *Let (A,σ) be a central simple k-algebra of degree $2m$ with a symplectic involution. Then after identifying $H^2(k,\mu_2)$ and $\mathrm{Br}_2(k)$, we have the equality*

$$w_2(T_\sigma^+) = \frac{m(m-1)}{2}([(-1,-1)] + [A]).$$

Proof. Let $B_0 = \begin{pmatrix} 0 & 1 & & & \\ -1 & 0 & & & \\ & & \ddots & & \\ & & & 0 & 1 \\ & & & -1 & 0 \end{pmatrix}$, and denote by σ_0 the involution σ_{B_0}.

We will write $\mathbf{PGSp}_{2m}$ and $\mathbf{Sp}_{2m}$ instead of $\mathbf{PGSp}(\mathrm{M}_{2m}(k),\sigma_0)$ and $\mathbf{Sp}(\mathrm{M}_{2m}(k),\sigma_0)$ respectively. It is known that the group $\mathbf{Sp}_{2m}$ is the universal cover of $\mathbf{PGSp}_{2m}$ (see [30] for example).

Let $\rho : \mathbf{PGSp}_{2m} \longrightarrow \mathbf{O}^+(\mathcal{T}_{\sigma_0}^+)$ be the morphism of algebraic groups defined earlier, and let $\tilde{\rho} : \mathbf{Sp}_{2m} \longrightarrow \mathbf{Spin}(\mathcal{T}_{\sigma_0}^+)$ the induced morphism on the universal covers. Therefore, we also get an induced group morphism $\tau : \mu_2 \longrightarrow \mu_2$.

For every field extension K/k, we then have a commutative diagram

$$\begin{array}{ccccccccc}
1 & \longrightarrow & \mu_2(K_s) & \longrightarrow & \mathbf{Sp}_{2m}(K_s) & \longrightarrow & \mathbf{PGSp}_{2m}(K_s) & \longrightarrow & 1 \\
 & & \downarrow \tau & & \downarrow \tilde{\rho} & & \downarrow \rho & & \\
1 & \longrightarrow & \mu_2(K_s) & \longrightarrow & \mathbf{Spin}(\mathcal{T}_{\sigma_0}^+)(K_s) & \longrightarrow & \mathbf{O}^+(\mathcal{T}_{\sigma_0}^+)(K_s) & \longrightarrow & 1
\end{array}$$

which induces a commutative diagram

$$\begin{array}{ccc}
H^1(K, \mathbf{PGSp}_{2m}) & \xrightarrow{\Delta_K^1} & H^2(K, \mu_2) \\
\downarrow \rho_{*,K} & & \downarrow \tau_{*,K} \\
H^1(K, \mathbf{O}^+(\mathcal{T}_{\sigma_0}^+)) & \xrightarrow{\delta_K^1} & H^2(K, \mu_2)
\end{array}$$

by Theorem III.7.39 (4).
Using Lemma VIII.21.34 and Corollary IV.11.10, we get the equality

$$\tau_{*,K}(\Delta_K^1(A, \sigma)) = w_2(\mathcal{T}_{\sigma_0,K}^+) + w_2(\mathcal{T}_\sigma^+)$$

for every K-algebra with a symplectic involution (A, σ). Let us now identify the map Δ_K^1. We have a commutative diagram

$$\begin{array}{ccccccccc}
1 & \longrightarrow & \mu_2(K_s) & \longrightarrow & \mathbf{Sp}_{2m}(K_s) & \longrightarrow & \mathbf{PGSp}_{2m}(K_s) & \longrightarrow & 1 \\
 & & \downarrow & & \downarrow & & \downarrow & & \\
1 & \longrightarrow & K_s^\times & \longrightarrow & \mathbf{GL}_{2m}(K_s) & \longrightarrow & \mathbf{PGL}_{2m}(K_s) & \longrightarrow & 1
\end{array}$$

which induces a commutative diagram

$$\begin{array}{ccc}
H^1(K, \mathbf{PGSp}_{2m}) & \xrightarrow{\Delta_K^1} & H^2(K, \mu_2) \\
\downarrow & & \downarrow \\
H^1(K, \mathbf{PGL}_{2m}) & \longrightarrow & H^2(K, \mathbb{G}_m)
\end{array}$$

by Theorem III.7.39 (4). Using Lemma VIII.21.20, Theorem VIII.20.6

and the commutativity of the diagram, we see that Δ^1_K maps the isomorphism class of (A,σ) onto $[A]$. Therefore, we get

$$w_2(\mathcal{T}_\sigma^+) = w_2(\mathcal{T}_{\sigma_0,K}^+) + \tau_{*,K}([A]).$$

By Example VIII.21.32 (4), we have

$$\mathcal{T}_{\sigma_0,K}^+ \simeq m \times \langle 2\rangle \perp m(m-1) \times \langle 1,-1\rangle.$$

Using Lemma IV.11.5, we have

$$w_2(\mathcal{T}_{\sigma_0,K}^+) = w_2(m\times\langle 2\rangle) + w_2(m(m-1)\times\langle 1,-1\rangle) + (2^m)\cup((-1)^{m(m-1)}).$$

Since $m(m-1)$ is even, the last term is 0. Moreover, $w_2(m\times\langle 2\rangle)$ is a multiple of $(2)\cup(2) = (2)\cup(-1)$. Since 2 is a sum of two squares, $(2)\cup(-1) = 0$. We then have

$$w_2(\mathcal{T}_{\sigma_0,K}^+) = w_2(m(m-1)\times\langle 1,-1\rangle) = w_2(m(m-1)\times\langle -1\rangle),$$

and therefore

$$w_2(\mathcal{T}_{\sigma_0,K}^+) = \frac{m(m-1)(m(m-1)-1)}{2}(-1)\cup(-1).$$

It is easy to check that $\dfrac{m(m-1)(m(m-1)-1)}{2}$ and $\dfrac{m(m-1)}{2}$ have the same parity. Therefore, after identification of $H^2(k,\mu_2)$ and $\mathrm{Br}_2(k)$, we get

$$w_2(\mathcal{T}_\sigma^+) = \frac{m(m-1)}{2}[(-1,-1)] + \tau_{*,K}([A])$$

for every field extension K/k.

Now we need to identify the map $\tau_{*,K}$. The map τ_K is either trivial or the identity map. Moreover, since $\mu_2(K_s) = \mu_2(k_s)$ and since the diagram

$$\begin{array}{ccc} \mu_2(k_s) & \xrightarrow{\tau_k} & \mu_2(k_s) \\ \downarrow & & \downarrow \\ \mu_2(K_s) & \xrightarrow{\tau_K} & \mu_2(K_s) \end{array}$$

commutes, we have $\tau_K = \tau_k$. Therefore, either τ_K is trivial for all K/k or equal to the identity for all K/k. Therefore, the induced map $\tau_{*,K}$ will be either trivial for every field extension K/k or equal to the identity map of $\mathrm{Br}_2(K)$ for every field extension K/k.

Set $K = k(t_1, t_2)$. Then the quaternion K-algebra $Q = (t_1, t_2)$ is not trivial. Let us apply the previous equality to the K-algebra

$$(A, \sigma) = (\mathrm{M}_m(K) \otimes_K Q, t \otimes \gamma).$$

By Example VIII.21.32 (3), we have

$$\mathcal{T}_\sigma^+ \simeq m \times \langle 2 \rangle \perp \frac{m(m-1)}{2} \times \langle 1, -t_1, -t_2, t_1 t_2 \rangle.$$

Similar computations to those performed before show that we have

$$w_2(\mathcal{T}_\sigma^+) = w_2\left(\frac{m(m-1)}{2} \times \langle -t_1, -t_2, t_1 t_2 \rangle\right).$$

Applying Lemma IV.11.5 several times, we get

$$w_2(\mathcal{T}_\sigma^+) = \frac{m(m-1)}{2} w_2(\langle -t_1, -t_2, t_1 t_2 \rangle).$$

The reader will check that we have

$$w_2(\langle -t_1, -t_2, t_1 t_2 \rangle) = (-1) \cup (-1) + (t_1) \cup (t_2).$$

We then get $w_2(\mathcal{T}_\sigma^+) = \dfrac{m(m-1)}{2}([(-1,-1)] + [Q])$ for this particular algebra with involution, and it follows that we have

$$\tau_{*,K}([Q]) = \frac{m(m-1)}{2}[Q].$$

If $\dfrac{m(m-1)}{2}$ is even, we get $\tau_{*,K}([Q]) = 0$. Thus $\tau_{*,K}$ is not the identity, and $\tau_{*,K}$ is trivial. If $\dfrac{m(m-1)}{2}$ is odd, we get $\tau_{*,K}([Q]) = [Q] \neq 0$. Thus $\tau_{*,K}$ is not trivial, and therefore is equal to the identity map. The previous considerations show that we have $\tau_{*,K} = 0$ for **all** K/k if $\dfrac{m(m-1)}{2}$ is even, and equal to the identity for all K/k if $\dfrac{m(m-1)}{2}$ is odd. Apply this to $K = k$, we get the desired result. □

Exercises

1. Let k be a field. Let $\mathbb{Z}/2\mathbb{Z}$ act by group automorphisms on $\mathbf{PGL}_n(k_s)$ as follows: if $f = \mathrm{Int}(M)$, set

$$\overline{1} \cdot \mathrm{Int}(M) = \mathrm{Int}((M^{-1})^t).$$

Let G be the corresponding semi-direct product.

(a) Show that G is a $\mathcal{G}_{k_s}$-group.

(b) Show that $H^1(k, G)$ classifies isomorphisms classes of central simple algebras (B, τ) with a unitary involution, whose center K is a quadratic étale k-algebra.

(c) Consider the exact sequence of $\mathcal{G}_{k_s}$-groups

$$1 \longrightarrow \mathbf{PGL}_n(k_s) \longrightarrow G \longrightarrow \mathbb{Z}/2\mathbb{Z} \longrightarrow 1.$$

Describe the induced maps $H^1(k, \mathbf{PGL}_n) \longrightarrow H^1(k, G)$ and $H^1(k, G) \longrightarrow H^1(k, \mathbb{Z}/2\mathbb{Z})$.

2. Let A be a central simple k-algebra. For any commutative k-algebra R, set

$$\mathbf{SL}_1(A)(R) = \{a \in A_R \mid \mathrm{Nrd}_{A_R}(a) = 1\}.$$

(a) For any field extension L/k, show that we have an exact sequence of $\mathcal{G}_{L_s}$-modules

$$1 \longrightarrow \mathbf{SL}_1(A)(L_s) \longrightarrow \mathbf{GL}_1(A)(L_s) \longrightarrow L_s^\times \longrightarrow 1.$$

(b) Deduce that $H^1(L, \mathbf{SL}_1(A)) \simeq L^\times/\mathrm{Nrd}_{A_L}(A_L^\times)$ for every field extension L/k.

3. Let A be a central simple k-algebra. For $c \in k^\times$, set

$$X_c = \{a \in A_{k_s} \mid \mathrm{Nrd}_{A_{k_s}}(a) = c\}.$$

Keeping the definitions of Chapter II, Exercice 7, prove that

(a) X_c is a principal homogeneous space over $\mathbf{SL}_1(A)(k_s)$

(b) Any principal homogeneous space over $\mathbf{SL}_1(A)(k_s)$ is isomorphic to X_c for some $c \in k^\times$

(c) $X_c \simeq X_d$ if and only if $cd^{-1} \in \mathrm{Nrd}_A(A^\times)$.

4. Let $n \geq 1$ be an integer, and let k be a field of characteristic different from 2. Let us denote by $\mathcal{T}$ the trace form of $\mathrm{M}_n(k)$. Show that for every central simple k-algebra A of degree n, we have

$$\det(\mathcal{T}_A) = \det(\mathcal{T}).$$

5. Let $n \geq 1$ be an **even** integer, and let k be a field of characteristic different from 2. Let us denote by $\mathcal{T}$ the trace form of $\mathrm{M}_n(k)$. The main goal of this exercise is to compute the Hasse invariant of the trace form of a central simple algebra. We will follow the arguments given by Lewis and Morales in [34].

(a) For every field extension K/k, show that the diagram

$$\begin{array}{ccccccccc} 1 & \longrightarrow & \mu_n(K_s) & \longrightarrow & \mathbf{SL}_n(K_s) & \longrightarrow & \mathbf{PGL}_n(K_s) & \longrightarrow & 1 \\ & & \downarrow \tau & & \downarrow & & \downarrow & & \\ 1 & \longrightarrow & \mu_2(K_s) & \longrightarrow & \mathbf{Spin}(\mathcal{T})(K_s) & \longrightarrow & \mathbf{O}^+(\mathcal{T})(K_s) & \longrightarrow & 1 \end{array}$$

commutes.

(b) Let $d \in K^\times$, and let $M = \mathrm{diag}(d, 1, \dots, 1) \in \mathrm{M}_n(K)$. Check that we have

$$\mathrm{Int}(M) = \tau_{v_2} \circ \tau_{w_2} \circ \cdots \circ \tau_{v_n} \circ \tau_{w_n},$$

where $v_i = E_{i1} - dE_{1i}$ and $w_i = E_{i1} - E_{1i}$ for $i = 2, \dots, n$.

(c) Show that the spinor norm of $\mathrm{Int}(M)$ is $d \in K^\times/K^{\times 2}$ (see Chapter IV for a definition of the spinor norm).

(d) Deduce that there exists a field extension K/k for which τ_K is not trivial.

(e) Show that for every central simple k-algebra of degree n, we have

$$w_2(\mathcal{T}_A) = w_2(\mathcal{T}) + \frac{n}{2}[A] = \frac{n(n-2)}{8}[(-1,-1)] + \frac{n}{2}[A].$$

(f) Deduce the values of $w_2(\mathcal{T}_\sigma^-)$ and $w_2(\mathcal{T}_\sigma)$ for any k-algebra (A, σ) with a symplectic involution.

6. Let k be a field, and let K/k be a quadratic étale k-algebra. If $K = k \times k$, we say B is a central simple K-algebra if isomorphic to K and $A \simeq A_1 \times A_2$, where A_1 and A_2 are central simple k-algebras. We then extend the notion of a unitary involution on B in an obvious way.

(a) Let A be a central simple k-algebra. Show that the map

$$\varepsilon\colon \begin{array}{ccc} A \times A^{op} & \longrightarrow & A \times A^{op} \\ (a_1, a_2^{op}) & \longmapsto & (a_2, a_1^{op}) \end{array}$$

is a unitary involution on $A \times A^{op}$.

(b) Assume that $K = k \times k$ and that (B, τ) is a central simple K-algebra with a unitary involution. Show that there exists a central simple k-algebra A such that

$$(B, \tau) \simeq (A \times A^{op}, \varepsilon).$$

7. Let (B, τ) be a central simple K-algebra with a unitary involution. We say that τ is hyperbolic if there exists an idempotent $e \in B$ such that $\tau(e) = 1 - e$.

   (a) Assume that $K = k \times k$. Show that ε is hyperbolic.

   (b) Assume that K is a field, let k be the subfield of K consisting of τ-symmetric elements, and let ι be the non-trivial k-automorphism of K. If $B = \mathrm{M}_n(K)$, show that we have

$$\tau = \mathrm{Int}(H) \circ t,$$

   where $H = (m_{ij}) \in \mathrm{GL}_n(K)$ satisfies $H^* = (\iota(m_{ji})) = H$.

   Show that τ is hyperbolic if and only if the (K, ι)-hermitian form on K^n represented by H is hyperbolic.

   (c) Show that all hyperbolic unitary involutions on B are conjugate.

IX
Digression: a geometric interpretation of $H^1(\_, G)$

This chapter is a preamble to the following ones, where we will need the notion of a G-torsor to derive some interesting properties of the functor $H^1(\_, G)$. Roughly speaking, a G-torsor is a scheme-theoretic generalization of a principal homogeneous space over $G(k_s)$. We will introduce this notion, as well as the notion of a generic G-torsor, which is a G-torsor which specializes densely to any other torsor. For example, $k(t)(\sqrt{t})/k(t)$ is a generic $\mathbb{Z}/2\mathbb{Z}$-torsor. We will see in this chapter and the following ones that quite often generic G-torsors concentrate all the essential information on the functor $H^1(\_, G)$.

All of this will require the notion of a scheme. Since giving precise definitions would be too lengthy, and out of the scope of this book, we will just give an insight of what a scheme is, and leave the interested reader to refer to his/her favorite book on the subject.

§IX.22 Reminiscences on schemes

Let k be a field. By [17, Théorème de comparaison, p.18], a scheme X over k is uniquely determined by its functor of points, still denoted by X by abuse of notation, that is the functor

$$X\colon \begin{array}{l} \mathbf{Alg}_k \longrightarrow \mathbf{Sets} \\ \quad R \longmapsto X(R) = \mathbf{Mor}_{k-Sch}(\mathrm{Spec}(R), X). \end{array}$$

Moreover, morphisms of schemes $X \longrightarrow Y$ correspond to natural transformations between the corresponding functors of points. Elements of $X(R)$ are called R-points.

For example, if A is a commutative k-algebra, the topological space

$\mathrm{Spec}(A)$ of prime ideals of A (endowed with Zariski topology) is a k-scheme (in fact, we also have to specify the sheaf of functions, but we will not do it here). Recall that Zariski topology is the topology whose closed subsets are the sets

$$V(\mathfrak{A}) = \{\mathfrak{p} \in \mathrm{Spec}(A) \mid \mathfrak{p} \supset \mathfrak{A}\},$$

where $\mathfrak{A}$ is an ideal of A. The functor of points of $\mathrm{Spec}(A)$ is given by

$$X : \begin{array}{rcl} \mathbf{Alg}_k & \longrightarrow & \mathbf{Sets} \\ R & \longmapsto & X(R) = \mathrm{Hom}_{k-alg}(A, R). \end{array}$$

Moreover, a morphism of k-schemes $\mathrm{Spec}(R) \longrightarrow \mathrm{Spec}(A)$ corresponds to a morphism of k-algebras $A \longrightarrow R$, in view of Yoneda's lemma. Notice that the k-algebra structure on A then gives rise to a morphism of k-schemes $\mathrm{Spec}(A) \longrightarrow \mathrm{Spec}(k)$. If moreover $A \simeq k[X_1, \ldots, X_n]/I$ is a finitely generated k-algebra, the functor of points of $\mathrm{Spec}(A)$ is $h_A \simeq V(I)$, where $V(I)$ is the functor introduced in Lemma III.7.14.

Such a k-scheme is called **affine**. If A is finitely generated, we say that $\mathrm{Spec}(A)$ is a k-scheme of finite type.

If L/k is a field extension, an L-point corresponds to a morphism of k-algebras $A \longrightarrow L$. Since L is an integral domain, the kernel $\mathfrak{p}$ of this morphism is a prime ideal, and we have a morphism of k-algebras $A_{\mathfrak{p}} \longrightarrow L$, where $A_{\mathfrak{p}}$ is the localization of A at $\mathfrak{p}$. We then have a commutative diagram

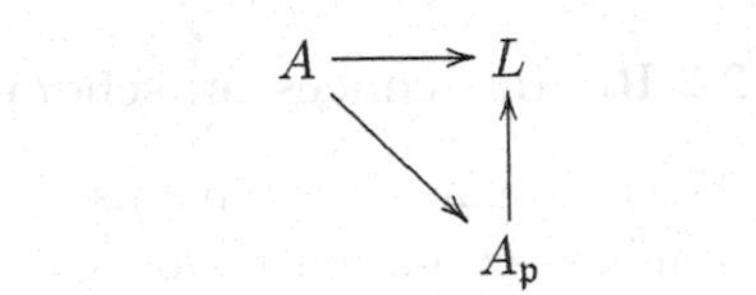

and thus the diagram

$$\begin{array}{ccc} \mathrm{Spec}(L) & \longrightarrow & \mathrm{Spec}(A) \\ \downarrow & \nearrow & \\ \mathrm{Spec}(A_{\mathfrak{p}}) & & \end{array}$$

commutes. If $\kappa(\mathfrak{p}) = A_{\mathfrak{p}}/\mathfrak{p}A_{\mathfrak{p}} \simeq \mathrm{Frac}(A/\mathfrak{p})$, the morphism $A_{\mathfrak{p}} \longrightarrow L$

gives rise to an injection $\kappa(\mathfrak{p}) \hookrightarrow L$, and the two diagrams

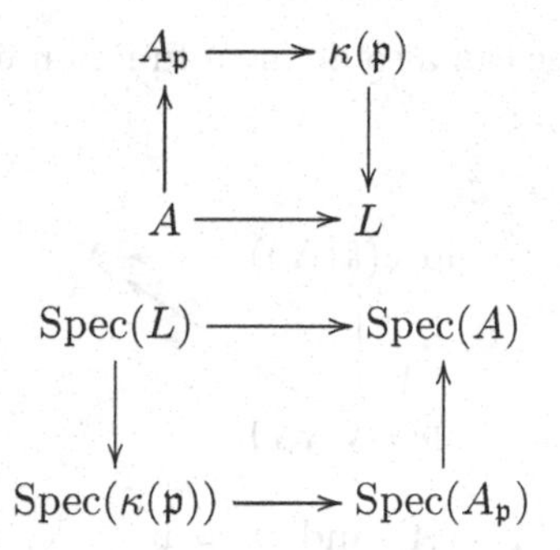

are commutative.

Assume now that A is an integral domain, with field of fractions $k(A)$. In this case, the topological space $\text{Spec}(A)$ is irreducible, and the diagrams

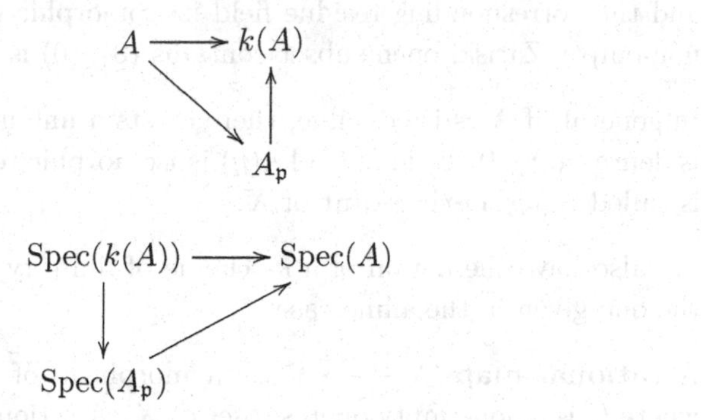

are commutative.

In general, given a k-scheme X, for any $x \in X$, one can define a local k-algebra $\mathcal{O}_{X,x}$, with residue field $\kappa(x)$. If now L/k is a field extension and $a : \text{Spec}(L) \longrightarrow X$ is an L-point, there exists a point $x \in X$ such that we have an injective morphism $\kappa(x) \hookrightarrow L$, and the two diagrams

$$\begin{array}{ccc} \text{Spec}(L) & \longrightarrow & X \\ \downarrow & \nearrow & \\ \text{Spec}(\mathcal{O}_{X,x}) & & \end{array}$$

$$\begin{array}{ccc} \text{Spec}(L) & \longrightarrow & X \\ \downarrow & & \uparrow \\ \text{Spec}(\kappa(x)) & \longrightarrow & \text{Spec}(\mathcal{O}_{X,x}) \end{array}$$

are commutative.

If X is irreducible, one can also define a function field $k(X)$, and for any $x \in X$, the diagram

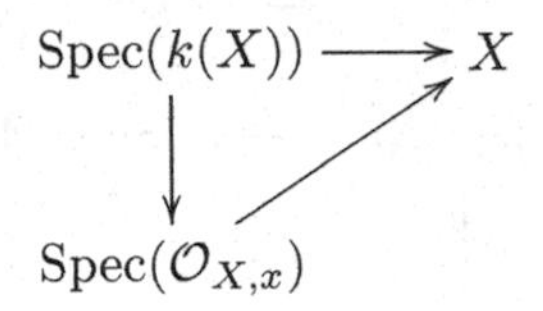

commutes. If $X = \mathrm{Spec}(A)$ and $x = \mathfrak{p} \in X$, we have $\mathcal{O}_{X,x} \simeq A_{\mathfrak{p}}$, $\kappa(x) \simeq \kappa(\mathfrak{p})$ and $k(X) \simeq k(A)$.

If $X = \mathrm{Spec}(A)$ and A is an integral domain, the ideal (0) is prime, and the corresponding residue field is isomorphic to $k(A)$. Since every non-empty Zariski open subset contains (0), (0) is dense in X.

In general, if X is irreducible, there exists a unique point $\eta \in X$ which is dense in X. Its residue field $\kappa(\eta)$ is isomorphic to $k(X)$. Such a point is called the **generic point** of X.

We also have the notion of a k-scheme of finite type, which generalizes the one given in the affine case.

A **rational map** $X \dashrightarrow Y$ is a morphism of k-schemes $U \longrightarrow Y$ where U is a non-empty open subset of X. A rational map is **dominant** if its image is dense in Y.

If Y is a k-scheme, a Y-scheme is a pair (X, f), where X is a k-scheme and $f : X \longrightarrow Y$ is a morphism of k-schemes. If X and X' are Y-schemes, a morphism of Y-schemes is a morphism of schemes $X \longrightarrow X'$ such that the diagram

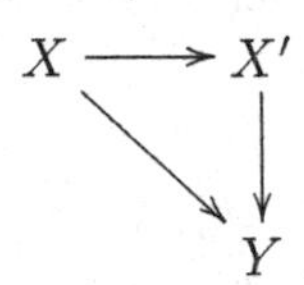

commutes.

We end this section by introducing the fibered product. If $f : X \longrightarrow Y$ and $f' : X' \longrightarrow Y$ are two morphisms of k-schemes, the **fibered product** $X \times_Y X'$ (or pullback) is the k-scheme corresponding to the functor of points

$$\begin{aligned} \mathbf{Alg}_k &\longrightarrow \mathbf{Sets} \\ R &\longmapsto X(R) \times_{Y(R)} X'(R), \end{aligned}$$

where $X(R) \times_{Y(R)} X'(R) = \{(a, a') \in X(R) \times X'(R) \mid f_R(a) = f'_R(a')\}$.

The fibered product has the following universal property: if $g : T \longrightarrow X$ and $g' : T \longrightarrow X'$ are two morphisms of schemes such that $f \circ g = f' \circ g'$, there exists a unique morphism $\varphi : T \longrightarrow X \times_Y X'$ satisfying

$$\pi_X \circ \varphi = g \text{ and } \pi_{X'} \circ \varphi = g',$$

where π_X and $\pi_{X'}$ are projections onto X and X' respectively.

If $X = \mathrm{Spec}(A), X' = \mathrm{Spec}(A')$ and $Y = \mathrm{Spec}(B)$ are affine k-schemes, then $X \times_Y X' = \mathrm{Spec}(A \otimes_B A')$, where the tensor product is formed with respect to the maps $B \longrightarrow A$ and $B \longrightarrow A'$.

If $f : X \longrightarrow Y$ is a morphism of k-schemes and $y \in Y$, we have a morphism $f' : \mathrm{Spec}(\kappa(y)) \longrightarrow Y$. The corresponding fibered product is called the **fiber** of y along f, and is denoted by f. If Y is irreducible, the **generic fiber** is $f^{-1}(\eta)$, where $\eta \in Y$ is the generic point of Y.

§IX.23 Torsors

We first extend the notion of a group-scheme.

Definition IX.23.1. Let Y be a scheme. A **group-scheme** over Y is a functor $G : \mathbf{Alg}_k \longrightarrow \mathbf{Grps}$ which is also the functor of points of a Y-scheme.

We now give the definition of a G-torsor.

Definition IX.23.2. Let G be an affine group-scheme over Y which is flat and locally of finite type over Y. We say that a morphism of schemes $X \longrightarrow Y$ is a **(flat) G-torsor over Y** if G acts on X (on the right), the morphism $X \longrightarrow Y$ is faithfully flat and locally of finite type, and the map $\varphi : G \times_Y X \longrightarrow X \times_Y X$ defined by

$$\begin{aligned} G \times_Y X &\longrightarrow X \times_Y X \\ (g, x) &\longmapsto (x, x \cdot g) \end{aligned}$$

is an isomorphism.

A morphism of G-torsors is just a G-equivariant morphism of Y-schemes.

One can show that a morphism between two Y-torsors is an isomorphism. If $Y' \longrightarrow Y$ is a morphism of schemes and $X \longrightarrow Y$ is a G-torsor, then $X \times_Y Y'$ is a G-torsor (in fact a $G \times_Y Y'$-torsor) over Y'.

The torsor $Y \times G_Y \longrightarrow Y$ is called the **split G-torsor**.

Remark IX.23.3. We will not define the notion of a (faithfully) flat morphism locally of finite type. Let us just say that if X and Y are affine k-schemes of finite type, any morphism $X \longrightarrow Y$ is locally of finite type. If $X = \mathrm{Spec}(A)$ and $Y = \mathrm{Spec}(B)$, then the morphism $X \longrightarrow Y$ is (faithfully) flat if and only if the corresponding morphism $B \longrightarrow A$ endows A with a structure of (faithfully) flat B-module. The reader will refer to [41] for the missing definitions and proofs on G-torsors.

Example IX.23.4. If G is a linear algebraic group over k, and K/k is a field extension, then $H^1(K, G)$ classifies isomorphism classes of G-torsors over $\mathrm{Spec}(K)$, as well as isomorphism classes of principal homogeneous spaces over $G(K_s)$ (see Chapter II, Exercise 5). One direction of the correspondence between these two sets is given as follows (the other direction is given by Exercice 2): if $X \longrightarrow \mathrm{Spec}(K)$ is a G-torsor, since G is affine and smooth, X is affine and smooth by faithfully flat descent, hence $X(K_s)$ is not empty, and is a principal homogeneous space over $G(K_s)$.

Let k be a field, and let G be a linear algebraic group over k. Since every k-algebra morphism $R \longrightarrow S$ gives rise to a morphism of k-schemes $\mathrm{Spec}(S) \longrightarrow \mathrm{Spec}(R)$, we get a functor $H^1(\_, G) : \mathbf{Alg}_k \longrightarrow \mathbf{Sets}^*$ by setting

$$H^1(R, G) = \{ \text{ isomorphism classes of } G_R - \text{torsors over } \mathrm{Spec}(R)\},$$

where the base point is the isomorphism class of the split G-torsor and the map $H^1(R, G) \longrightarrow H^1(S, G)$ is given by

$$H^1(R, G) \longrightarrow H^1(S, G), T \longmapsto T_S = T \times_{\mathrm{Spec}(R)} \mathrm{Spec}(S).$$

We would like now to introduce the notion of a generic torsor. We first explain how to specialize elements of a functor.

Definition IX.23.5. Let k be a field, and let $K/k, L/k$ be two field extensions. A pseudo **k-place** $f : K \rightsquigarrow L$ is a local morphism of k-algebras $\varphi_f : R_f \longrightarrow L$, where R_f is a local ring of K containing k.

Let $\mathbf{F} : \mathbf{Alg}_k \longrightarrow \mathbf{Sets}$ be a functor. Let $K/k, L/k$ be two field extensions. Let $a \in \mathbf{F}(K)$ and $b \in \mathbf{F}(L)$. We say that b is a **specialization** of a if there exists a pseudo k-place $f : K \rightsquigarrow L$ and $c \in \mathbf{F}(R_f)$ such that $c_K = a$ and $b = \mathbf{F}(\varphi_f)(c)$.

Example IX.23.6. Let us illustrate the previous definition when $\mathbf{F}$ is the functor $\mathbf{Quad}_n$.

Let $K = k(X_1, \ldots, X_n)$ and let $q = \langle X_1, \ldots, X_n \rangle$. Now let $R_0 = \mathrm{Spec}(k[X_1, \ldots, X_n][\frac{1}{X_1 \cdots X_n}])$ and let $Y = \mathrm{Spec}(R_0)$. Since $X_1 \cdots X_n \in R_0^\times$, the quadratic form $Q = \langle X_1, \ldots, X_n \rangle$ over R_0 is regular. Moreover, we have $Q_K \simeq q$.

Now if L/k is a field extension and if q' is a regular quadratic form over L, we have

$$q' \simeq \langle a_1, \ldots, a_n \rangle,$$

for some $a_i \in L^\times$. Let $\mathfrak{p} \in R_0$ be the kernel of the map

$$\begin{aligned} R_0 &\longrightarrow L \\ X_i &\longmapsto a_i. \end{aligned}$$

This is a prime ideal of R_0, that is a point y of Y. The corresponding local ring $\mathcal{O}_{Y,y}$ is $(R_0)_\mathfrak{p}$, and the map $\varphi : \mathcal{O}_{Y,y} \longrightarrow L$ which sends X_i onto a_i defines a pseudo k-place $K \rightsquigarrow L$.

Let $Q' = Q_{\mathcal{O}_{Y,y}}$. Then Q' is a regular quadratic form on $\mathcal{O}_{Y,y}$ satisfying $Q'_K \simeq q$. By definition of scalar extension for quadratic forms, we have

$$Q'_L = \langle \varphi(X_1), \ldots, \varphi(X_n) \rangle = \langle a_1, \ldots, a_n \rangle \simeq q'.$$

Hence q' is a specialization of q in the sense of the previous definition. Notice that, all in all, q' is the quadratic form obtained from q after performing the substitution $X_1 = a_1, \ldots, X_n = a_n$, which agrees with the intuitive notion of specialization. Notice also that if L is infinite, replacing a_i by $a_i\lambda_i^2$ provides a dense subset of Y such that the corresponding specialization is isomorphic to q'.

This example motivates the following definition.

Definition IX.23.7. Let $\mathbf{F} : \mathbf{Alg}_k \longrightarrow \mathbf{Sets}$ be a functor. An element $a \in \mathbf{F}(K)$ is **generic** if for every field extension L/k with L infinite, every element $b \in \mathbf{F}(L)$ is a specialization of a (hence (a, K) is a versal pair in the sense of [3].)

Example IX.23.8. The previous example shows that $\langle X_1, \ldots, X_n \rangle$ is a generic element of $\mathbf{Quad}_n$.

One can show that the functors $H^1(\_, \mathbf{O}_n)$ and $\mathbf{Quad}_n$ are isomorphic as functors from $\mathbf{Alg}_k$ to $\mathbf{Sets}^*$ (and not only as functors from $\mathfrak{C}_k$ to $\mathbf{Sets}^*$ as we already know from Galois descent). We are going to see that, more generally, the functor $H^1(\_, G)$ has a generic element.

Definition IX.23.9. Let k be a field and let G be a linear algebraic group over k. Let $f : X \longrightarrow Y$ be a G-torsor with Y irreducible. We say that it is **classifying for** G if, for any field extension k'/k with k' infinite and for any G-torsor P' over $\mathrm{Spec}(k')$, the set of points $y \in Y(k')$ such that P' is isomorphic to the fiber $f^{-1}(y)$ is dense in Y.

A **generic** G-torsor is the generic fiber of a classifying torsor.

Examples IX.23.10.

(1) If $\rho : G \hookrightarrow \mathbf{GL}(V)$, then $\mathbf{GL}(V) \longrightarrow \mathbf{GL}(V)/\rho(G)$ is a classifying G-torsor.

(2) Let V be a finite dimensional k-vector space. If G acts linearly and freely on a G-stable open subset of V, then by [64, Proposition 4.7] there exists a G-stable open subset U of V and a k-scheme Y such that $U \longrightarrow Y$ is a G-torsor.

For example, assume G is a finite group acting linearlly and faithfully on V, and set

$$U = V \setminus \ker_{g \neq 1}(g - 1).$$

Then $\mathrm{Spec}(k[U]) \longrightarrow \mathrm{Spec}(k[U]^G)$ is a classifying torsor.

Notice that if $k(V)$ denotes the field of rational functions on the affine space V, the action of G on V induces an action of G on $k(V)$. If $f : k(V) \longrightarrow k$ is such a rational function, then we set

$$g.f(v) = f(g^{-1} \cdot v) \text{ for all } g \in G, v \in V.$$

One can check that we get a linear faithful action of G on $k(V)$. The generic torsor corresponding to the above classifying torsor is then $k(V)/k(V)^G$.

See [25], Section 5.3 and Example 5.4. for a proof of these two facts.

Proposition IX.23.11. *Let G be a linear algebraic group defined over k, and let $P \longrightarrow \mathrm{Spec}(K)$ be a generic G-torsor. Then for every field extension L/k with L infinite and every G-torsor $T \longrightarrow \mathrm{Spec}(L)$, there exist a regular locar ring R of K containing k, a morphism of local k-algebras $R \longrightarrow L$ and a G-torsor $P' \longrightarrow \mathrm{Spec}(R)$ such that*

$$P'_K \simeq P \text{ and } P'_L \simeq T.$$

In particular, a generic G-torsor is a generic element of $H^1(\_, G)$.

Proof. Let $f : X \longrightarrow Y$ be a classifying torsor whose generic fiber is isomorphic to P. Replacing Y by an open subset U, and X by $f^{-1}(U)$ if necessary, one can assume that Y is smooth since $f^{-1}(U) \longrightarrow U$ is also a classifying torsor.

Take $T \longrightarrow \mathrm{Spec}(L)$ any torsor defined over L/k, L infinite. Since $X \longrightarrow Y$ is a classifying torsor, there exists a L-rational point $y : \mathrm{Spec}(L) \longrightarrow Y$ such the diagram

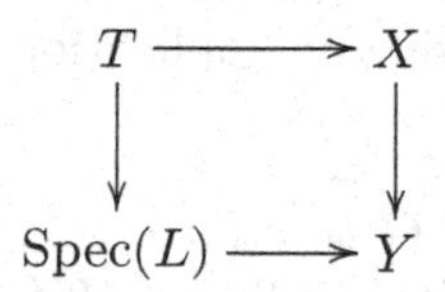

commutes, that is T is the pullback $X \times_Y \mathrm{Spec}(L)$.
Let $\mathcal{O}_{Y,y}$ be the local ring at the point y and let $\varphi : \mathrm{Spec}(\mathcal{O}_{Y,y}) \longrightarrow Y$ the canonical morphism. Notice that $\mathcal{O}_{Y,y}$ is a regular local ring, since Y is smooth. Consider $P' \longrightarrow \mathrm{Spec}(\mathcal{O}_{Y,y})$ the torsor obtained by pulling-back $X \longrightarrow Y$ along φ. The local ring $\mathcal{O}_{Y,y}$ is naturally a k-subalgebra of $k(Y)$ and using the universal property of the fibered product, we see that we have a diagram

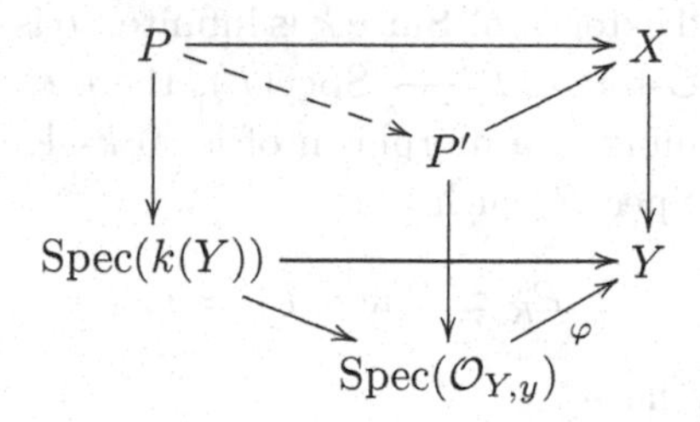

showing that $P \longrightarrow \mathrm{Spec}(k(Y))$ is nothing but the generic fiber of $P' \longrightarrow \mathrm{Spec}(\mathcal{O}_{Y,y})$. Moreover the morphism $y : \mathrm{Spec}(L) \longrightarrow Y$ factors through $\mathrm{Spec}(\kappa(y))$. Let us denote by $P'' \longrightarrow \mathrm{Spec}(\kappa(y))$ the torsor obtained by pulling-back $P' \longrightarrow \mathrm{Spec}(\mathcal{O}_{Y,y})$ along the morphism

$\mathrm{Spec}(\kappa(y)) \longrightarrow \mathrm{Spec}(\mathcal{O}_{Y,y})$. Then $P'' \longrightarrow \mathrm{Spec}(\kappa(y))$ is the pull-back of $P' \longrightarrow \mathrm{Spec}(\mathcal{O}_{Y,y})$ along the morphism $\mathrm{Spec}(\kappa(y)) \longrightarrow Y$. Using the universal property of the fibered product, one gets the existence of a map $T \to P''$ such that the diagram

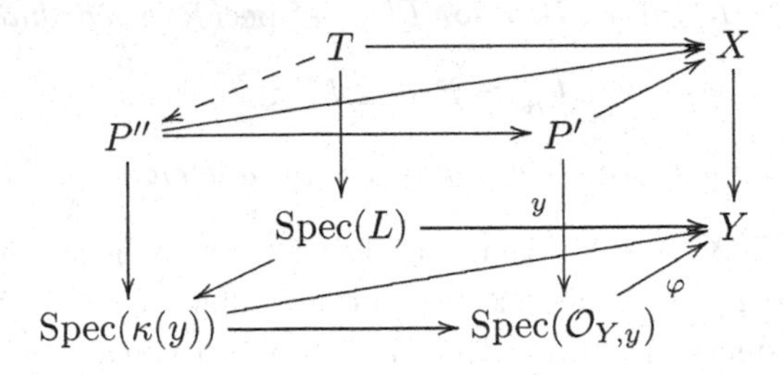

commutes.
This shows that $T \longrightarrow \mathrm{Spec}(L)$ comes from $P'' \longrightarrow \mathrm{Spec}(\kappa(y))$. We deduce that $T \longrightarrow \mathrm{Spec}(L)$ is the pullback of $P' \longrightarrow \mathrm{Spec}(\mathcal{O}_{Y,y})$ along $\mathrm{Spec}(L) \longrightarrow \mathrm{Spec}(\mathcal{O}_{Y,y})$. This concludes the proof. $\square$

We end this section by giving an application of generic torsors (see [25, Theorem 12.3]):

Theorem IX.23.12. *Let G be a linear algebraic group defined over k ($\mathrm{char}(k) \neq 2$). Let $\iota, \iota' : H^1(\_, G) \longrightarrow H^d(\_, \mu_2)$ be two cohomological invariants, and let $P \longrightarrow \mathrm{Spec}(K)$ be a generic G-torsor. If $\iota_K(P) = \iota'_K(P)$, then $\iota = \iota'$.*

Proof. We only give a partial proof in the case where k is infinite. The reader will refer to [25] to fill the gaps. Clearly, we can assume that $\iota' = 0$ without any loss of generality. Therefore, the assumption reads $\iota(P) = 0$.

Let L/k be any field extension. Since k is infinite, so is L. By Proposition IX.23.11, for any G-torsor $T \longrightarrow \mathrm{Spec}(L)$, there exist a regular locar ring R of K containing k, a morphism of local k-algebras $R \longrightarrow L$ and a G-torsor $P' \longrightarrow \mathrm{Spec}(R)$ such that

$$P'_K \simeq P \text{ and } P'_L \simeq T.$$

By assumption, we have

$$\iota_K(P'_K) = \iota_K(P) = 0.$$

Let $\mathfrak{m}$ be the maximal ideal of R, and set $k' = R/\mathfrak{m}$. By the Specialization Theorem [25, Theorem 12.2], we have $\iota_{k'}(P'_{k'}) = 0$. Since $R \longrightarrow L$

is a morphism of local rings, it factors through $R \longrightarrow k'$, so we have

$$\iota_L(T) = \iota_L(P'_L) = ((\iota_{k'}(P'_{k'}))_L = 0.$$

Hence $\iota = 0$, and this concludes the proof. □

Using this result, Serre computed the group of all cohomological invariants of $H^1(\_, G)$ for some algebraic groups G. See [25] for more details. We will give other applications of this result in the next chapter.

EXERCISES

1. Let $\mathbf{F}, \mathbf{F}' : \mathbf{Alg}_k \longrightarrow \mathbf{Sets}$ be two functors, and let

$$\varphi : \mathbf{F} \longrightarrow\!\!\!\!\!\rightarrow \mathbf{F}'$$

be a surjective natural transformation. Show that if $a \in \mathbf{F}(K)$ is a generic element of $\mathbf{F}$, then $\varphi_K(a)$ is a generic element of $\mathbf{F}'$.

2. Let k be a field, let G be an algebraic group over k represented by A, and let X be an affine k-scheme represented by B endowed with a right G-action.

   (a) Express the condition '$X \longrightarrow \mathrm{Spec}(k)$ is a G-torsor' in terms of A and B.

   (b) Let G be an abstract finite group, and let L be a Galois G-algebra over k. Using the previous question, show that $\mathrm{Spec}(L) \longrightarrow \mathrm{Spec}(k)$ is a G-torsor.

3. Let k be a field, let G be an algebraic group over k represented by A, and let P be a principal homogeneous space over $G(k_s)$.

   (a) Define a natural left $G(k_s)$-action on A_{k_s}, and check that $G(k_s)$ act on the right on the set $S = P \times A_{k_s}$ by

$$(x, a)\cdot g = (x\cdot g, g^{-1}\cdot a).$$

   (b) Fix an element $x_0 \in P$. Show that any orbit of $G(k_s)$ in S is represented in a unique way by an element of the form (x_0, a) for some $a \in A_{k_s}$.

(c) Let C the orbit set of $G(k_s)$ in S. We will denote by ξ_a the orbit of $(x_0, a) \in S$. Show that the operations

$$C \times C \longrightarrow C$$
$$(\xi_{a_1}, \xi_{a_2}) \longmapsto \xi_{a_1+a_2}$$

$$C \times C \longrightarrow C$$
$$(\xi_{a_1}, \xi_{a_2}) \longmapsto \xi_{a_1 a_2}$$

$$k_s \times C \longrightarrow C$$
$$(\lambda, \xi_a) \longmapsto \xi_{\lambda a}$$

endows C with a structure of a k_s-algebra.

(d) Check that the diagonal action of $\mathcal{G}_{k_s}$ on S induces an action of $\mathcal{G}_{k_s}$ on C.

(e) Let $B = C^{\mathcal{G}_{k_s}}$, and let $X = \mathrm{Spec}(B)$. Show that G acts on X on the right, that $X \longrightarrow \mathrm{Spec}(k)$ is a G-torsor, and that $X(k_s) \simeq P$.

4. Let $X \longrightarrow \mathrm{Spec}(k)$ be a G-torsor. Show that X is isomorphic to the split G-torsor if and only if $X(k) \neq \emptyset$.

X

Galois cohomology and Noether's problem

In this chapter, we give applications of Galois cohomology to Noether's problem: given an infinite field k and a finite group G acting linearly and faithfully on a finite dimensional k-vector space V, is the field extension $k(V)^G/k$ purely transcendental? We will show that the answer is negative in general by constructing a non-zero cohomological obstruction.

§X.24 Formulation of Noether's problem

Let k be an infinite field, and let G be a finite abstract group. Recall that a linear faithful representation of G is a finite dimensional k-vector space V on which G acts linearly and faithfully. Now, one can ask the following question:

does there exist a linear faithful representation such that $k(V)^G/k$ rational (i.e. a purely transcendental extension) ?

We will refer to this question as **Noether's problem** for G over k, and we will denote it by $(Noeth_{G,k})$.

If we choose a basis $\mathbf{e} = (e_1, \ldots, e_n)$ of V, this may be reinterpreted in a more explicit way. The faithful action of G on V gives rise to an injective group morphism $\rho : G \hookrightarrow \mathrm{GL}_n(k)$. Moreover, $k(V)$ identifies to $k(X_1, \ldots, X_n)$ via the isomorphism

$$k(X_1, \ldots, X_n) \xrightarrow{\sim} k(V)$$
$$F \longmapsto \left[\sum_{i=1}^{n} v_i e_i \longmapsto F(v_1, \ldots, v_n)\right].$$

If $g \in G$ and $\rho(g)^{-1} = (a_{ij})_{1 \leq i,j \leq n}$, the action of g on $F \in k(X_1, \ldots, X_n)$

is then given by

$$g \cdot F = F\left(\sum_{j=1}^{n} a_{1j} X_j, \dots, \sum_{j=1}^{n} a_{nj} X_j\right).$$

Noether's problem then reformulates as follows: does there exist an injective group morphism $\rho : G \hookrightarrow \mathrm{GL}_n(k)$ such that the field extension $k(X_1, \dots, X_n)^G/k$ is rational ?

The answer is known to be positive for $k = \mathbb{C}$ when G is abelian by a theorem of Fischer [22]. However, it becomes false for $k = \mathbb{Q}$ and for various cyclic groups, for example $G = \mathbb{Z}/8\mathbb{Z}$ (which is the smallest possible counterexample).
In this section, we are going to prove the following theorem, due to Serre [25, Theorem 33.16]:

Theorem X.24.1. *Let G be a finite group with a 2-Sylow subgroup which is cyclic of order ≥ 8. Then Noether's problem for G over $\mathbb{Q}$ has a negative answer.*

The case where G is abelian was proved by Endo-Miyata [21], Lenstra[33] and Voskrensenskiĭ [67]. Later on, Voskrensenskiĭ [68] and Saltman [51] proved the stronger property that there is no generic Galois extension of group G over a rational extension of $\mathbb{Q}$. Notice that Saltman also proved in [52] that Noether's problem has a negative answer over $\mathbb{C}$ in general when G is not abelian. In fact, he exhibits examples of finite groups G for which there is no generic Galois extension of group G over a rational extension of $\mathbb{C}$, by considering unramified cohomology classes in the Brauer group (see the next section for the notion of an unramified cohomology class). We refer the reader to the survey of Colliot-Thélène and Sansuc [15] on rationality problems for more results on Noether's problem and similar questions.

§X.25 The strategy

The main problem is to find a way to detect the non-rationality of a finitely generated field extension K/k. The answer is (sometimes) given by looking at the unramified cohomology of K/k. We will not be very precise here, since we will give details in the following paragraphs, but we would like to give a rough idea of the strategy of the proof. We assume that the reader is familiar with the language of valuations, and refer to [12], [44] or [71] for the missing definitions and basic properties of valued fields.

Assume that $\mathrm{char}(k) \neq 2$. Let v be a discrete valuation of K which is trivial on k, and let ∂_v be the map defined by

$$\partial_v : \begin{array}{rcl} H^1(K, \mu_2) & \longrightarrow & \mathbb{Z}/2\mathbb{Z} \\ (a) & \longmapsto & v(a) \end{array}$$

where a is any representative of the square class corresponding to (a) via the isomorphism $H^1(K, \mu_2) \simeq K^\times/K^{\times 2}$. This map is a well-defined group morphism (if $(a) = (b)$ in $H^1(K, \mu_2)$, then a and b differ by a non-zero square and $v(b) \equiv v(a) \pmod 2$), called the **residue at** v.

Say that a class $(a) \in H^1(K, \mu_2)$ is **unramified at** v if $\partial_v((a)) = 0$, and denote by $H^1_{nr}(K/k, \mu_2)$ the subgroup of all elements of $H^1(K, \mu_2)$ which are unramified at all discrete valuations of K which are trivial on k. Notice that if $a \in k^\times$, then $(a) \in H^1_{nr}(K/k, \mu_2)$ since v is trivial on k. Hence the restriction map induces a group morphism

$$H^1(k, \mu_2) \longrightarrow H^1_{nr}(K, \mu_2)$$

Elements lying in the image of this map are called **constant**.

Assume that $K = k(t)$ where t is an indeterminate over k. What is $H^1_{nr}(K/k, \mu_2)$ in this case ? Let $(a) \in H^1_{nr}(K/k, \mu_2)$. Multiplying by a square if necessary, we may assume that $a \in k[t]$. Now let π be an arbitrary monic irreducible polynomial dividing a, and let v_π the corresponding valuation on $k(t)$. By assumption, we will have $\partial_{v_\pi}((a)) = v_\pi(a) = 0 \in \mathbb{Z}/2\mathbb{Z}$. It implies that $a = cP^2$ for some $c \in k^\times$ and $P \in k[t]$. It follows that $(a) = (c)$, so $H^1_{nr}(K/k, \mu_2)$ consists of constant elements.

We can even say a bit more: assume that $a \in k^\times$ and that $\mathrm{Res}_{K/k}(a) = 0 \in H^1(K, \mu_2)$. It means that $a \in k(t)^{\times 2}$. Now if $a = F(t)^2, F(t) \in k(t)$, then write $F = \dfrac{P}{Q}$, where $P, Q \in k[t], Q \neq 0$ are relatively prime. We then have $aQ(t)^2 = P(t)^2$. It easily implies that $P(0) \neq 0$ and $Q(0) \neq 0$. Therefore $a = \dfrac{P(0)^2}{Q(0)^2} \in k^{\times 2}$, so $(a) = 0$. Consequently, we have proved

$$H^1_{nr}(k(t)/k, \mu_2) \simeq H^1(k, \mu_2).$$

In fact, if K/k is **any** finitely generated extension, one can define residue morphisms

$$\partial_v : H^d(K, \mu_2) \longrightarrow H^{d-1}(\kappa(v), \mu_2)$$

for all $d \geq 1$ and every discrete valuation of K which is trivial on k.

In particular, we can define $H^d_{nr}(K,\mu_2)$ as above for all $d \geq 1$. It follows from [25, Theorem 10.1] that the restriction map induces a group isomorphism

$$H^d(k,\mu_2) \simeq H^d_{nr}(k(X_1,\ldots,X_n),\mu_2) \text{ for all } d \geq 1.$$

In particular, the restriction map

$$H^d(k,\mu_2) \longrightarrow H^d(k(X_1,\ldots,X_n),\mu_2)$$

is injective.

How can it be useful for our problem ? To explain this, we first need the notion of an unramified cohomological invariant.

Definition X.25.1. Let k be a field, let G be an algebraic group-scheme defined over k. We say that a cohomological invariant

$$\iota : H^1(\_\,,G) \longrightarrow H^d(\_\,,\mu_2)$$

is **unramified** if for all finitely generated field extensions K/k and all $[\alpha] \in H^1(K,G)$ we have $\iota_K([\alpha]) \in H^d_{nr}(K/k,\mu_2)$.

We now come to the key idea of the proof of Theorem X.24.1, which is given by the following proposition.

Proposition X.25.2. *Let k be a field, and let G be a finite group. If $(Noeth)_{G,k}$ has a positive answer, then every unramified normalized cohomological invariant of G is trivial.*

Proof. Let V be a linear faithful representation of G such that $k(V)^G/k$ is rational, and let $\iota : H^1(\_\,,G) \longrightarrow H^d(\_\,,\mu_2)$ be an unramified normalized cohomological invariant. By Example IX.23.10 (2), $k(V)/k(V)^G$ is a generic G-torsor. By assumption $\iota_{k(V)^G}(k(V)/k(V)^G)$ is an element of $H^d_{nr}(k(V)^G,\mu_2)$, and since $k(V)^G/k$ is rational, it means that there exists $a \in H^d(k,\mu_2)$ such that

$$\iota_{k(V)^G}(k(V)/k(V)^G) = \mathrm{Res}_{k(V)^G/k}(a).$$

Let $\iota' : H^1(\_\,,G) \longrightarrow H^d(\_\,,\mu_2)$ be the constant cohomological invariant corresponding to a. By Theorem IX.23.12, we deduce that $\iota = \iota'$, that is

$$\iota_K(L/K) = \mathrm{Res}_{K/k}(a)$$

for every field extension K/k and every Galois G-algebra L/K. Applying this equality to the split Galois G-algebra L_0, we get $a = \iota_k(L_0) = 0$, since ι is normalized. It follows that $\iota = 0$. □

Therefore, to prove that $(Noeth_{G,k})$ has a negative answer, it is enough to produce a non-zero unramified normalized cohomological invariant of G. This is exactly the strategy we are going to use.

Remark X.25.3. In fact, the conclusion of the previous proposition remains true if we only assume that $H^1(\_, G)$ has a rational generic G-torsor, that is a generic G-torsor $T \longrightarrow \mathrm{Spec}(K)$, where K/k is rational. Hence producing a non-zero unramified normalized cohomological invariant of G leads to the stronger conclusion that $H^1(\_, G)$ has no rational generic G-torsor.

§X.26 Residue maps

In this section, we introduce the so-called residue maps

$$H^n(K, \mu_2) \longrightarrow H^{n-1}(\kappa(v), \mu_2),$$

where (K, v) is a valued field. All the valuations are supposed to be discrete and normalized.

Let (K, v) be a complete valued field. Since K is complete, the valuation v extends uniquely to a valuation v_s on K_s. The following result is certainly well-known, but we give a proof by lack of an appropriate reference.

Lemma X.26.1. *Keeping the previous notation, the residue field of v_s is an algebraic closure of $\kappa(v)$.*

Proof. Let us denote by k and ℓ the residue fields of (K, v) and (K_s, v_s) respectively. If K_{nr}/K is the maximal unramified extension of K in K_s and v_{nr} the unique extension of v to K_{nr}, then the residue field of (K_{nr}, v_{nr}) is k_s by [44, Chapter II, Prop.7.5], since a complete valued field is henselian. Let us also denote by $\mathcal{O}_v, \mathcal{O}_{nr}$ and $\mathcal{O}_s$ the valuation rings of v, v_{nr} and v_s respectively and by $\mathfrak{m}_v, \mathfrak{m}_{nr}$ and $\mathfrak{m}_s$ the corresponding maximal ideals.

Since $K \subset K_{nr} \subset K_s$, ℓ/k is algebraic and we have $k_s \subset \ell \subset k_{alg}$. If $\mathrm{char}(k) = 0$, then we have $k_{alg} = k_s = \ell$, and we are done. Now assume that $\mathrm{char}(k) = p > 0$. Since ℓ/k is algebraic, it remains to prove that ℓ is algebraically closed.

Let $\alpha \in \ell_{alg} = k_{alg}$. Then α is algebraic over k, and therefore over k_s.

Since k_s is separably closed, the minimal polynomial μ_{α,k_s} of α over k_s has the form

$$\mu_{\alpha,k_s} = X^{p^r} - a \in k_s[X],$$

for some $r \geq 1$ and some $a \in k_s \subset \ell$. Let $b \in \mathcal{O}_s$ any lifting of a. Let $c \in \mathcal{O}_s$ be any nonzero element, and consider the polynomial

$$P = X^{p^r} - cX - b \in \mathcal{O}_s[X] \subset K_s[X].$$

Since $P' = -c \neq 0$, P and P' are relatively prime, so P is separable over K_s. Since K_s is separably closed, P splits in K_s. Let $\beta \in K_s$ be any root of P. Since a valuation ring is integrally closed, we have $\beta \in \mathcal{O}_s$. Reducing the equality $\beta^{p^r} = c\beta + b$ modulo $\mathfrak{m}_s$, we get $\overline{\beta}^{p^r} = a \in \ell$ and we have

$$\overline{\beta}^{p^r} = \alpha^{p^r} \in \ell_{alg}.$$

Hence $(\alpha - \overline{\beta})^{p^r} = 0$ and thus $\alpha = \overline{\beta} \in \ell$. This concludes the proof. □

For all $\sigma \in \mathcal{G}_{K_s}$, $v_s \circ \sigma$ is a valuation on K_s extending v, and therefore $v_s \circ \sigma = v_s$. In particular, σ restricts to a K-automorphism

$$\mathcal{O}_{v_s} \longrightarrow \mathcal{O}_{v_s},$$

which maps $\mathfrak{m}_{v_s}$ onto $\mathfrak{m}_{v_s}$. Thus it induces by Lemma X.26.1 a $\kappa(v)$-automorphism

$$\overline{\sigma} : \kappa(v)_{alg} \longrightarrow \kappa(v)_{alg},$$

and therefore by restriction a $\kappa(v)$-automorphism

$$\overline{\sigma} : \kappa(v)_s \longrightarrow \kappa(v)_s.$$

We get in this way a morphism of profinite groups

$$\varphi_v : \begin{array}{c} \mathcal{G}_{K_s} \longrightarrow \mathcal{G}_{\kappa(v)} \\ \sigma \longmapsto \overline{\sigma}, \end{array}$$

whose kernel is by definition the inertia group I_v. We recall now the following result (see [25], Lemma 7.6):

Lemma X.26.2. *We have a split exact sequence*

$$1 \longrightarrow I_v \longrightarrow \mathcal{G}_{K_s} \xrightarrow{\varphi_v} \mathcal{G}_{\kappa(v)_s} \longrightarrow 1\,.$$

From now on, we will assume that K and $\kappa(v)$ both have characteristic different from 2. For $d \geq 0$, we will denote by

$$j_v : H^d(\kappa(v), \mu_2) \longmapsto H^d(K, \mu_2)$$

the group morphism induced by φ_v and the identity.

Lemma X.26.3. *For $d \geq 1$ and $p, q \geq 0$, the following properties hold:*

(1) *The map $j_v : H^d(\kappa(v), \mu_2) \longrightarrow H^d(K, \mu_2)$ is injective.*

(2) *If $(L, w)/(K, v)$ is an extension of complete valued fields, then $\kappa(v) \subset \kappa(w)$ and we have*

$$j_w \circ \mathrm{Res}_{\kappa(w)/\kappa(v)} = \mathrm{Res}_{L/K} \circ j_v.$$

(3) *For all $[\alpha] \in H^p(\kappa(v), \mu_2), [\beta] \in H^q(\kappa(v), \mu_2)$, we have*

$$j_v([\alpha] \cup [\beta]) = j_v([\alpha]) \cup j_v([\beta]).$$

(4) *For all $u_1, \ldots, u_d \in \mathcal{O}_v^\times$, we have*

$$j_v((\overline{u}_1) \cup \cdots \cup (\overline{u}_d)) = (u_1) \cup \cdots \cup (u_d).$$

Proof. By Lemma X.26.2, the sequence

$$1 \longrightarrow I_v \longrightarrow \mathcal{G}_{K_s} \xrightarrow{\varphi_v} \mathcal{G}_{\kappa(v)_s} \longrightarrow 1$$

splits. Let $\psi_v : \mathcal{G}_{\kappa(v)_s} \longrightarrow \mathcal{G}_{K_s}$ be a morphism of profinite groups such that

$$\varphi_v \circ \psi_v = \mathrm{Id}_{\mathcal{G}_{\kappa(v)_s}}.$$

Denoting by $\rho_v : H^d(K, \mu_2) \longrightarrow H^d(\kappa(v), \mu_2)$ the group morphism induced by ψ_v and the identity of μ_2, it is not difficult to see that we have

$$\rho_v \circ j_v = \mathrm{Id}_{H^d(\kappa(v), \mu_2)}.$$

In particular, j_v is injective. This proves (1).

Point (2) simply follows from the fact that we have a commutative diagram

$$\begin{array}{ccc} \mathcal{G}_{L_s} & \longrightarrow & \mathcal{G}_{K_s} \\ \downarrow & & \downarrow \\ \mathcal{G}_{\kappa(w)} & \longrightarrow & \mathcal{G}_{\kappa(v)} \end{array}$$

and from the definitions of the various maps involved.

Point (3) being an immediate consequence of Example II.6.7, it remains to prove (4). By (3), it is enough to prove it for $d = 1$. Let $u \in \mathcal{O}_v^\times$, and let $y \in \kappa(v)_s^\times$ such that $y^2 = \overline{u}$. By Hensel's lemma, y may be lifted to a square root $x \in \mathcal{O}_{v_s}^\times$ of u. We then have

$$x^2 = u \text{ and } \overline{x}^2 = \overline{u}.$$

A cocycle representing $(\overline{u})$ is given by

$$\alpha: \begin{aligned} \mathcal{G}_{\kappa(v)_s} &\longrightarrow \mu_2(\kappa(v)_s) \\ \tau &\longmapsto \frac{\tau(\overline{x})}{\overline{x}}, \end{aligned}$$

and therefore a cocycle representing $j_v((\overline{u}))$ is

$$\beta: \begin{aligned} \mathcal{G}_{K_s} &\longrightarrow \mu_2(K_s) \\ \sigma &\longmapsto \frac{\overline{\sigma}(\overline{x})}{\overline{x}}. \end{aligned}$$

We now claim that

$$\overline{\sigma}(\overline{x}) = \overline{x} \iff \sigma(x) = x \text{ for all } \sigma \in \mathcal{G}_{K_s}.$$

Let $\sigma \in \mathcal{G}_{K_s}$. If $\sigma(x) = x$, then $\sigma(x) \equiv x \mod \mathfrak{m}_v$, that is $\overline{\sigma}(\overline{x}) = \overline{x}$. Conversely, assume that $\overline{\sigma}(\overline{x}) = \overline{x}$. Suppose that $\sigma(x) = -x$. Reducing modulo $\mathfrak{m}_v$, we would get $\overline{x} = -\overline{x}$, and thus $1 = -1 \in \kappa(v)$, which is not possible since $\text{char}(\kappa(v)) \neq 2$. Hence $\sigma(x) = x$ as claimed. It follows that β is in fact the cocycle

$$\beta: \begin{aligned} \mathcal{G}_{K_s} &\longrightarrow \mu_2(K_s) \\ \sigma &\longmapsto \frac{\sigma(x)}{x}, \end{aligned}$$

which is known to represent (u). This concludes the proof. □

We continue with the following theorem:

Theorem X.26.4. *Let (K, v) be a complete valued field, and let π be a local parameter. Then for all $d \geq 1$, every $[\alpha] \in H^d(K, \mu_2)$ can be written in a unique way as*

$$[\alpha] = j_v([\alpha_0]) + (\pi) \cup j_v([\alpha_1])$$

where $[\alpha_i] \in H^{d-i}(\kappa(v), \mu_2)$. The class $[\alpha_1]$ does not depend on the choice of π.
Moreover, if $[\alpha_1] = 0$, the class $[\alpha_0]$ does not depend on the choice of π.

The reader will refer to [25], Chapter II for a proof of this result.

Remark X.26.5. If $d = 1$, then $H^{d-1}(\kappa(v), \mu_2)$ is canonically identified with $\mathbb{Z}/2\mathbb{Z}$, so $[\alpha_1] = \overline{m}, m = 0$ or 1, and $(\pi) \cup j_v([\alpha_1])$ has to be understood as the class $m(\pi)$.

Assume now that (K, v) is a valued field which is not necessarily complete, and let K_v be the completion of K with respect to v. Let $[\alpha] \in H^d(K, \mu_2)$. By Theorem X.26.4, we may write

$$\mathrm{Res}_{K_v/K}([\alpha]) = j_v([\alpha_0]) + (\pi) \cup j_v([\alpha_1]).$$

Definition X.26.6. Keeping the notation above, the **residue** of $[\alpha]$ at v is the class

$$\partial_v([\alpha]) = [\alpha_1] \in H^{d-1}(\kappa(v), \mu_2).$$

We say that $[\alpha]$ is **unramified** at v if $\partial_v([\alpha]) = 0$ In this case, we define the **specialization** $s_v([\alpha])$ of $[\alpha]$ at v by

$$s_v([\alpha]) = [\alpha_0] \in H^{d-1}(\kappa(v), \mu_2).$$

Clearly, the maps

$$\partial_v : H^d(K, \mu_2) \longrightarrow H^{d-1}(\kappa(v), \mu_2) \text{ and } s_v : \ker(\partial_v) \longrightarrow H^d(\kappa(v), \mu_2)$$

are group morphisms.

The following lemma immediately follows from the description of the map j_v in Lemma X.26.3 (4) and from the definition of the residue map:

Lemma X.26.7. *Let (K, v) be a valued field. Let $\pi \in \mathcal{O}_v$ be a local parameter, and assume that $u_1, \ldots, u_d \in \mathcal{O}_v^\times$. Then we have*

$$\partial_v((u_1) \cup \cdots \cup (u_d)) = 0,$$

$$\partial_v((u_1) \cup \cdots \cup (u_{d-1}) \cup (\pi)) = (\overline{u}_1) \cup \cdots \cup (\overline{u}_{d-1})$$

and

$$s_v((u_1) \cup \cdots \cup (u_d)) = (\overline{u}_1) \cup \cdots \cup (\overline{u}_d).$$

The next proposition shows how ∂_v and s_v behave with respect to scalar extensions.

Proposition X.26.8. *Let $(L, w)/(K, v)$ be an extension of valued fields, and let $d \geq 1$. Then for every $[\alpha] \in H^d(K, \mu_2)$, we have*

$$\partial_w(\mathrm{Res}_{L/K}([\alpha])) = e(w|v)\mathrm{Res}_{\kappa(w)/\kappa(v)}(\partial_v([\alpha])).$$

Moreover, if $[\alpha]$ *is unramified at* v, $\mathrm{Res}_{L/K}([\alpha])$ *is unramified at* w, *and we have*

$$s_w(\mathrm{Res}_{L/K}(\alpha)) = \mathrm{Res}_{\kappa(w)/\kappa(v)}(s_v([\alpha])).$$

Proof. To compute the residue of $\mathrm{Res}_{L/K}([\alpha])$ at w, we need first to extend scalars to the completion L_w of L with respect to w. Let π be a local parameter for v, and let π' be a local parameter for w. By assumption, we have $\pi = u\pi'^e$, where $e = e(w|v)$ and $u \in \mathcal{O}_w^\times$. We have

$$\mathrm{Res}_{L_w/L}(\mathrm{Res}_{L/K}([\alpha])) = \mathrm{Res}_{L_w/K}([\alpha]) = \mathrm{Res}_{L_w/K_v}(\mathrm{Res}_{K_v/K}([\alpha])).$$

To simplify notation, set $k = \kappa(v), \ell = \kappa(w)$ and $[\alpha'] = \mathrm{Res}_{L/K}([\alpha])$. We have

$$\mathrm{Res}_{K_v/K}([\alpha]) = j_v([\alpha_0]) + (\pi) \cup j_v([\alpha_1])$$

where $[\alpha_i] \in H^{d-i}(\kappa(v), \mu_2)$, and thus

$$\mathrm{Res}_{L_w/L}([\alpha']) = \mathrm{Res}_{L_w/K_v}(j_v([\alpha_0]) + (\pi) \cup j_v([\alpha_1])).$$

Using Lemma X.26.3 (2) and Remark III.9.20, we see that

$$\mathrm{Res}_{L_w/L}([\alpha']) = j_w(\mathrm{Res}_{\ell/k}([\alpha_0])) + \mathrm{Res}_{L/K}((\pi)) \cup j_w(\mathrm{Res}_{\ell/k}([\alpha_1])).$$

Now we have

$$\begin{aligned} \mathrm{Res}_{L/K}((\pi)) &= (u\pi'^e) \\ &= (u) + e(\pi') \\ &= j_w((\overline{u})) + e(\pi'), \end{aligned}$$

the last equality coming from Remark X.26.3 (4). Using Remark X.26.3 (3), we get that $\mathrm{Res}_{L_w/L}([\alpha'])$ is equal to

$$j_w(\mathrm{Res}_{\ell/k}([\alpha_0]) + (\overline{u}) \cup \mathrm{Res}_{\ell/k}([\alpha_1])) + (\pi') \cup j_w(e\mathrm{Res}_{\ell/k}([\alpha_1])).$$

By definition of the residue map, we then obtain

$$\partial_w([\alpha']) = e\mathrm{Res}_{\kappa(w)/\kappa(v)}([\alpha_1]) = e(w|v)\mathrm{Res}_{\kappa(w)/\kappa(v)}(\partial_v([\alpha])).$$

This equality shows in particular that if $[\alpha]$ is unramified at v, then $\mathrm{Res}_{L/K}([\alpha])$ is unramified at w. Moreover, we get in this case

$$\mathrm{Res}_{L_w/L}([\alpha']) = j_w(\mathrm{Res}_{\kappa(w)/\kappa(v)}([\alpha_0])),$$

that is

$$s_w([\alpha']) = \mathrm{Res}_{\kappa(w)/\kappa(v)}([\alpha_0]) = \mathrm{Res}_{\kappa(w)/\kappa(v)}(s_v([\alpha])).$$

This concludes the proof. □

Remark X.26.9. We have cheated a bit here. In fact, the residue map is defined in the hard way using spectral sequences and/or non-trivial results on cohomology of complete valued fields, and Theorem X.26.4 is a consequence of the properties of the residue map. The definitions above may be extended to cohomology classes with values in an arbitrary discrete $\mathcal{G}_{K_s}$-module. See [1],[25] or [26] for more details.

§X.27 An unramified cohomological invariant

Let G be a finite group, and let k be a field of characteristic different from 2. For every field extension K and every Galois G-algebra L/K, we set

$$\iota_K(L/K) = (2) \cup (d_L) \in H^2(K, \mu_2),$$

where d_L is the discriminant of L/K viewed as an étale algebra. It is not difficult to see that the cup-product commutes with scalar extensions. Moreover, since the same property holds for the discriminant, we get a cohomological invariant $\iota : H^1(\_, G) \longrightarrow H^2(\_, \mu_2)$, and this invariant is normalized since the discriminant of the split Galois G-algebra is trivial.

Lemma X.27.1. *If $k = \mathbb{Q}$ and $G = \mathbb{Z}/2^m\mathbb{Z}$, $m \geq 3$, the invariant ι is non-zero.*

Proof. Let $n = 2^{2^m} - 1$ and $L = \mathbb{Q}_2(\zeta_n)$. This extension is cyclic of group G by [57], Chapter IV, §4, Proposition 16 and Corollaire 1. Notice that $3 \mid n$, and therefore $\zeta_3 \in L$. In particular, $\sqrt{-3} \in L$ and $\mathbb{Q}_2(\sqrt{-3})/\mathbb{Q}_2$ is the unique quadratic subextension of $L/\mathbb{Q}_2$, since -3 is not a square in $\mathbb{Q}_2$. On the other hand, the discriminant of $L/\mathbb{Q}_2$ is not trivial by Corollary VII.17.10 and Remark VII.17.11. Since d_L is a polynomial expression in some elements of L by Lemma VII.17.9, we conclude that $K(\sqrt{d_L})$ is a quadratic subfield of L. Hence $\mathbb{Q}_2(\sqrt{d_L}) = \mathbb{Q}_2(\sqrt{-3})$. Thus $d_L = -3 \in \mathbb{Q}_2^\times/\mathbb{Q}_2^{\times 2}$ and $\iota_{\mathbb{Q}_2}(L/\mathbb{Q}_2) = (2) \cup (-3)$. We now check that $(2) \cup (-3) \neq 0 \in H^2(\mathbb{Q}_2, \mu_2)$.

Assume that $(2) \cup (-3) = 0$. Then there exist $a, b \in \mathbb{Q}_2$ such that $-3 = a^2 - 2b^2$, or equivalently there exists $x, y, z \in \mathbb{Z}_2, z \neq 0$ such that $-3z^2 = x^2 - 2y^2$. Dividing by a suitable power of 2, we may assume that x, y, z are not all lying in $2\mathbb{Z}_2$. Reducing modulo $2\mathbb{Z}_2$ shows that $x \equiv z$ mod $2\mathbb{Z}_2$. If $x, z \in 2\mathbb{Z}_2$, we easily get that $2y^2 \in 4\mathbb{Z}_2$ and thus $y \in 2\mathbb{Z}_2$, contradicting our assumption. Hence we have $x = 1 + 2x', z = 1 + 2z'$ for some $x', z' \in \mathbb{Z}_2$. We then get

$$-3(1 + 4z' + 4z'^2) = 1 + 4x' + 4x'^2 - 2y^2.$$

Thus $-3(4z'+4z'^2) = 4+4x'+4x'^2-2y^2$, so $2y^2 \in 4\mathbb{Z}_2$ and thus $y \in 2\mathbb{Z}_2$. If we set $y = 2y', y' \in \mathbb{Z}_2$, we get $-3(z'+z'^2) = 1+x'+x'^2-2y'^2$. Now we get a contradiction, since $z'+z'^2$ and $x'+x'^2$ both lie in $2\mathbb{Z}_2$. □

Lemma X.27.2. *If $G = \mathbb{Z}/2^m\mathbb{Z}$, $m \geq 3$, the invariant ι is unramified.*

Proof. Let K/k be a finitely generated field extension, let v be a valuation on K which is trivial on k, and let L/K be a Galois G-algebra. Then $\mathrm{Res}_{K_v/K}(\iota_K(L/K)) = \iota_{K_v}(L_{K_v}/K_v)$.

Set $L' = L_{K_v}$. This is a Galois G-algebra on K_v. If L' is not a field, then by Theorem V.14.20 $L' \simeq M^r$ (as an étale algebra) for some Galois extension M/K_v of group $H \subset G$, where H is a proper subgroup of G of index r. Since H is proper, r is a non-trivial power of 2, so we have $d_{L'} = (d_M)^r = 1 \in K_v^\times/K_v^{\times 2}$. Therefore $(2) \cup (d_{L'}) = 0$ and $\iota_K(L/K)$ is unramified at v.

Assume now that L' is a field. Then L'/K_v is a Galois extension of group G. Notice that by assumption, $\mathrm{char}(\kappa(v)) = \mathrm{char}(k) \neq 2$. Thus, if L'/K_v is totally ramified, it is tamely ramified. Hence K_v contains μ_{2^m} by [12, Chapter I, §8, Proposition 1], so it contains a primitive 8th root of 1 since $m \geq 3$, as well as a square root of 1. In particular 2 is a square in $K_v^\times$, and we conclude as before.

If L' is not totally ramified, then the maximal unramified subfield L_0 of L_{K_v} is a non-trivial subfield and therefore contains $K_v(\sqrt{d_{L'}})$. Thus, $K_v(\sqrt{d_{L'}})/K_v$ is unramified. It implies that the valuation of $d_{L'}$ is even, since the extension $K_v(\sqrt{u\pi_v})/K_v$ is totally ramified for any $u \in \mathcal{O}_v^\times$. Hence $(2) \cup (d_{L'}) = (2) \cup (u)$ for some $u \in \mathcal{O}_v^\times$. Notice that since v is trivial on k, we have $2 \in \mathcal{O}_v^\times$. It follows from Lemma X.26.7 that $\partial_v((2) \cup (d_{L'})) = 0$. We then get that $\iota_{K_v}(L'/K_v)$ is also unramified at v in this case. This concludes the proof. □

§X.28 Proof of Theorem X.24.1

We are now ready to conclude. Let G be a group with a cyclic 2-Sylow subgroup S of order at least 8. By [55, 6.2.11], S has a normal complement in G. In other words, there exists a surjective group morphism $\pi : G \longrightarrow S$ and a group morphism $s : S \longrightarrow G$ such that $\pi \circ s = \mathrm{Id}_S$. In particular, $\pi_* \circ s_* = \mathrm{Id}_{H^1(\_,S)}$ and

$$\pi_* : H^1(\_,G) \longrightarrow H^1(\_,S)$$

is a surjective natural transformation of functors.

Let $\alpha : H^1(\_, G) \longrightarrow H^2(\_, \mu_2)$ be the normalized cohomological invariant of G defined by

$$\alpha = \iota \circ \pi_*.$$

Since ι is unramified by Lemma X.27.2, so is α. By Lemma X.27.1, ι is non-zero, and since π_* is surjective, α is non-zero as well. By Proposition X.25.2, we conclude that $(Noeth_{G,\mathbb{Q}})$ has a negative answer.

Remark X.28.1. The arguments exposed here are those used by Serre in [25] to prove Theorem X.24.1. Using the same method, Serre also proved that $(Noeth_{G,\mathbb{Q}})$ has a negative answer for subgroups of odd index of $\widetilde{A}_7$.

XI

The rationality problem for adjoint algebraic groups

Let G be a connected linear algebraic group defined over k. In this chapter, we are interested in the following question: is G rational as an affine k-variety ? In other words, does there exist an open subset of G which is isomorphic to an open subset of an affine space ?

This question may be reformulated in more algebraic terms as follows. Let A be the k-algebra representing G. Since G is connected, A is an integral domain (see [69]), so we can consider its field of fractions $k(A)$. The problem now translates as: is $k(A)/k$ a rational extension ?

This is known to be true when k is algebraically closed, by a theorem of Chevalley [14, Cor.2]. When k is not algebraically closed, this is not true any more. We refer to the introduction of [38] and the associated references to have a brief account on the history of the problem.

In [38], Merkurjev studies the rationality problem for classical adjoint algebraic groups. These groups may be viewed as the connected component of the identity element of automorphism groups of some algebras with involutions. He shows that these groups are not rational in general. To do so, he computes the R-equivalence group $G(k)/R$ of such a group G, which measures in some sense how far G is from being rational. More precisely, a rational algebraic group satisfies $G(L)/R = 1$ for all field extensions L/k (see Section XI.29 for the definition of R-equivalence and a proof of this fact). He then exhibits infinite families of adjoint groups which have a non-trivial equivalence group over some field extension. The same method was applied by Chernousov and Merkurjev in [13] to construct infinite families of simply connected group which are not rational.

In [5], Monsurrò, Tignol and the author constructed new families of non-rational adjoints groups. To do so, they constructed a non-zero cohomogical invariant $G \longrightarrow H^4(\_, \mu_2)$, which turns out to factor through R-equivalence, giving rise to an invariant

$$\iota : G(\_)/R \longrightarrow H^4(\_, \mu_2)$$

satisfying $\iota_k \neq 0$. In particular, $G(k)/R \neq 1$ and G is not rational.

In this chapter, we will explain their construction in the case where G is the automorphism group of an algebra with a symplectic involution, in order to simplify the arguments. The non-zero cohomological invariant will be constructed using the restricted trace forms introduced in Chapter VIII.

§XI.29 R-equivalence groups

We start by defining the notion of R-equivalence for algebraic groups.

Let G be a group-scheme defined over k, let K/k be a field extension and let $K[t]_{0,1}$ be the localization of $K[t]$ with respect to the multiplicative subset generated by t and $(t-1)$. In other words, $K[t]_{0,1}$ is the subalgebra of $K(t)$ consisting of rational functions which are defined at 0 and 1. For $i = 0, 1$, we denote by $ev_i : K[t]_{0,1} \longrightarrow K$ the evaluation map at i. We then get two induced group morphisms

$$G(ev_i) : G(K[t]_{0,1}) \longrightarrow G(K).$$

If $g(t) \in G(K[t]_{0,1})$, we will denote by $g(i)$ the image of $g(t)$ under $G(ev_i)$. Since $G(ev_i)$ is a group morphism, for every $g_1(t), g_2(t) \in G(K[t]_{0,1})$, we get

$$(g_1(t)g_2(t)^{-1})(i) = g_1(i)g_2(i)^{-1}.$$

Notice also that any element $g \in G(K)$ defines an element $g(t) \in G(K[t]_{0,1})$ satisfying $g(0) = g(1) = g$, since the composition

$$K \subset K[t]_{0,1} \xrightarrow{ev_i} K$$

is the identity map. Such an element is called **constant**.

Definition XI.29.1. We say that an element $g \in G(K)$ is R-**trivial** if there exists $g(t) \in G(K[t]_{0,1})$ such that $g(0) = 1$ and $g(1) = g$.

Remark XI.29.2. This definition may be reformulated in more geometric terms: $g \in G(K)$ is R-trivial if there exists a rational map $f : \mathbb{A}^1_K \dashrightarrow G$ defined at 0 and 1 such that $f(0) = 1$ and $f(1) = g$.

Lemma XI.29.3. *The subset $RG(K)$ of $G(K)$ consisting of R-trivial elements is a normal subgroup of $G(K)$.*

Proof. Clearly $1 \in RG(K)$ since we can take $g(t) = 1$. Now if $g_j \in RG(K), j = 1, 2$, then there exists $g_j(t) \in G(K[t]_{0,1})$ such that $g_j(0) = 1$ and $g_j(1) = g_j$. The element $g(t) = g_1(t)g_2(t)^{-1} \in G(K[t]_{0,1})$ then satisfies $g(0) = 1$ and $g(1) = g_1 g_2^{-1}$. Hence $g_1 g_2^{-1}$ is R-trivial, and $RG(K)$ is a subgroup of $G(K)$.

Now let $g \in RG(K)$ and let $h \in G(K)$. Let $h(t)$ be the image of h under the morphism induced by $K \subset K[t]_{0,1}$, and let $g(t) \in G(K[t]_{0,1})$ such that $g(0) = 1$ and $g(1) = g$. Set $g'(t) = h(t)g(t)h(t)^{-1}$. Then $g'(0) = hh^{-1} = 1$, and $g'(1) = hgh^{-1}$. Hence hgh^{-1} is R-trivial. This shows that $RG(K)$ is a normal subgroup of $G(K)$. $\square$

Definition XI.29.4. The R-**equivalence group** of $G(K)$ is the group

$$G(K)/R = G(K)/RG(K).$$

Two elements $g, g' \in G(K)$ which differ by an element of $RG(K)$ will be called R-**equivalent**. If L/K is a field extension, the diagram

$$\begin{array}{ccc} K[t]_{0,1} & \longrightarrow & L[t]_{0,1} \\ \downarrow{\scriptstyle ev_i} & & \downarrow{\scriptstyle ev_i} \\ K & \longrightarrow & L \end{array}$$

commutes. It easily follows that the map $K \longrightarrow L$ induces a map $RG(K) \longrightarrow RG(L)$, so we get a group morphism

$$G(K)/R \longrightarrow G(L)/R.$$

We then obtain a functor $G(\_)/R : \mathfrak{C}_k \longrightarrow \mathbf{Grps}$.

Clearly, if G, G' are two group-schemes defined over k and K/k is a field extension, an element $(g, g') \in G(K) \times G'(K)$ is R-trivial if and only if g and g' are R-trivial. Therefore, we have a canonical isomorphism of functors

$$(G \times G')(\_)/R \simeq G(\_)/R \times G'(\_)/R.$$

We say that a group-scheme G defined over k is R-**trivial** if $G(\_)/R = 1$.

We now relate the notions of R-triviality and rationality. First, we need a definition.

Definition XI.29.5. Let G be a connected algebraic group-scheme defined over k. We say that G is **rational** if there exists a non-empty open subset U of G which is isomorphic (as an affine variety) to an open subset of some affine space. We say that G is **stably rational** if there exists $n \geq 0$ such that $G \times \mathbb{A}_k^n$ is rational.

The following result is the key ingredient of this chapter.

Proposition XI.29.6. *Let k be an infinite field, and let G be a connected algebraic group-scheme defined over k. If G is stably rational, then G is R-trivial.*

Proof. First, assume that G is rational. Let A be the k-algebra representing G, and let K/k be a field extension. The algebraic group-scheme

$$G_K\colon \begin{array}{c} \mathbf{Alg}_K \longrightarrow \mathbf{Grps} \\ R \longmapsto G(R) \end{array}$$

is represented by A_K, and therefore is also connected and rational. The definitions imply that $G_K(K)/R = G(K)/R$. Therefore, replacing G by G_K, we may assume that $K = k$.

Let U be an open subset of an affine space $\mathbb{A}_k^n$ and let U' be an open subset of G such that we have an isomorphism $f : U \xrightarrow{\sim} U'$. Since k is infinite, $U'(k)$ is dense in $G(k)$ since G is connected (hence irreducible as an affine k-variety). In particular, $U'(k)$ is not empty. Let $g_0 \in U'(k)$. Replacing $U'(k)$ by $g_0^{-1}U'(k)$ and $U(k)$ by $f_k^{-1}(g_0^{-1}U'(k))$, one may assume that $1 \in U'(k)$.

We denote by $u_0 \in U(k)$ the preimage of 1 under f_k. Now let $g \in G(k)$. Since $U'(k)$ and $gU'(k)$ are non-empty open subsets of $G(k)$, we get $U'(k) \cap gU'(k) \neq 0$. Hence there exist $g_1, g_2 \in U'(k)$ such that $g = g_1 g_2^{-1}$. Let $v_i = f_k^{-1}(g_i) \in U(k)$. Let $U_P(k)$ be an elementary open subset contained in $U(k)$, where $P \in k[X_1, \dots, X_n]$ is a non-zero polynomial, so we have

$$U_P(k) = \{(a_1, \dots, a_n) \in k^n \mid P(a_1, \dots, a_n) \neq 0\}.$$

It is easy to see that there exists $w_i \in k^n$ such that the polynomial

$$P((1-t)u_0 + tv_i + t(1-t)w_i) \in k[t]$$

is not zero. Consequently, we have

$$u_i(t) = (1-t)u_0 + tv_i + t(1-t)w_i \in U(k[t]_{0,1}).$$

Moreover, we have by construction $u_i(0) = u_0$ and $u_i(1) = v_i$. Let $g_i(t) = f_{k[t]_{0,1}}(u_i(t)) \in U'(k[t]_{0,1})$. Since we have a commutative diagram

$$\begin{array}{ccc} U(k[t]_{0,1}) & \longrightarrow & U'(k[t]_{0,1}) \\ \downarrow{\scriptstyle ev_i} & & \downarrow{\scriptstyle ev_i} \\ U(k) & \longrightarrow & U'(k) \end{array}$$

we get $g_i(0) = f_k(u_i(0)) = f_k(u_0) = 1$ and $g_i(1) = f_k(u_i(1)) = f_k(v_i) = g_i$. Therefore the element $g(t) = g_1(t)g_2^{-1}(t) \in G(k[t]_{0,1})$ satisfies

$$g(0) = 1 \text{ and } g(1) = g_1 g_2^{-1} = g.$$

This shows that $RG(k) = G(k)$, that is $G(k)/R = 1$. This proves that $G(\_)/R = 1$.

Assume now that G is stably rational. Then there exists $n \geq 0$ such that $G \times \mathbb{A}_k^n$ is rational. Notice that $\mathbb{A}_k^n$ is a rational connected algebraic group-scheme. Applying the previous point, we get

$$G(\_)/R \simeq G(\_)/R \times \mathbb{A}_k^n(\_)/R \simeq (G \times \mathbb{A}_k^n)(\_)/R = 1.$$

This concludes the proof. □

§XI.30 The rationality problem for adjoint groups

We start this section by giving a description of the R-equivalence group of $\mathbf{PGSp}(A, \sigma)$. Let (A, σ) be a central simple k-algebra with a symplectic involution. Let $\mathrm{Hyp}(A, \sigma)$ be the multiplicative subgroup of $k^\times$ generated by the elements of the form

$$N_{L/k}(z), z \in L^\times,$$

where L/k runs over all finite field extensions such that σ_L is hyperbolic. Let also $G(A, \sigma)$ be the multiplicative subgroup of $k^\times$ defined by

$$G(A, \sigma) = \{\mu(g) \mid g \in \mathbf{GSp}(A, \sigma)(k)\}.$$

In [38], Merkurjev proved the following result:

Theorem XI.30.1. *We have a group isomorphism*

$$\mathbf{PGSp}(A, \sigma)(k)/R \simeq G(A, \sigma)/k^{\times 2}\mathrm{Hyp}(A, \sigma).$$

As explained at the beginning of this chapter, we are going to construct a cohomological invariant of $\mathbf{PGSp}(A,\sigma)$ which factors through R-equivalence. We start with a lemma.

Lemma XI.30.2. *Let A be a central simple k-algebra of degree $2m$. Let σ, σ_0 be two symplectic involutions on A. If m is even, $\mathcal{T}_\sigma^+ \perp -\mathcal{T}_{\sigma_0}^+ \in I^3k$.*

Proof. By a celebrated theorem of Merkurjev ([39]), we have to prove that $e_i(\mathcal{T}_\sigma^+ \perp -\mathcal{T}_{\sigma_0}^+) = 0$ for $i = 0,1,2$. Since $\mathcal{T}_\sigma^+ \perp -\mathcal{T}_{\sigma_0}^+$ is even-dimensional, we have $e_0([\mathcal{T}_\sigma^+ \perp -\mathcal{T}_{\sigma_0}^+]) = 0$. By Lemma IV.12.13, we have

$$e_1([\mathcal{T}_\sigma^+ \perp -\mathcal{T}_{\sigma_0}^+]) = (\det(\mathcal{T}_\sigma^+)\det(-\mathcal{T}_{\sigma_0}^+)),$$

since m is even. Moreover, we have

$$\det(-\mathcal{T}_{\sigma_0}^+) = (-1)^{m(2m-1)}\det(\mathcal{T}_{\sigma_0}^+) = \det(\mathcal{T}_{\sigma_0}^+) = \det(\mathcal{T}_\sigma^+),$$

the last equality following from Corollary VIII.21.35. Therefore, we get

$$e_1([\mathcal{T}_\sigma^+ \perp -\mathcal{T}_{\sigma_0}^+]) = 0.$$

By Remark IV.12.14, we have

$$e_2([\mathcal{T}_\sigma^+ \perp -\mathcal{T}_{\sigma_0}^+]) = c([\mathcal{T}_\sigma^+ \perp -\mathcal{T}_{\sigma_0}^+]).$$

Set $r = m(2m+1)$.

The dimension of $\mathcal{T}_\sigma^+ \perp -\mathcal{T}_{\sigma_0}^+$ is $2r = 2m(2m+1)$. Since m is even, we have $2r \equiv 0$ or $4 \mod 8$. Moreover, we have $\det(\mathcal{T}_\sigma^+ \perp -\mathcal{T}_{\sigma_0}^+) = \det(\mathcal{T}_\sigma^+)\det(-\mathcal{T}_{\sigma_0}^+) = 1$, so we get

$$e_2([\mathcal{T}_\sigma^+ \perp -\mathcal{T}_{\sigma_0}^+]) = \begin{cases} w_2(\mathcal{T}_\sigma^+ \perp -\mathcal{T}_{\sigma_0}^+) & \text{if} \quad r \equiv 0[4] \\ w_2(\mathcal{T}_\sigma^+ \perp -\mathcal{T}_{\sigma_0}^+) + (-1)\cup(-1) & \text{if} \quad r \equiv 2[4]. \end{cases}$$

In other words, we have

$$e_2([\mathcal{T}_\sigma^+ \perp -\mathcal{T}_{\sigma_0}^+]) = w_2(\mathcal{T}_\sigma^+ \perp -\mathcal{T}_{\sigma_0}^+) + \frac{r}{2}(-1)\cup(-1).$$

Now using Lemma IV.11.5, we get

$$w_2(\mathcal{T}_\sigma^+ \perp -\mathcal{T}_{\sigma_0}^+) = w_2(\mathcal{T}_\sigma^+) + w_2(-\mathcal{T}_{\sigma_0}^+) + (\det(\mathcal{T}_\sigma^+))\cup(\det(-\mathcal{T}_{\sigma_0}^+)).$$

Applying this equality to σ_0, and substracting the two equations, we get

$$w_2(\mathcal{T}_\sigma^+ \perp -\mathcal{T}_{\sigma_0}^+) = w_2(\mathcal{T}_{\sigma_0}^+ \perp -\mathcal{T}_{\sigma_0}^+),$$

taking into account that the determinant and the Hasse invariant of $\mathcal{T}_\sigma^+$ do not depend of σ, by Corollary VIII.21.35 and Theorem VIII.21.36. Now, $\mathcal{T}_{\sigma_0}^+ \perp -\mathcal{T}_{\sigma_0}^+$ is a hyperbolic quadratic form of dimension $2r$, so we have

$$\mathcal{T}_{\sigma_0}^+ \perp -\mathcal{T}_{\sigma_0}^+ \simeq r \times \langle 1, -1 \rangle,$$

and thus we get

$$w_2(\mathcal{T}_{\sigma_0}^+ \perp -\mathcal{T}_{\sigma_0}^+) = w_2(r \times \langle -1 \rangle) = \frac{r(r-1)}{2}(-1) \cup (-1).$$

Since m is even, $r-1$ is odd, and we get

$$w_2(\mathcal{T}_\sigma^+ \perp -\mathcal{T}_{\sigma_0}^+) = \frac{r}{2}(-1) \cup (-1).$$

Hence we obtain

$$e_2([\mathcal{T}_\sigma^+ \perp -\mathcal{T}_{\sigma_0}^+]) = 0,$$

and this concludes the proof. □

We are now ready to construct the cohomological invariant we are looking for.

Proposition XI.30.3. *Let A be a central simple k-algebra carrying a hyperbolic symplectic involution σ_0. For any symplectic involution σ on A, and every field extension K/k, the map*

$$\iota_K : \begin{array}{l} \mathbf{GSp}(A,\sigma)(K) \longrightarrow H^4(K,\mu_2) \\ \qquad g \longmapsto (\mu(g)) \cup e_3([\mathcal{T}_\sigma^+ \perp -\mathcal{T}_{(\sigma_0)_K}^+]) \end{array}$$

induces a well-defined group morphism

$$\theta_K : \mathbf{PGSp}(A,\sigma)(K)/R \longrightarrow H^4(K,\mu_2).$$

These maps give rise to cohomological invariants

$$\iota : \mathbf{GSp}(A,\sigma) \longrightarrow H^4(\_,\mu_2) \text{ and } \theta : \mathbf{PGSp}(A,\sigma)(-)/R \longrightarrow H^4(\_,\mu_2).$$

Proof. If $g \in \mathbf{GSp}(A,\sigma)(K)$ and $\lambda \in K^\times$, we have $\mu(\lambda g) = \lambda^2\mu(g)$. Thus ι induces a well-defined group morphism

$$\overline{\iota}_K : \mathbf{PGSp}(A,\sigma)(K) \longrightarrow H^4(K,\mu_2).$$

We now prove that $\overline{\iota}_K$ factors through R-equivalence. By Theorem XI.30.1, it is enough to prove that $\iota_K(\lambda^2) = 0$ for all $\lambda \in K^\times$, and that for every finite field extension L/K such that σ_L is hyperbolic and all $z \in L^\times$, we have

$$\iota_K(N_{L/K}(z)) = 0.$$

The first point is clear. Now let L/K and $z \in L^\times$ as above. Since σ_L and $\sigma_{0,L}$ are hyperbolic involutions on A_K, they are conjugate by Theorem VIII.21.15. Therefore, the corresponding trace forms are isomorphic by Proposition VIII.21.30 (1). The second part of the same proposition then implies that $(\mathcal{T}_\sigma^+ \perp -\mathcal{T}_{(\sigma_0)_K}^+)_L$ is hyperbolic. By Corollary IV.12.16, we get

$$(N_{L/K}(z)) \cup e_3([\mathcal{T}_\sigma^+ \perp -\mathcal{T}_{(\sigma_0)_K}^+]) = 0,$$

which is what we wanted to prove. The functoriality part is left to the reader. □

Corollary XI.30.4. *Assume that there exists a field extension K/k such that ι_K is not identically zero. Then $\mathbf{PGSp}(A,\sigma)$ is not R-trivial, and in particular not (stably) rational.*

§XI.31 Examples of non-rational adjoint groups

In order to use the previous corollary, we need some examples of algebras with involutions for which it is possible to compute this invariant explicitly. We will need some intermediate results.

Proposition XI.31.1. *Let (A,ρ) be a central simple k-algebra of even degree with an orthogonal involution, let $u \in \mathrm{Sym}(A,\rho)^\times$, and let $\rho_u = \mathrm{Int}(u) \circ \rho$. Then we have*

$$\det(\mathcal{T}_{\rho_u}^+) = \mathrm{Nrd}_A(u)\det(\mathcal{T}_\rho^+).$$

Proof. Notice first that the map

$$\begin{aligned} \mathbf{O}(A,\rho)(k_s) &\longrightarrow \mathbf{O}(\mathcal{T}_\rho^+)(k_s) \\ a &\longmapsto \mathrm{Int}(a)_{|\mathrm{Sym}(A_{k_s},\rho_{k_s})} \end{aligned}$$

factors through $\mathbf{PGO}(A,\rho)(k_s)$. By Lemma VIII.21.21 and Lemma VIII.21.34, the induced map

$$H^1(k,\mathbf{O}(A,\rho)) \longrightarrow H^1(k,\mathbf{O}(\mathcal{T}_\rho^+))$$

maps a class $u/_\sim$ onto the isomorphism class of $\mathcal{T}_{\rho_u}^+$.

Claim: the map

$$H^1(k,\mathbf{O}(A,\rho)) \longrightarrow H^1(k,\mu_2)$$

induced by the reduced norm sends $u/_\sim$ onto the square class of $\mathrm{Nrd}_A(u)$.

Indeed, by Galois descent, a cocycle α corresponding to the class $u/_\sim$ is given by

$$\alpha\colon \begin{array}{l} \mathcal{G}_{k_s} \longrightarrow \mathbf{O}(A,\rho)(k_s) \\ \tau \longmapsto a\,\tau\!\cdot\! a^{-1}, \end{array}$$

where $a \in A_{k_s}^\times$ satisfies

$$a u_{k_s} \rho_{k_s}(a) = 1.$$

The image of $u/_\sim$ under the map induced by the reduced norm is then represented by the cocycle

$$\beta\colon \begin{array}{l} \mathcal{G}_{k_s} \longrightarrow \mu_2(k_s) \\ \tau \longmapsto \dfrac{\mathrm{Nrd}_{A_{k_s}}(a)}{\mathrm{Nrd}_{A_{k_s}}(\tau\!\cdot\! a)}. \end{array}$$

Notice now that we have

$$\mathrm{Nrd}_{A_{k_s}}(\tau\!\cdot\! a) = \tau\!\cdot\!\mathrm{Nrd}_{A_{k_s}}(a).$$

Indeed, let $\varphi : A_{k_s} \xrightarrow{\sim} \mathrm{M}_n(k_s)$ be an isomorphism of k_s-algebras. Using the definition of the action of $\mathcal{G}_{k_s}$ on A_{k_s}, we see that we have

$$\varphi(\tau\!\cdot\! a) = \tau\!\cdot\!\varphi(a).$$

The desired equality follows from properties of the determinant (details are left to the reader). Hence we have

$$\beta_\tau = \frac{\mathrm{Nrd}_{A_{k_s}}(a)}{\tau\!\cdot\!\mathrm{Nrd}_{A_{k_s}}(a)} = \frac{\tau\!\cdot\!\mathrm{Nrd}_{A_{k_s}}(a^{-1})}{\mathrm{Nrd}_{A_{k_s}}(a^{-1})} \text{ for all } \tau \in \mathcal{G}_{k_s}.$$

The equality $a u_{k_s} \rho_{k_s}(a) = 1$ and the properties of the reduced norm imply that we have

$$\mathrm{Nrd}_{A_{k_s}}(a^{-1})^2 = \mathrm{Nrd}_{A_{k_s}}(u_{k_s}) = \mathrm{Nrd}_A(u).$$

Hence β represents the square-class of $\mathrm{Nrd}_A(u)$, and this proves the claim.

Since the map $\det_* : H^1(k, \mathbf{O}(\mathcal{T}_\rho^+)) \longrightarrow H^1(k, \mu_2)$ maps the isomorphism class of a quadratic form q onto the square class $\dfrac{\det(q)}{\det(\mathcal{T}_\rho^+)}$ by Proposition IV.11.2, it is therefore enough to prove that the diagram

$$\begin{array}{ccc} \mathbf{O}(A,\rho)(k_s) & \xrightarrow{\mathrm{Nrd}_{A_{k_s}}} & \mu_2(k_s) \\ & \searrow & \uparrow \mathrm{det} \\ & & \mathbf{O}(\mathcal{T}_\rho^+)(k_s) \end{array}$$

commutes to get the desired equality. By Lemma VIII.21.17, there exists an isomorphism

$$f : (A_{k_s}, \sigma_{k_s}) \xrightarrow{\sim} (\mathrm{M}_n(k_s), t)$$

of k_s-algebras with involution. Let $a \in \mathbf{O}(A,\sigma)(k_s)$. We have

$$\mathrm{Nrd}_{A_{k_s}}(a) = \det(f(a)).$$

Moreover, let us denote by $\mathrm{Int}^+(a)$ and $\mathrm{Int}^+(f(a))$ the restriction of $\mathrm{Int}(a)$ and $\mathrm{Int}(f(a))$ to $\mathrm{Sym}(A_{k_s}, \rho_{k_s})$ and $\mathrm{Sym}(\mathrm{M}_n(k_s), t)$ respectively. Now conjugation by f induces an isomorphism between the endomorphism ring of these vector spaces, which maps $\mathrm{Int}^+(a) \in \mathbf{O}(\mathcal{T}_\rho^+)(k_s)$ onto $\mathrm{Int}^+(f(a)) \in \mathbf{O}(\mathrm{M}_n(k), t)(k_s)$. Since this isomorphism preserves determinants, we have $\det(\mathrm{Int}^+(a)) = \det(\mathrm{Int}^+(f(a)))$, and the commutativity of the diagram reads

$$\det(\mathrm{Int}^+(f(a))) = \det(f(a)) \text{ for all } a \in A_{k_s}.$$

Notice that we have an equality

$$\mathbf{O}(\mathrm{M}_n(k), t)(k_s) = \mathbf{O}_n(k_s).$$

Since reflections span $\mathbf{O}_n(k_s)$ by Proposition IV.10.5, it is enough to prove the equality

$$\det(\mathrm{Int}^+(\tau_x)) = \det(\tau_x) = -1$$

for all $x \in k_s^n$ which are anisotropic for the unit quadratic form $n \times \langle 1 \rangle$.

Since $\tau_{\lambda x} = \tau_x$ for all $\lambda \in k_s$, one may assume that x is a unit vector. Let T be the matrix of τ_x in the canonical basis of k_s^n. Let $e_1 = x, e_2, \ldots, e_n$ be an orthonormal basis of k_s^n with respect to the unit form (which exists since k_s is separably closed), and let P be the corresponding base change matrix. We then have the equality $T = PT'P^{-1}$, where $T' = \begin{pmatrix} -1 & \\ & I_{n-1} \end{pmatrix}$.

Moreover, the matrix P is orthogonal by construction, so $P^{-1} = P^t$, and $\mathrm{Int}(P)$ induces an automorphism $\mathrm{Int}^+(P)$ of $\mathrm{Sym}(\mathrm{M}_n(k_s), t)$. Now we have

$$\mathrm{Int}^+(T) = \mathrm{Int}^+(P) \circ \mathrm{Int}^+(T') \circ \mathrm{Int}^+(P)^{-1}.$$

It easily implies that we are reduced to check that

$$\det(\mathrm{Int}^+(T')) = -1.$$

Set $\varepsilon_{ij} = \dfrac{E_{ij} + E_{ji}}{2}$ for all $1 \leq i \leq j \leq n$. It is easy to check that we have $\mathrm{Int}^+(T')(\varepsilon_{ij}) = -\varepsilon_{ij}$ if $i = 1$ and $j \geq 2$, and $\mathrm{Int}^+(T')(\varepsilon_{ij}) = \varepsilon_{ij}$ otherwise. Hence we get

$$\det(\mathrm{Int}^+(T')) = (-1)^{n-1} = -1,$$

since n is even. This concludes the proof. □

Proposition XI.31.2. *Let B be a central simple k-algebra of even degree, let ρ, ρ_0 be two orthogonal involutions on B, and write*

$$\rho = \mathrm{Int}(u) \circ \rho_0, u \in \mathrm{Sym}(B, \rho_0)^\times.$$

Let $Q = (a, b)$ be a quaternion k-algebra, and let γ be the corresponding symplectic involution. Let $A = B \otimes_k Q$. Then the involutions $\sigma = \rho \otimes \gamma$ and $\sigma_0 = \rho_0 \otimes \gamma$ are symplectic, and we have

$$e_3(\mathcal{T}_\sigma^+ \perp -\mathcal{T}_{\sigma_0}^+) = (\mathrm{Nrd}_B(u)) \cup (a) \cup (b).$$

Proof. By Lemma VIII.21.31, σ and σ_0 are symplectic, since ρ and ρ_0 are orthogonal and γ is symplectic. According to the same lemma, we have the following equality in the Witt group of k:

$$[\mathcal{T}_\sigma^+] = [\mathcal{T}_\rho^+][\mathcal{T}_\gamma^+] + [\mathcal{T}_\rho^-][\mathcal{T}_\gamma^-].$$

Using Proposition VIII.21.30, we get

$$\begin{aligned} [\mathcal{T}_\sigma^+] &= [\mathcal{T}_\rho^+][\mathcal{T}_\gamma^+] + ([\mathcal{T}_\rho^+] - [\mathcal{T}_B])[\mathcal{T}_\gamma^-] \\ &= [\mathcal{T}_\rho^+]([\mathcal{T}_\gamma^+] + [\mathcal{T}_\gamma^-]) - [\mathcal{T}_B][\mathcal{T}_\gamma^-] \\ &= [\mathcal{T}_\rho^+][\mathcal{T}_\gamma] - [\mathcal{T}_B][\mathcal{T}_\gamma^-]. \end{aligned}$$

Since the same equality is true if we replace ρ by ρ_0, we get

$$[\mathcal{T}_\sigma^+ \perp -\mathcal{T}_{\sigma_0}^+] = [\mathcal{T}_\rho^+ \perp -\mathcal{T}_{\rho_0}^+][\mathcal{T}_\gamma].$$

By Example VIII.21.32 (2), we have

$$\mathcal{T}_\gamma \simeq 2\langle\langle a, b\rangle\rangle.$$

Hence $\mathcal{T}_\gamma \in I^2(k)$. By Remark IV.12.9, we have

$$e_2([\mathcal{T}_\gamma]) = e_2([\langle\langle a, b\rangle\rangle]) = (a) \cup (b).$$

This may also be seen using the fact that e_2 coincide with the Clifford invariant on $I^2(k)$. Since $\mathcal{T}_\rho^+ \perp -\mathcal{T}_{\rho_0}^+$ is even dimensional, we have $\mathcal{T}_\rho^+ \perp -\mathcal{T}_{\rho_0}^+ \in I(k)$. By Lemma IV.12.12, we get

$$e_3([\mathcal{T}_\sigma^+ \perp -\mathcal{T}_{\sigma_0}^+]) = e_1([\mathcal{T}_\rho^+ \perp -\mathcal{T}_{\rho_0}^+]) \cup (a) \cup (b).$$

Notice that we have $\dim(\mathcal{T}_\rho^+ \perp -\mathcal{T}_{\rho_0}^+) = n(n-1)$, where n is the degree of B, and that $\dfrac{n(n-1)(n(n-1)-1)}{2}$ and $\dfrac{n(n-1)}{2}$ have same parity. By Lemma IV.12.13, we then get

$$e_1([\mathcal{T}_\rho^+ \perp -\mathcal{T}_{\rho_0}^+]) = ((-1)^{\frac{n(n-1)}{2}} \det(\mathcal{T}_\rho^+ \perp -\mathcal{T}_{\rho_0}^+)).$$

Now by Proposition XI.31.1, we have

$$\begin{aligned} \det(\mathcal{T}_\rho^+ \perp -\mathcal{T}_{\rho_0}^+) &= \det(\mathcal{T}_\rho^+)\det(-\mathcal{T}_{\rho_0}^+) \\ &= (-1)^{\frac{n(n-1)}{2}} \det(\mathcal{T}_\rho^+)\det(\mathcal{T}_{\rho_0}^+) \\ &= (-1)^{\frac{n(n-1)}{2}} \mathrm{Nrd}_B(u) \in k^\times/k^{\times 2}. \end{aligned}$$

The desired result follows. □

We are now ready to construct our family of examples.

Theorem XI.31.3. *Assume that $-1 \in k^{\times 2}$, and that there exist elements $a, b, c, d \in k^\times$ such that*

$$(a) \cup (b) \cup (c) \cup (d) \neq 0 \text{ in } H^4(k, \mu_2).$$

Let $H = (a, b)$ and $Q = (c, d)$. Let $m, s \geq 1$ be two integers, with s odd and let ρ be the orthogonal involution on

$$\mathrm{M}_{2ms}(H) \simeq (\mathrm{M}_{2m}(k) \otimes_k H) \otimes_k \mathrm{M}_s(k)$$

defined by

$$\rho = [\mathrm{Int}(\mathrm{diag}(j, i, i, \ldots, i) \otimes 1)] \circ ((t \otimes \gamma_H) \otimes \sigma_B),$$

where B is any symmetric invertible matrix of $\mathrm{M}_s(k)$ and $1, i, j, ij$ is the standard basis of H. Finally, let $A = \mathrm{M}_{2ms}(H) \otimes Q$ and let $\sigma = \rho \otimes \gamma_Q$. Then $\mathbf{PGSp}(A, \sigma)$ is not R-trivial, and in particular not stably rational.

Proof. Let $\rho_0 = (t \otimes \gamma_H) \otimes \sigma_B$ and let $\sigma_0 = \rho_0 \otimes \gamma_Q$. Then ρ_0 is an orthogonal hyperbolic involution on $\mathrm{M}_{2m}(H)$, and σ_0 is a hyperbolic involution on A. Indeed, since $-1 \in k^{\times 2}$, the unit quadratic form $2m \times \langle 1 \rangle$ is isomorphic to $m \times \langle 1, -1 \rangle$, which is hyperbolic. Hence $t = \sigma_{I_n}$ is a hyperbolic involution, and therefore so are ρ_0 and σ_0, since it is easy to see that the tensor product of an involution by a hyperbolic one is still hyperbolic. Now we have

$$\rho = \mathrm{Int}(u_0 \otimes 1) \circ \rho_0,$$

where $u_0 = \mathrm{diag}(j, i, i, \ldots, i)$. Since we have

$$\begin{aligned} \mathrm{Nrd}_{\mathrm{M}_{2ms}(H)}(u_0 \otimes 1) &= \mathrm{Nrd}_{\mathrm{M}_{2m}(H) \otimes \mathrm{M}_s(k)}(u_0 \otimes 1) \\ &= \mathrm{Nrd}_{\mathrm{M}_{2m}(H)}(u_0)^s \\ &= (-b)^s(-a)^{(2m-1)s} \\ &= b^s a^{(2m-1)s}, \end{aligned}$$

we get

$$e_3(T_\sigma^+ \perp -T_{\sigma_0}^+) = (a^{(2m-1)s} b^s) \cup (c) \cup (d) = (ab) \cup (c) \cup (d)$$

by the previous proposition, since s and $(2m-1)s$ are odd. Now let

$$g' = \mathrm{diag}(j, \omega j, \ldots, \omega j),$$

where $\omega \in k^\times$ satisfies $\omega^2 = -1$, and let $g = g' \otimes 1$. Then we have

$$\rho(g') = \mathrm{diag}(j, \omega i j i^{-1}, \ldots, \omega i j i^{-1}) = \mathrm{diag}(j, -\omega j, \ldots, -\omega j),$$

and therefore $g\sigma(g) = g'\rho(g') = b$. Hence $g \in \mathbf{GSp}(A, \sigma)(k)$ and $\mu(g) = b$. Therefore, we get

$$\iota_k(g) = (b) \cup (ab) \cup (c) \cup (d).$$

But we have

$$(b) \cup (ab) = (b) \cup (a) + (b) \cup (b) = (a) \cup (b) + (b) \cup (-1) = (a) \cup (b),$$

since -1 is a square. Thus, we get

$$\iota_k(g) = (a) \cup (b) \cup (c) \cup (d) \neq 0,$$

and we conclude using Corollary XI.30.4. □

To have explicit examples, we need to find fields satisfying the conditions of the theorem above. This will be provided by the following lemma.

Lemma XI.31.4. *Let k_0 be any field of characteristic different from 2, and let $k = k_0(t_1, \dots, t_n)$, where $t_1, \dots, t_n$ are independent indeterminates over k_0. Then we have*

$$(t_1) \cup \cdots \cup (t_n) \neq 0 \text{ in } H^n(k, \mu_2).$$

Proof. We prove it by induction. If $n = 1$, this is clear since t_1 is not a square in $k(t_1)^\times$. Assume the result is true for some $n \geq 1$, and let us prove that

$$(t_1) \cup \cdots \cup (t_{n+1}) \neq 0 \text{ in } H^{n+1}(k, \mu_2).$$

Let v be the t_{n+1}-adic valuation on $k_0(t_1, \dots, t_n)(t_{n+1})$. Then we have

$$\partial_v((t_1) \cup \cdots \cup (t_{n+1})) = (t_1) \cup \cdots \cup (t_n) \neq 0 \text{ in } H^n(k_0(t_1, \dots, t_n), \mu_2).$$

In particular, $(t_1) \cup \cdots \cup (t_{n+1}) \neq 0$ and we are done. □

We may then take $k = k_0(t_1, t_2, t_3, t_4)$ and $a = t_1, b = t_2, c = t_3, d = t_4$ in the previous theorem, where k_0 is any field of characteristic different from 2 such that $-1 \in k_0^{\times 2}$. More examples of non-rational automorphism groups of algebra with involutions constructed using similar arguments may be found in [5].

Exercises

1. Let K/k be a quadratic étale algebra. In this exercise, as well as the following ones, we denote by ι the non-trivial k-automorphism of K, (B, τ) will denote a central simple K-algebra with a unitary involution (see Chapter VIII, Exercise 7 for a definition in the case where $K = k \times k$) such that $\tau_{|K} = \iota$.

   (a) Check that if $x \in \mathrm{Sym}(B, \tau)$, then $\mathrm{Trd}_B(x) \in k$.

   We define a quadratic form over k by

   $$\mathcal{T}_\tau : \begin{array}{l} B \longrightarrow k \\ x \longmapsto \mathrm{Trd}_B(\tau(x)x). \end{array}$$

   We denote by $\mathcal{T}_\tau^+$ and $\mathcal{T}_\tau^-$ its restriction to $\mathrm{Sym}(B, \tau)$ and $\mathrm{Skew}(B, \tau)$ respectively.

(b) Let $\alpha \in K^\times$ such that $\tau(\alpha) = -\alpha$, so that $d = \alpha^2 \in k^\times$. Check that $\mathrm{Skew}(B,\tau) = \alpha\mathrm{Sym}(B,\tau)$, and deduce that we have

$$\mathcal{T}_\tau^- \simeq -d\mathcal{T}_\tau^+.$$

(c) If $(B,\tau) = (A \times A^{op}, \varepsilon)$, show that $\mathcal{T}_\tau^+ \simeq \mathcal{T}_A$.

(d) Deduce that $\mathcal{T}_\tau, \mathcal{T}_\tau^+$ and $\mathcal{T}_\tau^-$ are regular quadratic forms.

2. Let τ' be another unitary involution on B such that $\tau'_{|K} = \iota$, show that $\det(\mathcal{T}_{\tau'}^+) = \det(\mathcal{T}_\tau^+)$ and deduce that $\mathcal{T}_{\tau'}^+ \perp -\mathcal{T}_\tau^+ \in I^2k$.

Hint : Let us denote by $\mathbf{PGU}(B,\tau)$ the automorphism group of (B,τ). This a connected affine algebraic group-scheme defined over k.

3. Let A be a central simple k-algebra. Show that $\mathbf{PGL_1}(A)$ is rational.

As for the case of involutions of the first kind, we may define the notion of a similitude for τ. We then obtain an affine algebraic group-scheme $\mathbf{GU}(B,\tau)$ defined over k. We then define a subgroup $G(B,\tau)$ of $k^\times$ by

$$G(B,\tau) = \{\mu(g) \mid g \in \mathbf{GU}(B,\tau)(k)\}.$$

In [38], Merkurjev proved that

$$\mathbf{PGU}(B,\tau)(k)/R \simeq G(B,\tau)/N_{K/k}(K^\times)\mathrm{Hyp}(B,\tau),$$

where $\mathrm{Hyp}(B,\tau)$ is the subgroup of $k^\times$ generated by the elements $N_{L/k}(z), z \in L^\times$, where L/k describes the finite field extensions such that τ_L is hyperbolic.

4. Let (A,ρ) be a central simple k-algebra with an involution of the first kind. If $\alpha \in K^\times$ satisfies $\tau(\alpha) = -\alpha$, set $d = \alpha^2 \in k^\times$. Let $B = A \otimes_k K$ and let $\tau = \rho \otimes_k \iota$.

   (a) If ρ_0 is another involution of the first kind of same type as ρ and $\tau_0 = \rho_0 \otimes \iota$, show that we have

$$\mathcal{T}_\tau^+ \perp -\mathcal{T}_{\tau_0}^+ \simeq \langle 1, -d\rangle \otimes (\mathcal{T}_\rho^+ \perp -\mathcal{T}_{\rho_0}^+).$$

(b) Assume that ρ_0 is hyperbolic. For every field extension L/k, define a map $\iota_L : \mathbf{GU}(B,\tau)(L) \longrightarrow H^3(L,\mu_2)$ by

$$\iota_L(g) = (\mu(g)) \cup e_2([\mathcal{T}_\tau^+ \perp -\mathcal{T}_{(\tau_0)_L}^+]).$$

Show that the maps ι_L induce a well-defined cohomological invariant

$$\theta : \mathbf{PGU}(B,\tau)(\_)/R \longrightarrow H^3(k,\mu_2).$$

(c) Using the previous questions, find a field K and a central simple K-algebra with a unitary involution (B,τ) such that $\mathbf{PGU}(B,\tau)$ is not stably rational.

Hint: Assume that $-1 \in k^{\times 2}$. Set $A = \mathrm{M}_{2m}(H)$, where H is a quaternion algebra, and take ρ to be the orthogonal involution $\rho = \mathrm{Int}(\mathrm{diag}(j,i,i,\ldots,i)) \circ (t \otimes \gamma_H)$.

XII
Essential dimension of functors

When studying a class of mathematical objects, it is natural to ask how many independent parameters are needed to define them up to isomorphism, in order to measure their 'degree of complication'. The notion of essential dimension has been defined to give a precise meaning to this number of parameters. Essential dimension was first introduced by Buhler and Reichstein in [10] for finite groups. It has then been extended to arbitrary algebraic groups by Reichstein [47]. Recently, Merkurjev [37] generalized this notion to arbitrary functors in some private notes. In the meantime, Rost also proposed a valuative approach to essential dimension [49]. In [3], a systematic study of essential dimension is developed, based on Merkurjev's private notes. In particular, it is shown that the notions of essential dimension introduced by Buhler, Merkurjev, Reichstein and Rost all agree. In this chapter, we will follow the treatment of [3]. We will often state some results without proof.

§XII.32 Essential dimension: definition and first examples

Definition XII.32.1. Let $\mathbf{F} : \mathfrak{C}_k \longrightarrow \mathbf{Sets}$ be a covariant functor, let K/k be a field extension and let $a \in \mathbf{F}(K)$. Finally, let E/k be a subextension of K/k. We say that a is **defined** over E if a lies in the image of $\mathbf{F}(E) \longrightarrow \mathbf{F}(K)$. The **essential dimension** of a is the integer $\mathrm{ed}(a)$ defined by

$$\mathrm{ed}(a) = \min\{\mathrm{trdeg}(E/k) \mid E \subset K, a \text{ is defined over } E\}.$$

The **essential dimension of F** is the supremum of $\mathrm{ed}(a)$ for all $a \in \mathbf{F}(K)$ and for all K/k. The essential dimension of $\mathbf{F}$ will be denoted by $\mathrm{ed}_k(\mathbf{F})$.

Let us give some examples.

Examples XII.32.2.

(1) If K/k is algebraic, then $\mathrm{ed}(a) = 0$ for all $a \in \mathbf{F}(K)$.

(2) If $\mathrm{char}(k) \neq 2$, let q be a non-degenerate quadratic form of dimension n over K/k. Then $q \simeq \langle a_1, \ldots, a_n \rangle$, for some $a_i \in K^\times$. Set $K' = k(a_1, \ldots, a_n) \subset K$ and $q' = \langle a_1, \ldots, a_n \rangle$. Then $q \simeq q'_K$ and therefore

$$\mathrm{ed}(q) \leq \mathrm{trdeg}(K'/k) = n.$$

In particular,

$$\mathrm{ed}_k(\mathbf{Quad}_n) \leq n.$$

It takes a bit more effort to prove that equality holds.

(3) In the same spirit, if K/k is a field extension and L/K is an étale multiquadratic K-algebra of dimension $2n$, we have

$$L \simeq K[\sqrt{a_1}] \times \cdots \times K[\sqrt{a_n}] \simeq L' \otimes_{K'} K,$$

where $K' = k(a_1, \ldots, a_n)$ and $L' = K'[\sqrt{a_1}] \times \cdots \times K'[\sqrt{a_n}]$. Hence $\mathrm{ed}(L/K) \leq \mathrm{trdeg}(K'/k) = n$. In fact, the essential dimension of multiquadratic algebras of dimension $2n$ is equal to n if $\mathrm{char}(k) \neq 2$, as we will see later.

(4) If $\mathrm{char}(k) = 2$, the situation is quite different. In this case, every multiquadratic étale K-algebra of dimension $2n$ is isomorphic to

$$K[\wp^{-1}(a_1)] \times \cdots \times K[\wp^{-1}(a_n)]$$

for some $a_1, \ldots, a_n \in K$, where $\wp$ is the Weierstrass function

$$\wp\colon \begin{array}{l} K \longrightarrow K \\ x \longmapsto x^2 - x. \end{array}$$

Thus, we still have

$$\mathrm{ed}(L/K) \leq n.$$

However, if k is large enough, we can have a strict inequality. For example, assume that k contains $\mathbb{F}_4$, so there exists $j \in k$ satisfying $j^2 = j + 1$. If L/K is a biquadratic extension we have

$$L \simeq K[\wp^{-1}(a)] \times K[\wp^{-1}(b)] \simeq K[\wp^{-1}(j^2\lambda)] \times K[\wp^{-1}(\lambda)],$$

where $\lambda = \wp(a) + j\wp(b) \in K$. To prove this, notice first that we have

$$j^2\wp(c) = j^2(c^2 - c) = (jc)^2 - (1 + j)c = c + \wp(jc).$$

Then we get

$$j^2\lambda = j^2\wp(a) + \wp(b) = a + \wp(b + ja),$$

and therefore $K[\wp^{-1}(j^2\lambda)] \simeq K[\wp^{-1}(a)]$. We also have

$$\lambda = \wp(a) + j\wp(b) = \wp(a) + (1 + j^2)\wp(b) = \wp(a + b) + b + \wp(jb).$$

Hence $\lambda = b+\wp(a+b+jb)$, and therefore $K[\wp^{-1}(\lambda)] \simeq K[\wp^{-1}(b)]$. Thus L/K is defined over $k(\lambda)$, and we have

$$\mathrm{ed}(L/K) \leq 1.$$

We will see an explanation of this phenomenon later.

(5) Let X be a scheme over k. It gives rise to a functor

$$X : \mathfrak{C}_k \longrightarrow \textbf{Sets}, K/k \longmapsto X(K) = \mathrm{Mor}(\mathrm{Spec}(K), X).$$

If L/K is a field extension, the induced map $X(K) \longrightarrow X(L)$ is just composition on the right by $\mathrm{Spec}(L) \longrightarrow \mathrm{Spec}(K)$.

Let us compute $\mathrm{ed}_k(X)$. Let K/k be a field extension, let $a \in X(K)$ and let $x \in X$ be the corresponding point. Assume that K/k as minimal transcendence degree over k, so that $\mathrm{ed}(a) = \mathrm{trdeg}(K/k)$. The morphism $a : \mathrm{Spec}(K) \longrightarrow X$ factors through $a' : \mathrm{Spec}(\kappa(x)) \longrightarrow X$, so $a = a'_K$. Since K/k has minimal transcendence degree, we get

$$\mathrm{ed}(a) = \mathrm{trdeg}(K/k) = \mathrm{trdeg}(\kappa(x)/k).$$

Hence we get

$$\mathrm{ed}_k(X) = \sup_{x\in X} \mathrm{trdeg}\left(\kappa(x) : k\right) = \dim(X).$$

If G is an algebraic group-scheme defined over k, we will write $\mathrm{ed}_k(G)$ for the essential dimension of the functor $H^1(\_, G)$. Computing $\mathrm{ed}_k(G)$ is a particularly interesting problem because $H^1(\_, G)$ often classifies algebraic objects up to isomorphism, as we have already seen in the previous chapters.

§XII.33 First results

Let $\mathbf{F} : \mathfrak{C}_k \longrightarrow \textbf{Sets}$ be a covariant functor, and let k'/k be a field extension. If K/k' is a field extension of k', we can associate the field extension K/k. Then one can view $\mathbf{F}$ as a functor $\mathfrak{C}_{k'} \longrightarrow \textbf{Sets}$.

Lemma XII.33.1. *Let k'/k a field extension. Then*

$$\mathrm{ed}_{k'}(\mathbf{F}) \leq \mathrm{ed}_k(\mathbf{F}).$$

Proof. If $\mathrm{ed}_k(\mathbf{F}) = \infty$, the result is obvious. Let $\mathrm{ed}_k(\mathbf{F}) = n$. Take K/k' a field extension and $a \in \mathbf{F}(K)$. There is a subextension $k \subset E \subset K$ with $\mathrm{trdeg}(E/k) \leq n$ such that a is in the image of the map $\mathbf{F}(E) \longrightarrow \mathbf{F}(K)$. The composite extension $E' = Ek'$ then satisfies $\mathrm{trdeg}(E'/k') \leq n$ and clearly a is in the image of the map $\mathbf{F}(E') \longrightarrow \mathbf{F}(K)$. Thus $\mathrm{ed}(a) \leq n$ and $\mathrm{ed}_{k'}(\mathbf{F}) \leq n$. □

This lemma can be useful to give lower bounds (taking $k' = k_{alg}$ for example).

Remark XII.33.2. In general one does not have $\mathrm{ed}_k(\mathbf{F}) = \mathrm{ed}_{k'}(\mathbf{F})$ for any field extension k'/k. Indeed, let $\mathbf{F} : \mathfrak{C}_k \longrightarrow \mathbf{Sets}$ be the functor defined by

$$\mathbf{F}(K) = \begin{cases} \{0\} & \text{if } \mathrm{trdeg}(K/k) \leq 1 \\ \{0,1\} & \text{if } \mathrm{trdeg}(K/k) \geq 2 \end{cases}$$

the map induced by a morphism $K \longrightarrow K'$ being the inclusion of sets. Clearly, $\mathrm{ed}_k(\mathbf{F}) = 2$, but $\mathrm{ed}_{k'}(\mathbf{F}) = 0$ as soon as $\mathrm{trdeg}(k'/k) \geq 2$ (since $\mathbf{F}$ becomes constant over k').

Lemma XII.33.3. *Let $\varphi : \mathbf{F} \longrightarrow \mathbf{F}'$ be a natural transformation of functors. For all K/k and all $a \in \mathbf{F}(K)$, we have*

$$\mathrm{ed}(a) \geq \mathrm{ed}(\varphi_K(a)).$$

In particular, if φ is surjective, we have $\mathrm{ed}_k(\mathbf{F}') \leq \mathrm{ed}_k(\mathbf{F})$.

Proof. Let $K' \subset K$ such that $a = b_K$ for some $b \in \mathbf{F}(K')$ and $\mathrm{ed}(a) = \mathrm{trdeg}(K'/k)$. Then we have

$$\varphi_K(a) = \varphi_K(b_K) = (\varphi_{K'}(b))_K.$$

Hence $\varphi(a)$ is defined over K' and $\mathrm{ed}(\varphi_K(a)) \leq \mathrm{trdeg}(K'/k) = \mathrm{ed}(a)$. The last part is clear. □

The lemma above and Example XII.32.2 (5) then yield:

Corollary XII.33.4. *Let $\mathbf{F}$ be a functor, and let X be a k-scheme. Assume we have a surjective natural transformation*

$$X \twoheadrightarrow \mathbf{F}.$$

Then $\mathrm{ed}_k(\mathbf{F}) \leq \dim(X)$.

Examples XII.33.5.

(1) If $\mathrm{char}(k) \neq 2$, we have a surjection

$$K^{\times n} \longrightarrow \mathbf{Quad}_n(K)$$
$$(a_1, \ldots, a_n) \longmapsto \langle a_1, \ldots, a_n \rangle$$

for all K/k, so we get a surjective natural transformation

$$\mathbb{G}_m^n \longrightarrow \mathbf{Quad}_n.$$

We then recover the inequality $\mathrm{ed}_k(\mathbf{Quad}_n) \leq n$.

(2) If $\mathrm{char}(k) \neq 2$, we have a surjection

$$\mathbb{G}_m^n(K) \longrightarrow H^1(K, (\mathbb{Z}/2\mathbb{Z})^n),$$

which sends for all K/k the element $(a_1, \ldots, a_n) \in (K^\times)^n$ onto the K-algebra $K[\sqrt{a_1}] \times \ldots \times K[\sqrt{a_n}]$. We then get a surjective natural transformation

$$\mathbb{G}_m^n \longrightarrow H^1(\_, (\mathbb{Z}/2\mathbb{Z})^n).$$

Thus $\mathrm{ed}_k((\mathbb{Z}/2\mathbb{Z})^n) \leq n$.

(3) If $\mathrm{char}(k) = 2$, $k \supset \mathbb{F}_q, q = 2^n$, we have an exact sequence

$$0 \longrightarrow (\mathbb{Z}/2\mathbb{Z})^n \longrightarrow K_s \longrightarrow K_s \longrightarrow 0,$$

for every field extension K/k, where the last map is

$$K_s \longrightarrow K_s$$
$$x \longmapsto x^q - x.$$

Passing to cohomology, we get an exact sequence

$$\mathbb{G}_a \longrightarrow H^1(\_, (\mathbb{Z}/2\mathbb{Z})^n) \longrightarrow H^1(\_, \mathbb{G}_a) = 0.$$

Thus the map above is surjective, and we get

$$\mathrm{ed}_k((\mathbb{Z}/2\mathbb{Z})^n) \leq 1.$$

Definition XII.33.6. We say that a functor $\mathbf{F} : \mathfrak{C}_k \longrightarrow$ **Sets** acts over a functor $\mathbf{G} : \mathfrak{C}_k \longrightarrow$ **Sets** if we have a natural transformation

$$\varphi : \mathbf{F} \times \mathbf{G} \longrightarrow \mathbf{G}.$$

For every field extension $K/k, x \in \mathbf{F}(K)$ and $a \in \mathbf{G}(K)$, we will write $x \cdot a$ instead of $\varphi(x, a)$.

We will say that the action of $\mathbf{F}$ on $\mathbf{G}$ is **transitive** if for every field extension K/k and all $a, a' \in \mathbf{G}(K)$, there exists $x \in \mathbf{F}(K)$ such that $a' = x{\cdot}a$. If $\pi : \mathbf{G} \longrightarrow \mathbf{H}$ is a natural transformation of functors and K/k is an extension, each element $a \in \mathbf{H}(K)$ gives rise to a functor $\pi^{-1}(a)$, defined over the category $\mathfrak{C}_K$, by setting

$$\pi^{-1}(a)(L) = \pi_L^{-1}(a) = \{x \in \mathbf{G}(L) \mid \pi_L(x) = a_L\}$$

for every extension L/K.

Let $\pi : \mathbf{G} \twoheadrightarrow \mathbf{H}$ be a surjective natural transormation. We say that a functor $\mathbf{F}$ is in **fibration position** for π if $\mathbf{F}$ acts transitively on each fiber of π. More precisely, for every extension K/k and every $a \in \mathbf{H}(K)$, we require that the functor $\mathbf{F}$ (viewed over the category $\mathfrak{C}_K$) acts transitively on $\pi^{-1}(a)$. When $\mathbf{F}$ is in fibration position for π we simply write $\mathbf{F} \rightsquigarrow \mathbf{G} \xrightarrow{\pi} \mathbf{H}$ and call this a **fibration of functors**.

Proposition XII.33.7. *Let* $\mathbf{F} \rightsquigarrow \mathbf{G} \xrightarrow{\pi} \mathbf{H}$ *be a fibration of functors. Then*

$$\mathrm{ed}_k(\mathbf{H}) \leq \mathrm{ed}_k(\mathbf{G}) \leq \mathrm{ed}_k(\mathbf{F}) + \mathrm{ed}_k(\mathbf{H}).$$

Proof. Let K/k be a field extension and $a \in \mathbf{G}(K)$. By definition there is a field extension E with $k \subset E \subset K$, satisfying $\mathrm{trdeg}(E : k) \leq \mathrm{ed}_k(\mathbf{H})$, and an element $b' \in \mathbf{H}(E)$ such that $b'_K = \pi_K(a)$. Since π_E is surjective there exists $a' \in \mathbf{G}(E)$ such that $\pi_E(a') = b'$. Now $\pi_K(a'_K) = \pi_K(a)$ and thus a'_K and a are in the same fiber. By assumption there exists an element $c \in \mathbf{F}(K)$ such that $c{\cdot}a'_K = a$. Now there exists an extension E' with $k \subset E' \subset K$ and $\mathrm{trdeg}(E' : k) \leq \mathrm{ed}_k(\mathbf{F})$ such that c is in the image of the map $\mathbf{F}(E') \longrightarrow \mathbf{F}(K)$. Pick $c' \in \mathbf{F}(E')$ such that $c'_K = c$. Considering now the composite extension $E'' = EE'$ and setting $d = c'_{E''}{\cdot}a'_{E''} \in \mathbf{G}(E'')$ we have, since the action is functorial,

$$d_K = (c'_{E''}{\cdot}a'_{E''})_K = c'_K{\cdot}a'_K = c{\cdot}a'_K = a,$$

and thus

$$\mathrm{ed}(a) \leq \mathrm{trdeg}(E'' : k) \leq \mathrm{trdeg}(E : k) + \mathrm{trdeg}(E' : k) \leq \mathrm{ed}_k(\mathbf{H}) + \mathrm{ed}_k(\mathbf{F}).$$

Since this is true for an arbitrary element a the desired inequality follows. The other inequality is already known from Lemma XII.33.3. □

Corollary XII.33.8. *We have*

$$\max(\mathrm{ed}_k(\mathbf{F}), \mathrm{ed}_k(\mathbf{G})) \leq \mathrm{ed}_k(\mathbf{F} \times \mathbf{G}) \leq \mathrm{ed}_k(\mathbf{F}) + \mathrm{ed}_k(\mathbf{G}).$$

Proof. The reader will check easily that we have two fibrations of functors

$$\mathbf{F} \rightsquigarrow \mathbf{F} \times \mathbf{G} \xrightarrow{\pi} \mathbf{G}$$

and

$$\mathbf{G} \rightsquigarrow \mathbf{F} \times \mathbf{G} \xrightarrow{\pi} \mathbf{F}$$

The result then follows from the previous proposition. □

§XII.34 Cohomological invariants and essential dimension

Definition XII.34.1. Let $\mathbf{F} : \mathfrak{C}_k \longrightarrow \mathbf{Sets}^*$ be a functor, and M be a torsion $\mathcal{G}_{k_s}$-module M.

We say that a cohomological invariant $\iota : \mathbf{F} \longrightarrow H^d(\_, M)$ is **non-trivial** if for all K/k, there exists L/K such that $\iota_L \neq 0$.

Proposition XII.34.2. *If* $\mathbf{F}$ *admits a non-trivial cohomological invariant of degree d, then* $\mathrm{ed}_k(\mathbf{F}) \geq d$.

Proof. Assume to the contrary that $\mathrm{ed}_k(\mathbf{F}) < d$. Since $\mathrm{ed}_{k_{alg}}(\mathbf{F}) \leq \mathrm{ed}_k(\mathbf{F})$, one can assume that k is algebraically closed.

Let K/k be a field extension, and let L/K be a field extension of K. We are going to show that $\iota_L = 0$, contradicting the assumption. Let $a \in \mathbf{F}(L)$. By assumption, $a = b_L$ for some $b \in \mathbf{F}(K')$, where $K' \subset L$ and $\mathrm{trdeg}(K'/k) < d$. We then have $\iota_L(a) = (\iota_{K'}(b))_L$. But $H^d(K', M) = 0$ since $\mathrm{trdeg}(K'/k) < d$ and k is algebraically closed by [58, Proposition 11,p.93], and therefore $\iota_L(a) = 0$. Thus $\iota_L = 0$, and this concludes the proof. □

Examples XII.34.3.

(1) Let k be a field of characteristic different from 2. Then we have $\mathrm{ed}_k((\mathbb{Z}/2\mathbb{Z})^n) = n$.

Indeed, we already know that we have

$$\mathrm{ed}_k((\mathbb{Z}/2\mathbb{Z})^n) \leq n.$$

To prove the other inequality, we define a cohomological invariant $\iota : H^1(\_, (\mathbb{Z}/2\mathbb{Z})^n) \longrightarrow H^n(\_, \mu_2)$ as follows. For every field extension K/k, we set

$$\iota_K(K[\sqrt{a_1}] \times \cdots \times K[\sqrt{a_n}]) = (a_1) \cup \cdots \cup (a_n).$$

Using Lemma XI.31.4, it is easy to see that ι is non-trivial, and we are done.

(2) Let k be a field, $\mathrm{char}(k) \neq 2$. Then $\mathrm{ed}_k(\mathbf{Quad}_n) = n$. Indeed the nth Stiefel-Whitney class provides a cohomological invariant of $\mathbf{Quad}_n$ of degree n, which is non-trivial for the same reason as in the previous example. We conclude as before.

This method can be quite limited as the next proposition shows.

Proposition XII.34.4. *Let k be a field, $\mathrm{char}(k) \neq 2$, let Q be a division quaternion k-algebra, and let $\mathbf{SL}_1(Q)$ be the group of elements of reduced norm 1. Then $\mathrm{ed}_k(\mathbf{SL}_1(Q)) = 1$, but $H^1(\_, \mathbf{SL}_1(Q))$ has no non-trivial cohomological invariants of degree 1.*

Proof. From Chapter VIII, Exercise 2, we have a bijection

$$H^1(K, \mathbf{SL}_1(Q)) \simeq K^\times / \mathrm{Nrd}_{Q_K}(Q_K^\times)$$

for every field extension K/k. Let us show that $\mathrm{ed}_k(\mathbf{SL}_1(Q)) = 1$. From the previous description of $H^1(\_, \mathbf{SL}_1(Q))$, we have an obvious surjection

$$\mathbb{G}_m \longrightarrow\!\!\!\!\!\!\rightarrow H^1(\_, \mathbf{SL}_1(Q)) \,,$$

and therefore $\mathrm{ed}_k(\mathbf{SL}_1(Q)) \leq 1$ by Corollary XII.33.4. To prove the other equality, let t be an indeterminate over k.

We now prove that $\mathrm{ed}(\bar{t}) \geq 1$. Assume that $\bar{t}$ is defined over an algebraic subextension of $k(t)/k$. Since $k(t)/k$ is purely transcendental, it means that $\bar{t}$ is defined over k. Therefore there exist $u \in k$ and $z \in Q_{k(t)}^\times$ such that

$$t = u\mathrm{Nrd}_{Q_{k(t)}}(z).$$

Write $Q = (a, b)$ for some $a, b \in k^\times$. Then $\mathrm{Nrd}_{Q_{k(t)}}(z)$ is represented by the Pfister form $\langle\langle a, b\rangle\rangle$ We then get

$$\langle\langle t, a, b\rangle\rangle \simeq \langle\langle u, a, b\rangle\rangle$$

by the well-known multiplicativity property of Pfister forms (see [53] for more details). Computing the invariants e_3 on both sides, we get

$$(t) \cup (a) \cup (b) = (u) \cup (a) \cup (b) \in H^3(k(t), \mu_2).$$

Taking the residues corresponding to the t-adic valuation, we obtain

$$(a) \cup (b) = 0 \in H^2(k, \mu_2).$$

This is equivalent to say that $[Q] = 0 \in \mathrm{Br}(k)$ by Corollary VIII.20.10, which is a contradiction.

Finally, let us prove that $H^1(-, \mathbf{SL}_1(Q))$ has no non-trivial cohomological invariants. Indeed, let $\iota : H^1(\_, \mathbf{SL}_1(Q)) \longrightarrow H^d(\_, M)$ be a cohomological invariant, and let $K = k_{alg}$, so we have $Q_K \simeq \mathrm{M}_2(K)$. Let L/K be any field extension. Then we also have

$$Q_L \simeq \mathrm{M}_2(L).$$

We then have $H^1(L, \mathbf{SL}_1(Q)) = 1$, since $\mathrm{Nrd}_{Q_L}(Q_L^\times) = L^\times$. Therefore, $\iota_L = 0$ and ι is not non-trivial. □

§XII.35 Generic objects and essential dimension

Definition XII.35.1. Let $\mathbf{F} : \mathbf{Alg}_k \longrightarrow \mathbf{Sets}$ be a functor with a generic object. We say that a generic objet $a \in \mathbf{F}(K)$ is **nice** if

$$\text{for all } K' \subset K, a' \in \mathbf{F}(K'), a = a'_K \Rightarrow a' \text{ is a generic object.}$$

Proposition XII.35.2. *Assume that* $\mathbf{F} : \mathbf{Alg}_k \longrightarrow \mathbf{Sets}$ *has a nice generic object* $a \in \mathbf{F}(K)$. *Then* $\mathrm{ed}_k(\mathbf{F}) = \mathrm{ed}(a)$.

Proof. Since $\mathrm{ed}_k(\mathbf{F}) \geq \mathrm{ed}(a)$ by definition, it remains to prove that for all L/k and all $b \in \mathbf{F}(L)$, we have $\mathrm{ed}(b) \leq \mathrm{ed}(a)$.

If L is finite, then $\mathrm{ed}(b) = 0$, and the inequality is clear. Assume that L is infinite. Let $a' \in \mathbf{F}(K')$ such that $a'_K = a$ and $\mathrm{ed}(a) = \mathrm{trdeg}(K'/k)$. By assumption, a' is generic, so b is a specialization of a'. Therefore, there exists a pseudo k-place $f : K' \rightsquigarrow L$ and $c \in \mathbf{F}(R_f)$ such that $a' = c_{K'}$ and $b = \mathbf{F}(\varphi_f)(c)$. Since $\varphi_f : R_f \longrightarrow L$ factors through $\kappa(f) = R_f/\mathfrak{m}_f \hookrightarrow L$, we see that $b = c'_L$, where $c' \in \mathbf{F}(\kappa(f))$. Hence

$$\mathrm{ed}(b) \leq \mathrm{trdeg}(\kappa(f)/k) \leq \mathrm{trdeg}(K'/k) = \mathrm{ed}(a).$$

This concludes the proof. □

We now give examples of functors with nice generic objects.

Proposition IX.23.11 shows that a generic G-torsor $P \longrightarrow \mathrm{Spec}(K)$ is a generic object of $H^1(\_, G)$. In fact, we are going to see that a generic torsor is a nice generic object for $H^1(\_, G)$. We need a definition first.

Definition XII.35.3. Let $f : X \longrightarrow Y$ and $f' : X' \longrightarrow Y'$ be two G-torsors. We say that f' is a **compression** of f if there is a diagram

$$\begin{array}{ccc} X & \overset{g}{\dashrightarrow} & X' \\ {\scriptstyle f}\downarrow & & \downarrow{\scriptstyle f'} \\ Y & \underset{h}{\dashrightarrow} & Y' \end{array}$$

where g is a G-equivariant *rational* dominant morphism and h is a rational morphism too (necessarily dominant).

Proposition XII.35.4. *Let K be a field. Let $T \longrightarrow \mathrm{Spec}(K)$ be a G-torsor, and let $f : X \longrightarrow Y$ be a model for T, that is a G-torsor whose generic fiber is isomorphic to T. Let $K' \subset K$, and let $T' \longrightarrow \mathrm{Spec}(K')$ be a G-torsor such that $T \simeq T'_K$. Then there exists a compression $f' : X' \longrightarrow Y'$ of f whose generic fiber is isomorphic to T'.*

Proof. See [3] for a proof. □

Lemma XII.35.5. *Let $f' : X' \longrightarrow Y'$ be a compression of a classifying torsor $f : X \longrightarrow Y$. Then f' is also classifying.*

Proof. Let

$$\begin{array}{ccc} X & \overset{g}{\dashrightarrow} & X' \\ {\scriptstyle f}\downarrow & & \downarrow{\scriptstyle f'} \\ Y & \underset{h}{\dashrightarrow} & Y' \end{array}$$

be such a compression. Let k'/k be a field extension with k' infinite and let $P' \in H^1(k', G)$. Since f is classifying one can find a k'-rational point $y \in Y(k')$ which lies in U, the open set on which h is defined, such that $f^{-1}(y) \simeq P'$. Then the fiber of f' at $h(y)$ clearly gives a torsor isomorphic to P'. □

We now have the following result:

Theorem XII.35.6. *A generic G-torsor $T \longrightarrow \mathrm{Spec}(K)$ is a nice generic object of $H^1(\_, G)$. In particular, $\mathrm{ed}_k(G) = \mathrm{ed}_k(T)$.*

Proof. This follows from Proposition XII.35.4, Lemma XII.35.5 and Proposition XII.35.2. □

Remark XII.35.7. This theorem reduces the computation of $\mathrm{ed}_k(G)$ to the computation of a generic element. In particular, if G is an abstract finite group, Example IX.23.10 shows that $\mathrm{ed}_k(G)$ is the essential dimension of $k(V)/k(V)^G$, where V is any faithful linear representation of G.

Theorem XII.35.8. *Let G be an algebraic group and H a closed algebraic subgroup of G. Then*

$$\mathrm{ed}_k(H) + \dim(H) \leq \mathrm{ed}_k(G) + \dim(G).$$

In particular, if G is finite, we have

$$\mathrm{ed}_k(H) \leq \mathrm{ed}_k(G).$$

Proof. Once again, we refer to [3] for a proof. □

§XII.36 Generically free representations

Definition XII.36.1. Let G be an algebraic group defined over k. A linear representation V of G is **generically free** if there exists a G-stable open subset U of V on which G acts freely.

Example XII.36.2. Any faithful linear representation $V$of a finite group G is generically free, since one may take

$$U = V \setminus \bigcap_{g \neq 1} \ker(g-1).$$

Proposition XII.36.3. *Let V be a generically free linear representation of G. Then*

$$\mathrm{ed}_k(G) \leq \dim(V) - \dim(G).$$

In particular, if G is a finite group and V is a faithful linear representation of G, then

$$\mathrm{ed}_k(G) \leq \dim(V).$$

Proof. By Example IX.23.10 (2), one can find a G-stable open subset U such that $U \longrightarrow U/G$ is a classifying G-torsor. Thus its generic fiber T is a generic torsor, which is defined over $k(U/G)$. Since $\mathrm{trdeg}(k(U/G)/k) = \dim(V) - \dim(G)$, applying Theorem XII.35.6 yields the desired result. The last part follows from the previous point and Example XII.36.2. □

Proposition XII.36.4. *Let G be a finite constant group scheme over k acting linearly and faithfully on a k-vector space V. Then the essential* *dimension of G is the minimum of the integers* $\mathrm{trdeg}(E/k)$*, where E runs through the set of subfields of $k(V)$ on which G acts faithfully.*

Proof. By Theorem XII.35.6, $\mathrm{ed}_k(G)$ is equal to the essential dimension of $k(V)/k(V)^G$, since this latter extension is a generic G-torsor by Example IX.23.10 (2). Let $E \subset k(V)$. We are going to prove that G acts faithfully on E if and only if

$$k(V) \simeq E \otimes_{E^G} k(V)^G,$$

which will be enough to get the desired result. Assume first that the isomorphism above holds. Since G acts faithfully on $k(V)$, it follows that G acts also faithfully on E. Conversely, assume that G acts faithfully on $E \subset k(V)$. Since we have $E \cap k(V)^G = E^G$, then E/E^G and $k(V)^G/E^G$ are linearly disjoint and then $Ek(V)^G \simeq E \otimes_{E^G} k(V)^G$. We then have

$$(Ek(V)^G)^G \simeq (E \otimes_{E^G} k(V)^G)^G = E^G \otimes_{E^G} k(V)^G \simeq k(V)^G.$$

Hence $Ek(V)^G/k(V)^G$ is a Galois subextension of $k(V)/k(V)^G$ with Galois group G. We then get $Ek(V)^G = k(V)$. Thus

$$E \otimes_{E^G} k(V)^G \simeq Ek(V)^G = k(V),$$

and $k(V)/k(V)^G$ comes from E/E^G. This concludes the proof. □

Remark XII.36.5. This result shows that the notion of essential dimension of a finite group G coincides with the one defined originally by Buhler and Reichstein defined in [10].

Recently, Karpenko and Merkurjev [29] proved the following remarkable result:

Theorem XII.36.6. *Let p be a prime number, let k be a field such that* $\mathrm{char}(k) \neq p$ *and $\mu_p \subset k$. Finally, let G a finite p-group. Then* $\mathrm{ed}_k(G)$ *is the minimal dimension of a faithful linear representation V of G.*

Remark XII.36.7. This result is not true is G has composite order.

§XII.37 Some examples

In all the following examples, we suppose that $\mathrm{char}(k) \neq 2$. We first would like to estimate the essential dimension of S_n, that is the essential dimension of étale algebras of dimension n.

With the assumption on the ground field, S_n acts faithfully on the hyperplane $H = \{x \in k^n \mid x_1 + \cdots + x_n = 0\}$ and thus on $k(H) = k(X_1, \ldots, X_{n-1})$. Now $k^\times$ acts on $k(X_1, \ldots, X_{n-1})$ by $\lambda \cdot X_i = \lambda X_i$ for all $\lambda \in k^\times$ and all $i = 1, \ldots, n-1$. This action commutes with the action of S_n. We easily see that

$$k(X_1, \ldots, X_{n-1})^{\mathbb{G}_m} = k\left(\frac{X_1}{X_{n-1}}, \ldots, \frac{X_{n-2}}{X_{n-1}}\right).$$

Now, if $n \geq 3$, the group S_n acts faithfully on the latter field. The transcendence degree of $k\left(\frac{X_1}{X_{n-1}}, \ldots, \frac{X_{n-2}}{X_{n-1}}\right)$ being equal to $n-2$, one concludes that

$$\mathrm{ed}_k(S_n) \leq n-2 \text{ for all } n \geq 3$$

by Proposition XII.36.4.

In particular we find $\mathrm{ed}_k(S_3) = 1$ and $\mathrm{ed}_k(S_4) = 2$.

The inequality above may be easily viewed as follows in terms of equations if $\mathrm{char}(k) \nmid n$. Let K be an infinite field containing k and let $E = K[X]/(f)$ an étale K-algebra of rank n.

Write $f = X^n + a_{n-1}X^{n-1} + \cdots + a_0$, and let $\alpha = X + (f)$. The substitution $\alpha \leftrightarrow \alpha - \dfrac{a_{n-1}}{n}$ induces an isomorphism $E \simeq K[X]/(g)$, where g is a separable polynomial of the form

$$g = X^n + b_{n-2}X^{n-2} + \ldots + b_1X + b_0.$$

If $b_1 \neq 0$, and if we set $\beta = X + (g)$, the substitution $\beta \leftrightarrow \frac{b_0}{b_1}\beta$ induces an isomorphism $E \simeq K[X]/(h)$, where h is a separable polynomial of the form

$$g = X^n + c_{n-2}X^{n-2} + \ldots + c_1X + c_1.$$

Hence E is defined over $k(c_1, \ldots, c_{n-2})$, which has transcendence degree at most $n-2$ over k.

Assume now that $n \geq 5$, and let us prove that $\mathrm{ed}_k(S_n) \leq n-3$ in this case.

The group $\mathbf{PGL}_2(k)$ acts on $k(X_1, \ldots, X_n)$ in the following way:

$$\left[\begin{pmatrix} a & b \\ c & d \end{pmatrix}\right] \cdot X_i = \frac{aX_i + b}{cX_i + d} \qquad \forall\, i = 1, \ldots, n.$$

If now i, j, k, ℓ are distinct, the cross-sections

$$[X_i, X_j, X_k, X_\ell] = \frac{(X_i - X_k)(X_j - X_\ell)}{(X_j - X_k)(X_i - X_\ell)}$$

are $\mathbf{PGL}_2$-invariant. Hence we have

$$k([X_i, X_j, X_k, X_\ell]) \subset k(X_1, \ldots, X_n)^{\mathbf{PGL}_2(k)}$$

where $k([X_i, X_j, X_k, X_\ell])$ is a short notation for the field generated by the biratios $[X_i, X_j, X_k, X_\ell]$ for i, j, k, l all distinct. But it is easy to see that $k([X_i, X_j, X_k, X_\ell])$ is generated by the biratios $[X_1, X_2, X_3, X_i]$ with $i = 4, \ldots, n$. Hence $k([X_i, X_j, X_k, X_\ell]) \simeq k(Y_1, \ldots, Y_{n-3})$. If $n \geq 5$, every $\sigma \in S_n \setminus \{1\}$ moves at least one of the $[X_i, X_j, X_k, X_\ell]$'s. Consequently, since the above action commutes with the S_n-action, S_n acts faithfully on $k(Y_1, \ldots, Y_{n-3})$.

Proposition XII.36.4 then shows that we have

$$\mathrm{ed}(S_n) \leq n - 3 \text{ for all } n \geq 5.$$

Also, we have $(\mathbb{Z}/2\mathbb{Z})^{[n/2]} \subset S_n$, and therefore

$$\mathrm{ed}_k(S_n) \geq \mathrm{ed}_k((\mathbb{Z}/2\mathbb{Z})^{[n/2]}) = [n/2].$$

Compare with [10, Theorem 6.2].

In particular we have $\mathrm{ed}_k(S_5) = 2$ and $\mathrm{ed}_k(S_6) = 3$.

Remark XII.37.1. The question is still open concerning S_7. Do we have $\mathrm{ed}_k(S_7) = 3$ or 4 ?

We are now going to compute the essential dimension of twisted forms of μ_4, that is algebraic groups which are isomorphic to μ_4 after a suitable field extension. Since $\mathbf{Aut}(\mu_4) \simeq \mathbb{Z}/2\mathbb{Z}$, each twisted form corresponds to a quadratic étale algebra K/k. For any commutative k-algebra R, the set of R-points of the associated twisted form $\mu_{4[K]}$ can be described as

$$\mu_{4[K]}(R) = \{x \in (K \otimes_k R)^\times \mid x^4 = 1 \text{ and } N_{K \otimes_k R/R}(x) = 1\}.$$

If $K = k \times k$, we just obtain μ_4, and if $K = k(\sqrt{-1})$, we obtain $\mathbb{Z}/4\mathbb{Z}$.

The following result is due to Rost (see [50]).

Theorem XII.37.2. *We have* $\mathrm{ed}_k(\mu_{4[K]}) = 1$ *if* $K = k \times k$ *and* 2 *otherwise.*

Proof. Since the case $K = k \times k$ is clear, we will assume that K is a field from now on.

We first give a description of $H^1(\_, \mu_{4[K]})$. Denoting by $R_{K/k}(\mathbb{G}_{m,K})$ the algebraic group-scheme defined by

$$R_{K/k}(\mathbb{G}_{m,K})(R) = (K \otimes_k R)^\times$$

for any commutative k-algebra R, let $\mathbb{T}$ be the kernel of the map $\mathbb{G}_m \times R_{K/k}(\mathbb{G}_{m,K}) \longrightarrow \mathbb{G}_m$ given on the R-points by

$$(x,y) \longmapsto x^4 N_{K\otimes_k R/R}(y)^{-1}.$$

Now define a group-scheme morphism $\theta : R_{K/k}(\mathbb{G}_{m,K}) \longrightarrow \mathbb{T}$ by

$$\theta_R(z) = (N_{K\otimes_k R/R}(z), z^4).$$

We can see that, for every field extension L/k, we have an exact sequence

$$1 \longrightarrow \mu_{4[K]}(L_s) \longrightarrow R_{K/k}(\mathbb{G}_{m,K})(L_s) \longrightarrow \mathbb{T}(L_s) \longrightarrow 1.$$

Since $K \otimes_k L$ is semi-simple, taking cohomology and applying Hilbert 90 yield an exact sequence

$$(K \otimes_k L)^\times \longrightarrow \mathbb{T}(L) \longrightarrow H^1(L, \mu_{4[K]}) \longrightarrow 1.$$

In particular, we get

$$H^1(L, \mu_{4[K]}) \simeq \frac{\{(x,y) \in L^\times \times (K\otimes_k L)^\times \mid x^4 = N_{K\otimes_k L/L}(y)\}}{\{(N_{K\otimes_k L/L}(z), z^4) \mid z \in (K \otimes_k L)^\times\}}.$$

The same exact sequence proves the existence of a surjective natural transformation

$$\mathbb{T} \longrightarrow\!\!\!\!\!\rightarrow H^1(\_, \mu_{4[K]}) .$$

By Corollary XII.33.4, we get

$$\mathrm{ed}_k(\mu_{4[K]}) \leq \dim(\mathbb{T}) = 2.$$

It remains to prove the missing inequality. Let σ be the unique non-trivial automorphism of K/k, and let $\sigma_L = \sigma \otimes \mathrm{Id}_L$. The reader will check easily that the classical version of Hilbert 90 remains true if $K \otimes_k L \simeq L \times L$. Therefore, we get

$$x^4 = N_{K\otimes_k L/L}(y) \iff N_{K\otimes_k L/L}(yx^{-2}) = 1 \iff y = x^2 \frac{\sigma_L(\lambda)}{\lambda},$$

for some $\lambda \in (K \otimes_k L)^\times$. Thus, we have

$$H^1(L, \mu_{4[K]}) \simeq \frac{\{(x, x^2\frac{\sigma_L(\lambda)}{\lambda}) \mid x \in L^\times, \lambda \in (K \otimes L)^\times\}}{\{(N_{K\otimes_k L/L}(z), z^4) \mid z \in (K \otimes_k L)^\times\}}.$$

If $x \in L^\times$ and $\lambda \in (K \otimes_k L)^\times$, set

$$[x, \lambda] = \overline{(x, x^2 \frac{\sigma_L(\lambda)}{\lambda})}.$$

Since $\mu_{4[K]}$ is isomorphic to μ_4 over some field extension and μ_4 has essential dimension equal to 1, $\mu_{4[K]}$ does not have any non-trivial cohomological invariant of degree 2. However, we are going to use non-zero invariants of degree 1 or 2 to get the result.

Write $K = k(\sqrt{d})$, and set

$$\eta_{1,L}: \begin{array}{rcl} H^1(L, \mu_{4[K]}) & \longrightarrow & H^1(L, \mu_2) \\ [x, \lambda] & \longmapsto & (N_{K\otimes_k L/L}(\lambda)) \end{array}$$

and

$$\eta_{2,L}: \begin{array}{rcl} H^2(L, \mu_{4[K]}) & \longrightarrow & H^2(L, \mu_2) \\ [x, \lambda] & \longmapsto & (x) \cup (d). \end{array}$$

Let us check that these maps are well-defined. It follows from the definitions that we have

$$[x, \lambda] = [x', \lambda'] \iff x' = xN_{K\otimes_k L/L}(z), x'^2 \frac{\sigma_L(\lambda')}{\lambda'} = x^2 \frac{\sigma_L(\lambda)}{\lambda} z^4$$

for some $z \in (K \otimes_k L)^\times$. Since $(N_{K\otimes_k L/L}(z)) \cup (d) = 0$ by Proposition III.9.15 (2), it follows that $\eta_{2,L}$ is well-defined. Moreover, since we have $x' = x\sigma_L(z)z$, we get

$$1 = \frac{x^2 \sigma_L(z^2) z^2}{x'^2} = \frac{\sigma_L(\lambda'\lambda^{-1}z^2)}{\lambda'\lambda^{-1}z^2},$$

and therefore

$$\sigma_L(\lambda'\lambda^{-1}z^2) = \lambda'\lambda^{-1}z^2.$$

Hence there exists $u \in L^\times$ such that $\lambda' z^2 = \lambda u$. Taking norms on both sides, we obtain

$$N_{K\otimes_k L/L}(\lambda') = N_{K\otimes_k L/L}(\lambda) \in L^\times / L^{\times 2}.$$

It follows that the map $\eta_{1,L}$ is well-defined. Clearly, these maps give rise to two cohomological invariants η_1, η_2 (notice that η_2 is not non-trivial).

Let $L_0 = k(s, t)$ and let $\alpha = [s, 1 + t\sqrt{d}]$. To get the desired result, it is enough to show that $\mathrm{ed}(\alpha) \geq 2$. Let $\alpha' \in H^1(L, \mu_{4[K]})$ be a torsor such that $\alpha'_{L_0} = \alpha$ ($L \subset L_0$) with $\mathrm{trdeg}(L/k) = \mathrm{ed}(\alpha)$. We need to show that L/k has at least transcendence degree 2 over k.

Let v the s-adic valuation on $k(t)(s)$, let v' be its restriction to L and let $e_{v'|v}$ be the corresponding ramification index. We have $\eta_2(\alpha) = \eta_2(\alpha')_{L_0}$. Notice that

$$\partial_v(\eta_2(\alpha)) = \partial_v((s) \cup (d)) = (d) \neq 0 \in H^1(k(t), \mu_2),$$

since $d \in k^\times$ is not a square. Now we have

$$\partial_v(\eta_2(\alpha)) = e_{v'|v}(\partial_{v'}(\eta_2(\alpha')))_{k(t)}$$

by Proposition X.26.8. It follows that $e_{v'|v}$ is necessarily odd. In particular, v' is not trivial. Hence it is enough to prove that $\mathrm{trdeg}(\kappa(v')/k) \geq 1$ to get the desired conclusion. Indeed, we will have

$$\mathrm{trdeg}(L/k) > \mathrm{trdeg}(\kappa(v')) \geq 1,$$

and thus $\mathrm{trdeg}(L/k) \geq 2$.

Assume that $\kappa(v')/k$ is an algebraic extension. Since $\kappa(v') \subset k(t)$, we get $\kappa(v') = k$. Now we have

$$\partial_v(\eta_1(\alpha)) = \partial_v((1 - dt^2)) = 0 = e_{v'|v}(\partial_{v'}(\eta_1(\alpha')))_{k(t)},$$

so we get $e_{v'|v}(\partial_{v'}(\eta_1(\alpha')))_{k(t)} = 0$. Since $e_{v'|v}$ is odd, this yields

$$(\partial_{v'}(\eta_1(\alpha')))_{k(t)} = 0.$$

Since $\kappa(v') = k$ and the restriction from k to $k(t)$ is injective (see Chapter X, Section X.25), we get $\partial_{v'}(\eta_1(\alpha')) = 0$. Therefore, the specializations of $\eta_1(\alpha)$ and $\eta_1(\alpha')$ are well-defined and we have

$$s_v(\eta_1(\alpha)) = (1 - t^2 d) = (s_{v'}(\eta_1(\alpha')))_{k(t)}$$

by Proposition X.26.8. Let $u \in k^\times$ such that $s_{v'}(\eta_1(\alpha')) = (u)$. The previous equality shows that there exists $f(t) \in k(t)^\times$ such that

$$1 - t^2 d = u f(t)^2.$$

Let $\pi \in k[t]$ be an irreducible divisor of $1 - t^2 d$. Comparing π-adic valuations yields a contradiction. Hence $\mathrm{trdeg}(\kappa(v')/k) \geq 1$, and this concludes the proof. □

Corollary XII.37.3. *Assume that* $\mathrm{char}(k) \neq 2$. *Then* $\mathrm{ed}_k(\mathbb{Z}/4\mathbb{Z}) = 1$ *if* $-1 \in k^{\times 2}$ *and* 2 *otherwise.*

Proof. If $-1 \in k^{\times 2}$, $\mathbb{Z}/4\mathbb{Z} \simeq \mu_4$, and the result is clear. If $-1 \notin k^{\times 2}$, $\mathbb{Z}/4\mathbb{Z} \simeq \mu_{4[K]}$, where $K = k(\sqrt{-1})$. Now apply the previous result. □

§XII.38 Complements and open problems

We would like to end this chapter by giving some complements and stating some open problems concerning essential dimension.

First of all, we would like to mention the existence of a local version of essential dimension of a functor at a prime p, defined as follows:

Definition XII.38.1. Let $\mathbf{F} : \mathfrak{C}_k \longrightarrow$ **Sets** be a covariant functor, let p be a prime number, let K/k be a field extension and let $a \in \mathbf{F}(K)$. Finally, let E/k a subextension of K/k.

The **essential dimension** at p of a is the integer $\mathrm{ed}(a;p)$ defined by

$$\mathrm{ed}(a;p) = \min\{\mathrm{ed}(a_L) \mid [L:K] \text{ is prime to } p\}.$$

The **essential dimension at** p of $\mathbf{F}$ is the supremum of $\mathrm{ed}(a;p)$ for all $a \in \mathbf{F}(K)$ and for all K/k. The essential dimension of $\mathbf{F}$ at p will be denoted by $\mathrm{ed}_k(\mathbf{F};p)$.

If $\mathbf{F} = H^1(\_, G)$, we simply denote it by $\mathrm{ed}_k(G;p)$.

When G is a finite abstract group and $\mathrm{char}(k) \neq p$, one can show that

$$\mathrm{ed}_k(G;p) = \mathrm{ed}_k(G_p;p) = \mathrm{ed}_{k(\mu_p)}(G_p;p),$$

where G_p is a p-Sylow subgroup of G.

Therefore, the essential dimension at p of any finite group may be computed over an arbitrary field k satisfying $\mathrm{char}(k) \neq p$ in view of Theorem XII.36.6. See [29] for more details.

Clearly, we have

$$\mathrm{ed}_k(G) \geq \sup_p(\mathrm{ed}_k(G;p)).$$

This inequality is known to be strict for finite groups (see the exercices for a counterexample), but no examples of connected groups for which the inequality is strict are known (at least to the author).

Question 1: If G is a connected algebraic group defined over k, do we have $\mathrm{ed}_k(G) = \sup_p(\mathrm{ed}_k(G;p))$ (at least if $\mathrm{char}(k) = 0$) ?

In 'bad' characteristic, only few results are known. For example, we have the following result, proved in [28]:

Proposition XII.38.2. *Let p be a prime number, let $n \geq 1$, and let k be a field of characteristic p. Then we have*

$$\mathrm{ed}_k((\mathbb{Z}/p\mathbb{Z})^n) = \begin{cases} 1 & \text{if } |k| \geq p^n \\ 2 & \text{if } |k| < p^n. \end{cases}$$

However, the essential dimension $\mathbb{Z}/p^n\mathbb{Z}$ over a field of characteristic p is not known for $n \geq 3$.

Question 2: Let k be a field of characteristic $p > 0$, and let G be a finite p-group. Is $\mathrm{ed}_k(G)$ equal to the least dimension of a linear faithful representation of G?

The computation of the essential dimension of a connected algebraic group G is a very interesting problem, far from being solved, even in the case where G is split and $k = \mathbb{C}$.

Recently, Brosnan, Reichstein and Vistoli [9] proved that $\mathrm{ed}_k(\mathbf{Spin}_n)$ is equivalent to $2^{[\frac{n-1}{2}]}$ when $n \longrightarrow +\infty$.

The number $\mathrm{ed}_{\mathbb{C}}(\mathbf{PGL}_n)$ is known only for few values of n. It is equal to 1 if $n = 2, 3, 6$ (see for example [47]) and is equal to 5 for $n = 4$ (Rost[49]; see also [36]). The essential of dimension $\mathbf{PGL}_5$, or more generally of $\mathbf{PGL}_p$ (p prime) is not known. An old conjecture states that every central simple algebra of degree p whose center contains $\mathbb{C}$ is isomorphic to a symbol algebra $\{a, b\}_p$. If this is true, this would lead to the equality $\mathrm{ed}_{\mathbb{C}}(\mathbf{PGL}_p) = 2$. Conversely, proving that $\mathrm{ed}_{\mathbb{C}}(\mathbf{PGL}_p) \geq 3$ ($p \geq 5$) would immediately disprove this conjecture. It is interesting to note that Reichstein related the essential dimension of a given central simple algebra A to the essential dimension of higher trace forms. Later on, Florence generalized Reichstein's result to separable algebras. See [48] and [23] for more details.

Notes

The reader will find more computations of essential dimension of some functors in [4],[6] and [49]. Essential dimension of finite groups is also studied in [28]. In fact, in [28], the authors are focusing mainly on the existence of generic polynomials for a given finite group G, that is on the existence of generic G-torsors over a rational extension of k.

Exercises

1. Let k be a field and L be an étale algebra over k. Recall that the algebraic group-scheme $\mathbb{G}_{m,L}^{(1)}$ is defined by

$$\mathbb{G}_{m,L}^{(1)}(R) = \{z \in L_R^\times \mid N_{L_R/R}(z) = 1\}$$

for every commutative k-algebra R, and that we have

$$H^1(K, \mathbb{G}_{m,L}^{(1)}) \simeq K^\times / N_{L_K/K}(L_K^\times)$$

for every field extension K/k.

(a) Let t be a transcendental element over k. Show that t belongs to $N_{L_{k(t)}/k(t)}(L_{k(t)}^{\times})$ if and only if L is isomorphic to a product of some finite separable field extensions of relatively prime degrees.

(b) Deduce the value of $\mathrm{ed}_k(\mathbb{G}_{m,L}^{(1)})$.

2. Let k be a field of characteristic different from 2. Show that $\mathrm{ed}_k(\mathbf{PGL}_2) = 2$.

Hint: Use the fact that every central simple algebra of degree 2 is isomorphic to a quaternion algebra.

3. Keeping the notation of the proof of Theorem XII.37.2, show that $[s, 1 + t\sqrt{d}]$ is a generic $\mu_{4[K]}$-torsor.

4. Let G be a finite abstract group.

(a) Show that if $\mathrm{ed}_k(G) = 1$, then G identifies to a subgroup of $\mathbf{PGL}_2(k)$.

Hint: If E/k is a subextension of $k(V)/k$ of transcendence degree 1, then E/k is purely transcendental.

(b) Deduce that for all prime $p \geq 5$, we have $\mathrm{ed}_{\mathbb{Q}}(\mathbb{Z}/p\mathbb{Z}) \geq 2$.

(c) Show that $\mathrm{ed}_{\mathbb{Q}}(\mathbb{Z}/5\mathbb{Z}) = 2$.

(d) Assume that $\mathrm{char}(k) = p$. Show that $\mathrm{ed}_k(\mathbb{Z}/p^2\mathbb{Z}) = 2$.

5. Let G be a finite abstract group of order n. Show that for every prime integer $p \nmid n$, we have $\mathrm{ed}_k(G; p) = 0$.

Hint: Show that for any Galois G-algebra L/K, there exists a finite field extension M/K of degree dividing n such that L_M is isomorphic to the split Galois G-algebra over M.

6. Let k be a field, and let p be prime number.

(a) Assume that $p \nmid \mathrm{char}(k)$. Show that $\mathrm{ed}_k(\mathbb{Z}/p\mathbb{Z}; p) = 1$.

(b) Compare $\mathrm{ed}_{\mathbb{Q}}(\mathbb{Z}/p\mathbb{Z})$ and $\sup_{\ell}(\mathrm{ed}_{\mathbb{Q}}(\mathbb{Z}/p\mathbb{Z}; \ell))$.

References

[1] J. Kr. Arason, *Cohomologische Invarianten quadratischer Formen,* J. Algebra **36** (1975), 448–491.

[2] G. Berhuy, *An introduction to Galois cohomology and its applications.* Lecture notes (2009). Available at http://www-fourier.ujf-grenoble.fr/~berhuy/fichiers/NTUcourse.pdf.

[3] G. Berhuy, G. Favi, *Essential Dimension: a Functorial Point of View (After A. Merkurjev)* Doc. Math **8** (2003), 279–330.

[4] G. Berhuy, G. Favi, *Essential dimension of cubics.* J. of Algebra **278**, No. 1 (2004), 199-216.

[5] G. Berhuy, M. Monsurrò, J.-P. Tignol, *Cohomological invariants and R-triviality of adjoint classical groups.* Math. Z. **248**, No 2 (2004) 313–323.

[6] G. Berhuy, Z. Reichstein, *On the notion of canonical dimension for algebraic groups.* Advances in Maths 198, No. 1 (2005) 128–171.

[7] A. Borel, J. Tits, *Compléments à l'article: Groupes réductifs.* Pub. Math. IHES **41** (1972), 253–276.

[8] N. Bourbaki, *Algèbre, Chapitre V,* Hermann, Paris, 1959.

[9] P. Brosnan, Z. Reichstein, A. Vistoli, *Essential dimension, spinor groups and quadratic forms.* Preprint (2007). To appear in Annals of Maths.

[10] J. Buhler, Z. Reichstein, *On the essential dimension of a finite group.* Comp. Math. **106** (1997), 159–179.

[11] H. Cartan, S. Eilenberg, *Homological algebra.* Princeton Univ. Press, Princeton, N.J., 1956.

[12] J.W.S. Cassels, A. Fröhlich, *Algebraic number theory.* Academic Press, 1967.

[13] V.Chernousov, A. S. Merkurjev, *R-equivalence in spinor groups.* J. Amer. Math. Soc. **14** (2001), 509–534.

[14] C. Chevalley, *On algebraic group varieties.* J. Math. Soc. Jap. **6** (1954), 303–324.

[15] J.-L. Colliot-Thélène, J.-J. Sansuc, *The rationality problem for fields of invariants under linear algebraic groups (with special regards to the rationality problem).* Proceedings of the International Colloquium on Algebraic groups and Homogeneous Spaces (Mumbai 2004), ed. V. Mehta, TIFR Mumbai, Narosa Publishing House (2007), 113–186. Available online at http://www.math.u-psud.fr/~colliot/mumbai04.pdf.

[16] T. Crespo, *Galois representations, embedding problems and modular forms,* Collect. Math. **48**, 1-2 (1997), 63–83.

[17] M. Demazure, P. Gabriel, *Groupes algébriques. Tome I: Géométrie algébrique,*

généralités, groupes commutatifs. Masson & Cie Ed., Paris (1970).

[18] A. Delzant, *Définition des classes de Stiefel-Whitney d'un module quadratique sur un corps de caractéristique différente de* 2, C. R. Acad. Sci. Paris **255** (1962), 1366–1368.

[19] P. K. Draxl, *Skew fields,* L.M.S. Lecture Note Series **81**, Cambridge University Press, Cambridge, 1983.

[20] C. Drees, M. Epkenhans, M. Krüskemper, *On the computation of the trace form of some Galois extensions,* J. Algebra **192** (1997), no. 1, 209–234.

[21] S. Endo, T. Miyata,*Invariants of finite abelian groups.* J.Math.Soc.Japan **25** (1973), 7–26.

[22] E. Fischer, *Die Isomorphie der Invariantenk'orper der endlicher Abelschen Gruppen linearer Transformationen.* Nachr.Akad.Wiss.G'ottingen Math.-Phys. (1915), 77–80.

[23] M. Florence, *On higher trace forms of separable algebras.* Preprint (2006). Available at http://www.math.jussieu.fr/~florence/ed_alg.pdf.

[24] A. Fröhlich, *Orthogonal representations of Galois groups, Stiefel-Whitney classes and Hasse-Witt invariants,* J. Reine Angew. Math. **360** (1985), 84–123.

[25] R. S. Garibaldi, A. S. Merkurjev, J.-P. Serre, *Cohomological invariants in Galois cohomology,* University Lecture Series **28**, A.M.S., Providence, RI, 2003.

[26] P. Gille, T. Szamuely, *Central simple algebras and Galois cohomology.* Cambridge studies in advanced mathematics **101**, Cambridge University Press, 2006.

[27] B. Jacob, M. Rost, *Degree four cohomological invariants for quadratic forms.* Invent. Math. **96**, No. 3 (1989), 551–570.

[28] C. Jensen, A. Ledet, N. Yui, *Generic Polynomials,* MSRI Publ. **45**, Cambridge University Press, 2002.

[29] N. Karpenko, A. S. Merkurjev, *Essential dimension of finite p-groups.* Invent. Math. **172**, No. 3 (2008), 491-508.

[30] M.-A. Knus, A.S. Merkurjev, M. Rost, J.-P. Tignol, *The book of involutions,* A.M.S. Coll. Pub. **44**, 1998.

[31] M.-A. Knus, M. Ojanguren, *Sur le polynôme caractéristique et les automorphismes des algèbres d'Azumaya.* Annali della Scuola Normale Superiore di Pisa - Classe di Scienze Sér. 3, **26** No. 1 (1972), 225–231.

[32] S. Lang, *Algebra (Revised third edition),* Graduate Texts in Mathematics **211**, Springer-Verlag, New York, 2002.

[33] H. W. Lenstra, Jr., *Rational functions invariant under a finite abelian group.* Invent. Math. **25** (1974), 299–325.

[34] D. W. Lewis, J. Morales, *The Hasse invariant of the trace form of a central simple algebra.* Pub. Math. Besançon, Théorie des nombres, Année 1993-94, 1–6.

[35] S. Mac Lane, *Categories for the working mathematician,* Second edition. Graduate Texts in Mathematics **5**, Springer-Verlag, New York, 1998.

[36] A. S. Merkurjev, *Essential p-dimension of* $\mathbf{PGL}_{p^2}$. Preprint (2008). Available at http://www.math.uni-bielefeld.de/LAG/man/313.pdf.

[37] A. S. Merkurjev, *Essential dimension.* Private notes (1999), Lecture notes (2000).

[38] A. S. Merkurjev, *R-equivalence and rationality problem for semisimple adjoint classical algebraic groups.* Pub. Math. IHES **84** (1996), 189–213.

[39] A.S. Merkurjev, *On the norm residue symbol of degree* 2. Dokladi Akad.

Nauk. SSSR **261** (1981), 542–547. English translation: Soviet Math. Doklady **24** (1981), 546–551.

[40] J.-F. Mestre, *Extensions régulières de* $\mathbb{Q}(T)$ *de groupe de Galois* $\tilde{A}_n$, J. Algebra **131**, No. 2 (1990), 483–495.

[41] J.S. Milne, *Étale Cohomology.* Princeton Math. Series **33**, Princeton University Press, Princeton, N.J. (1980).

[42] P. Morandi, *Field and Galois theory,* Graduate Texts in Mathematics **167**, Springer-Verlag, New York, 1996.

[43] F. Morel, *Milnor's conjecture on quadratic forms and mod 2 motivic complexes.* www.math.uiuc.edu/K-theory/0684/. Preprint (2004)

[44] J. Neukirch, *Algebraic number theory,* Grundlehren der Math. Wiss. **322**, Springer-Verlag, Berlin, 1999.

[45] D. Orlov, A. Vishik, V. Voevodsky, *An exact sequence for* $K_*^M/2$ *with applications to quadratic forms.* Ann. of Math. (2) **165**, No. 1 (2007) 1–13.

[46] A. Quéguiner, *Invariants d'algèbres à involution.* Thèse de doctorat (1996).

[47] Z. Reichstein, *On the notion of essential dimension for algebraic groups.* Transformation Groups **5**, no. 3 (2000), 265–304.

[48] Z. Reichstein, *Higher trace forms and essential dimension in central simple algebras.* Archiv der Mathematik **88**, No. 1 (2007), 12–18.

[49] M. Rost, *Computation of some essential dimensions.* Preprint (2000). Available at http://www.math.uni-bielefeld.de/~rost/data/ed.pdf.

[50] M. Rost, *Essential dimension of twisted* C_4. Preprint (2000). Available at http://www.math.uni-bielefeld.de/~rost/data/ed1.pdf.

[51] D. Saltman, *Generic Galois extensions and problems in field theory.* Adv. in Math. **43** (1982), 250–283.

[52] D. Saltman, *Noether's problem over an algebraically closed field.* Invent. Math**77** (1984), 71–84.

[53] W. Scharlau, *Quadratic and Hermitian forms,* Grundlehren der Math. Wiss. **270**, Springer-Verlag, Berlin, 1985.

[54] I. Schur, *Über die Darstellung der symmetrischen und der alternierenden Gruppe durch gebrochene lineare Substitutionen,* J. reine angew. Math. **139** (1911), 155-250.

[55] W.R.Scoot,*Group theory, second ed.*, Dover Publications Inc., New York, 1987.

[56] E. S. Selmer, *On the irreducibility of certain trinomials.* Math. Scand. **4** (1956), 287–302.

[57] J-P. Serre, *Corps locaux,* Hermann, Paris, 1962; 3rd ed., 1968. English translation: *Local fields,* Graduate Texts in Mathematics **67**, Springer-Verlag, 1979.

[58] J-P. Serre, *Cohomologie galoisienne,* Lecture notes in Mathematics **5**, Springer-Verlag, 1965; 5th ed., révisée et complétée, 1994. English translation of the 5th edition: *Galois cohomology,* Springer-Verlag, 1997.

[59] J-P. Serre, *Cohomologie galoisienne: progrès et problèmes,* Astérisque (1995), no **227**, 229-257, Sém. Bourbaki 1993/1994, no. **783**.

[60] J.-P. Serre, *Topics in Galois Theory.* Research notes in Math. **1**, Jones and Bartlett Pib., Boston, MA (1992).

[61] J-P. Serre, *L'invariant de Witt de la forme* $\mathrm{Tr}(x^2)$, Comment. Math. Helv. **59** (1984), no. 4, 651–676.

[62] I. R. Shafarevich, *Construction of fields of algebraic numbers with given solvable Galois group.* Izv. Akad. Naukr. SSSR. Ser. Mat. **18** (1954) 523–578. Amer. Math. Soc. Transl. **4** (1960), 185–237.

[63] T. A. Springer, *On the equivalence of quadratic forms,* Indag. Math. **21** (1959), 241–253.

[64] R. W. Thomason, *Comparison of equivariant algebraic and topological K-theory*. Duke Math. J. **53**, No 3, (1986), 795–825.

[65] N. Vila, *On central extensions of A_n as Galois group over $\mathbb{Q}$,* Arch. Math. **44** (1985), 424–437.

[66] V. Voevodsky, *Motivic cohomology with $\mathbb{Z}/2\mathbb{Z}$-coefficients*. Pub. Math. IHES. **98** (2003), 59–104.

[67] V. E. Voskrensenskiĭ, *On the question of the structure of subfield of invariants of a cyclic group of automorphisms of the field $\mathbb{Q}(x_1, \dots, x_n)$*. Izv. Akad. Nauk SSSR Ser.Mat. **34** (1970), 366–375 (Russian). English translation: Math. USSR Izv. **4** (1970), 371–380.

[68] V. E. Voskrensenskiĭ, Algebraic tori, Nauka, Moscow, 1977 (Russian); revised and augmented English translation: Translations of Math. Monographs **179**, AMS, 1998.

[69] W. C. Waterhouse, *Introduction to affine group schemes,* Graduate Texts in Mathematics **66**, Springer-Verlag, New York-Berlin (1979).

[70] E. Witt, *Konstruktion von galoisschen Körpern der Charakteristik p zu vorgegebener Gruppe der Ordnung p^f*, J. reine angew. Math. **174** (1936), 237–245.

[71] O. Zariski, P. Samuel, *Commutative algebra II,* Graduate texts in Mathematics **28**, Springer-Verlag, Berlin, 1975.

Index

Printed in the United States
by Baker & Taylor Publisher Services